Contents

Chapter One: An Overview of the Human Body 1

Chapter Two: Molecular Behavior and the Chemistry of Water Environments 15

Chapter Three: Similarities and Differences Between Cells Living in the Society 35

Chapter Four: A Question of Priorities: An Organization Chart of the Body 49

Chapter Five: A Question of Pressure 67

Chapter Six: Barriers of the Body 83

Chapter Seven: Nutrition and Metabolism 105

Chapter Eight: The Brain in Control of it All 123

Chapter Nine: Reproduction 147

Chapter Ten: The Immune System 171

Chapter Eleven: Infections and Cancer 191

Chapter Twelve: Stress 205

Chapter Thirteen: Pain, A Fundamental Stress 219

Chapter Fourteen: Manipulating the Brain 233

Chapter Fifteen: Nature and Nurture 251

Index 269

Drugs, Stress, and Human Function

Richard Almon, PhD
Debra DuBois, PhD
Amanda Almon, MFA, CMI

Bassim Hamadeh, CEO and Publisher
Michael Simpson, Vice President of Acquisitions
Jamie Giganti, Managing Editor
Jess Busch, Graphic Design Supervisor
John Remington, Acquisitions Editor
Brian Fahey, Licensing Associate

First published in the United States of America in 2013 by Cognella, Inc.

Printed in the United States of America

ISBN: 978-1-62131-210-9 (pbk)

www.cognella.com 800.200.3908

chapter one

an overview of the human body

Humans are living: First and foremost humans are living organisms. Although this may sound like an absurdly simple statement, "living" governs the rules by which every human must play by, or they die and become simple matter (molecules and atoms). The fact is, being alive takes a great deal of energy. The cardinal rule of life is that organizational energy (enthalpy) must be put into the system to maintain a balance with the drive towards disorder (entropy) (Fig. 1.1).

You, as a human, get this organizational energy that maintains life by breaking down and processing food that you eat. However, the process of breaking down that food to form usable energy requires oxygen. Simply put, if you do not eat for a relatively short period of time, and you do not breathe for an even shorter period of time, you die. You die because humans produce almost all of their organizational energy by a process that uses oxygen to break down the molecules in food. Without usable energy, you are dead.

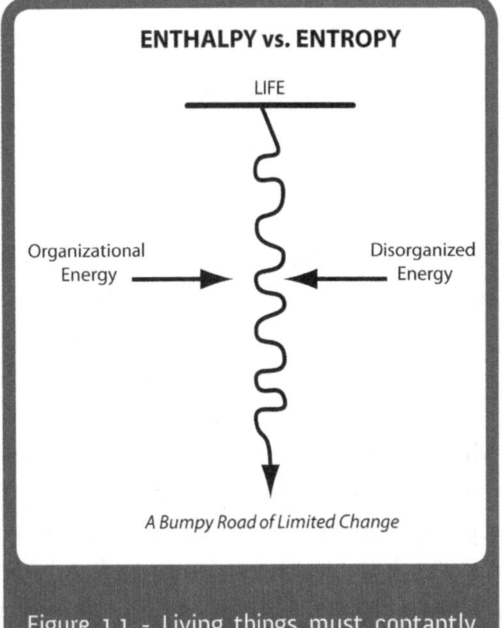

Figure 1.1 - Living things must contantly expend energy to stay alive.

All living organisms are made up of molecules, which themselves are formed from the fundamental building blocks of matter. These fundamental building blocks are called elements. The word atom means "one" of a particular element. You may remember having seen a periodic table of elements somewhere, especially if you've had a chemistry class. There are 92 different elements that exist naturally on earth, and a few dozen more that can be produced by man.

Some examples of different elements that you are probably familiar with include oxygen (O), carbon (C), and iron (Fe). These are the fundamental units of matter because you can't chemically break them down into other things: in other words, they are the smallest individual units of which everything in our world (including humans) is made. Atoms of different elements interact to form larger structures called molecules (Fig.1.2).

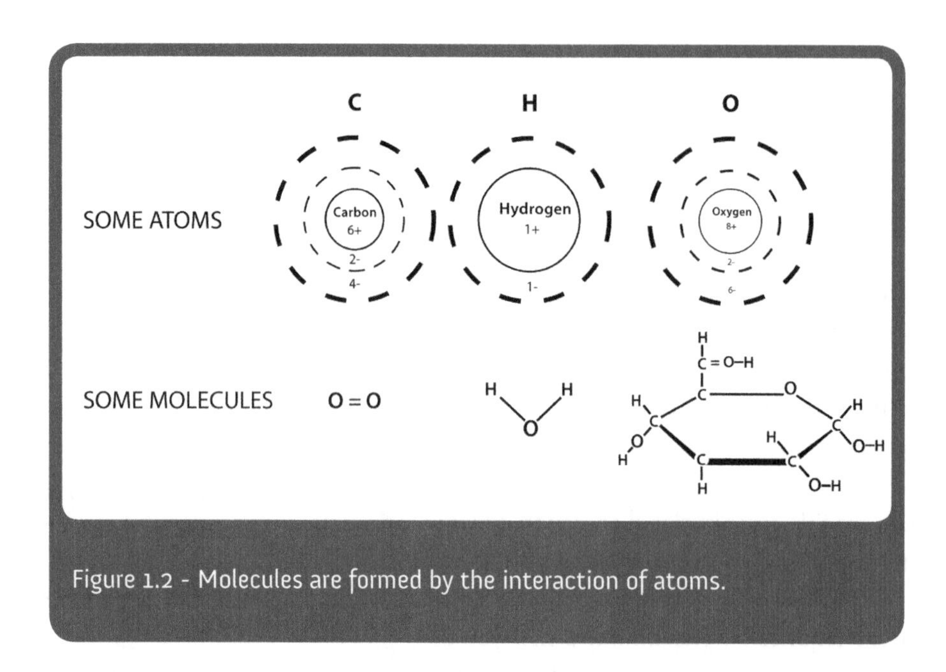

Figure 1.2 - Molecules are formed by the interaction of atoms.

The "rules" by which atoms interact to form molecules will be discussed in Chapter 2. However, what is important here is that because sticking atoms together to form molecules is organizational, energy must be used. This energy is not lost, but rather it is stored in the bonds between the atoms and will be released when the molecules are taken apart. Life works by building and taking apart molecules.

All energy for life on earth comes from the sun: Plants, through a process called photosynthesis, obtain their organizational energy by capturing it directly from the sun. A plant can then use this organizational energy captured from sunlight to make many different types of complex molecules. Non-photosynthetic living organisms (such as humans) consume (either directly by eating plants or indirectly by eating other animals that have eaten plants) and use the organizational energy stored in the bonds between atoms in molecules made by plants (Fig. 1.3).

Because there are many different types of molecules, there is a great variation in the amount of energy stored in the bonds between atoms. But in all cases the result is stored potential energy.

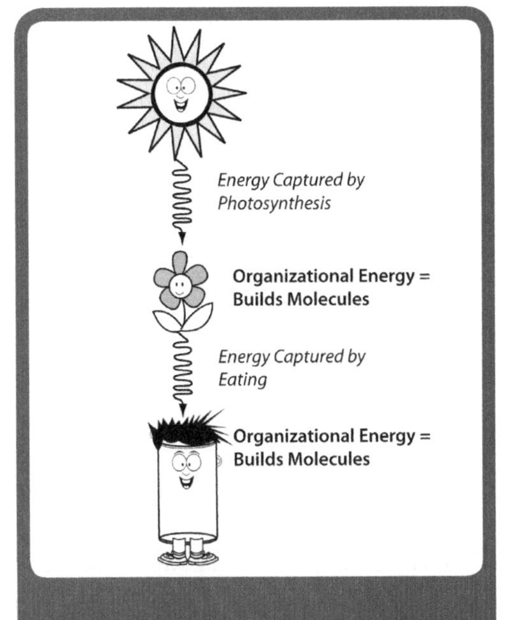

Figure 1.3 - Energy from the sun is used as organizational energy by other plants to build molecules. Humans and other animals use the organizational energy they get from breaking down food to build molecules.

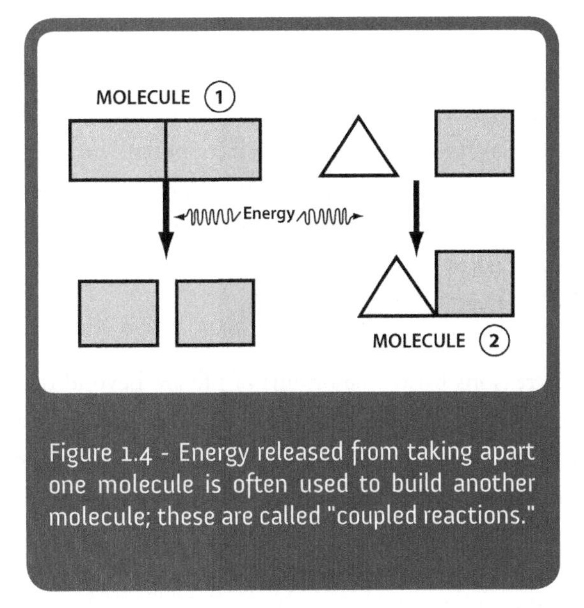

Figure 1.4 - Energy released from taking apart one molecule is often used to build another molecule; these are called "coupled reactions."

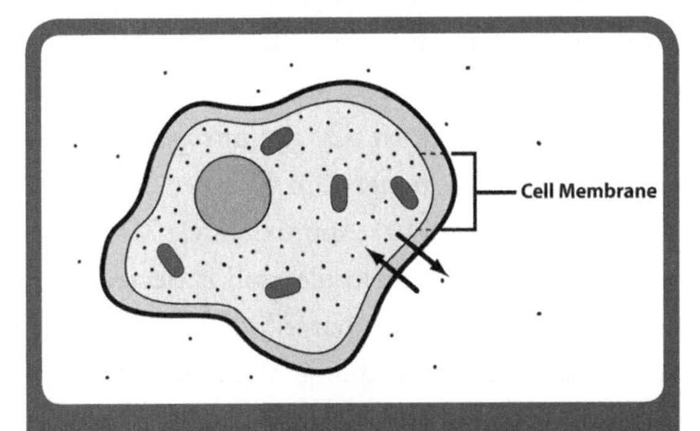

Figure 1.5 - The water environment inside the cell is different than the water environment outside the cell. The cell membrane separates these environments and controls the movement of molecules between the two environments.

The energy stored in chemical bonds is "potential" because it can be released. Life exists by coupling, or linking together, favorable energy releasing processes (more disorder) with unfavorable energy requiring processes (more order). For example, the bonds between two atoms can be taken apart to provide the energy to stick two other atoms together making a different molecule (Fig 1.4). The general requirement is that there must be more energy in the bonds between the two atoms you take apart than is needed to stick the other two atoms together. By analogy, if you drive a car to the top of a hill and put on the break you have stored potential energy. The potential is for the car to roll back down the hill. If you tie a rope to a car pointing down one side of a hill to a car at the bottom of the other side of the hill and release the brake, the first car rolling down the hill will accomplish the work of pulling the second car to the top of the hill as long as the first car weighs more than the second. Because life exists by coupling favorable energy releasing processes with unfavorable energy requiring processes (like the car analogy) some productive energy is often lost as heat in each coupling. The heat is energy that dissipates from the immediate life reaction going on in the cell. However, this energy dissipated as heat is still used by the complex society of cells. This energy provides heat. Hence, living organisms must always be acquiring additional energy storing molecules. In the case of humans, these energy containing molecules are obtained by eating food. More simply put, if you don't eat you can starve to death.

The reasons for or the origins of life are beyond the scope of this book. The fact is life does exist. Life is a complex balance between organizational energy and the natural drive towards disorganization. A life can perpetuate itself until parts wear out. Even then, life can remake and perpetuate itself in new individuals (reproduction) with enough flexibility that changes in the general environment can be adjusted for (evolution).

Water is necessary for life: There are many different kinds of molecules in the human body. Some are quite small, and others are quite large and often complex in structure (macromolecules) and are formed by joining together individual units of smaller molecules. Many of these molecules contain a great deal of the element carbon (C) and are referred to as "organic molecules." Other elements found a lot in our molecules include hydrogen (H) and oxygen (O) and nitrogen (N). We will discuss different types of molecules and what they do in the body in more detail in Chapter 2. However, the molecule that makes up the largest percentage of our bodies is water (H_2O). Approximately 70% of our bodies is water. An interesting and important feature of water is that the water molecule itself is electrically charged, or "polar": part of the molecule carries a positive charge and part a negative charge (why this occurs will be discussed in more detail in Chapter 2). Water is an environment of electrical charge. Water is necessary for life to work.

Life exists in water compartments which themselves are surrounded by water. The smallest compartment of life is the cell. This compartment is composed of a water environment sequestered inside a small "shell" of oil (the cell membrane) (Fig 1.5). And as we all know, "oil and water don't mix." There are other specialized water compartments within the cell itself (called "organelles" which carry out particular jobs inside the cell). However, the cell is considered the smallest unit of life because in the right circumstances it can continue living on its own and is capable of reproducing.

This shell, that we call a membrane, allows the small water environment inside the cell to be different than the larger water environment outside the cell. However, two water environments sitting side-by-side is organized (order). Fundamental to the distribution of molecules among compartments in the body is the drive towards disorganization (disorder). When the compositions of two water environments sitting side-by-side are different, there is potential. The potential is to succumb to the natural drive for disorder and to make them identical in all regards. Resisting the natural drive to make the inside of a cell no different than the water outside is one reason a living organism needs to acquire energy storing molecules from the outside. So what does life do? Life keeps itself ordered against the drive to disorder until parts wear out (aging). Life can also make reasonable facsimiles of itself (reproduction) so that it exists with the flexibility for adjustment to alterations in the environment (evolution).

Life exists in different forms: There are forms of life that are single cells. Bacteria are single-celled organisms. You can measure the generation time of a bacterium in minutes. Their lifespan is short. In that short lifespan they keep going long enough to create reasonable facsimiles of themselves (reproduce). They also at times trade genetic material among the members of their community to add adaptive flexibility. If you have some diversity in the community then adjustments to changes in the environment are possible. They don't reproduce using sex; they just at times get together and exchange chunks of genetic material (DNA) with other individuals in the same pond of water. This exchange creates enough diversity to facilitate survival of the general group of very similar but slightly diverse individuals that are the same form of life. For example, if you have a bacterial infection such as strep throat or bacterial meningitis, you can develop an antibiotic resistant form of the disease if a small number of individuals in the population of bacteria causing the disease are resistant to the antibiotic. This process of selection based on vulnerabilities and exchanging genetic information changes the vulnerabilities of the general population. This is why infections acquired in hospitals are particularly difficult to treat. Humans have a generation time in years, so change in the human population is much slower. Bacteria have a generation time measured in minutes, and therefore the population can change much faster. To their advantage, these very short-lived populations of individuals can change rapidly because of rapid adaptability of the community.

Humans are multicellular organisms: Unlike bacteria which are single cells (and can only be seen with a microscope), humans are made up of trillions of cells. Although all cells of an individual human body start out with the same "blueprint," (that is, they contain the same genetic information), during development different cells become different from each other. This happens because different parts of the genetic "blueprint" are read and expressed in different types of cells. Thus, cells become specialized. Different types of cells become organized together to form the different tissues and organs of the body, which are designed to carry out different tasks within the body (Fig 1.6).

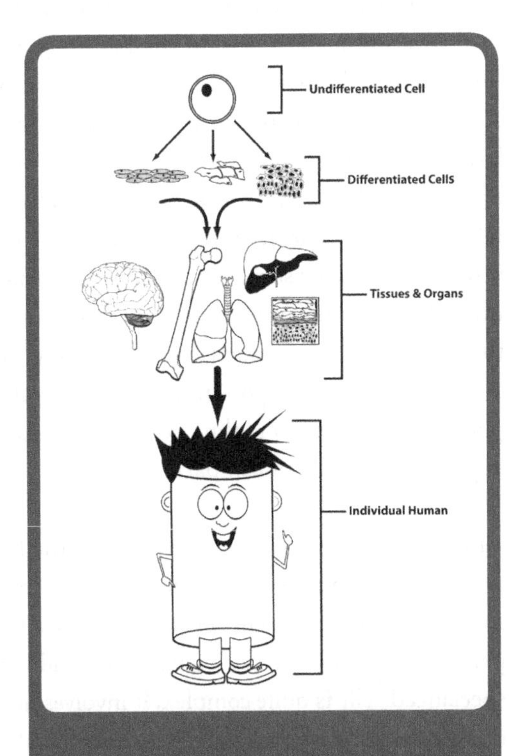

Figure 1.6 - Differentiated cells interact to form discrete structures called tissues and organs, which are organized to form a human.

The development of organized societies of cells (multicellular organisms) out of loosely linked communities of individual cells in the same pond of water provided many advantages and imparts new complexities to life.

Single-celled organisms, even when living in a community, must for the most part do everything for themselves, and the existence of any individual is measured in a lifespan of minutes. Forming organized societies of cells allowed for specialization and therefore increased complexity. In a society, individuals specialize in doing a small number of tasks that are important to the existence of the society as a whole. To do this they must give up the ability to do everything for themselves. A society is a group of cooperating individuals. Although none of them can take care of themselves independently, the existence of the society is enhanced by their specialized functional existence. The advantage is complexity and longevity. The lifespan of a cell is no longer the lifespan of the individual. The problem is that a society requires that individuals coordinate. The individuals in a society must "talk" to each other to coordinate the actions of all individuals for the good of the society.

Humans develop from a single cell: A human is a complex society made up of a diverse collection of specialized cells. However, this society of diverse individuals started as a single cell. Mom and dad got together and combined genetic information to produce a single cell with a unique genetic make-up (sexual reproduction, which will be discussed more fully in Chapter 9). This is how complex multi-cellular organisms create genetic diversity. This unique cell begins to divide as it wanders its way in a week-long trip down the Fallopian tube to the uterus. During this trip cell division continues, creating a ball of a few hundred cells with the only specialization being where in the ball they reside, inside or outside. The cells on the outside will become the placenta (the structure which connects mom and the baby) while the cells on the inside will become the embryo. The cells on the inside of this ball are what we call embryonic stem cells. These embryonic stem cells have the capability to become any type of specialized adult cell. Because of this potential to become any type of specialized cell, these cells are of great medical interest. For example, an insulin dependent diabetic must take insulin because his or her immune system has destroyed the cells of the pancreas that normally produce insulin. Embryonic stem cells have the potential to become and thus be used to replace insulin producing beta islet cells in the pancreas. Similarly, Parkinson's disease is caused by the loss of a population of cells in a part of the brain called the substantia nigra (literally, "black substance"). These cells produce a specific chemical called dopamine which is important for communications between cells of the brain. Embryonic stem cells have the potential to become and thus be used to replace these cells. Embryonic stem cells have the potential to grow any necessary replacement part.

After a week of wandering its way down the Fallopian tube, the ball of stem cells implants in the uterus and pre-natal (pre-birth) development begins. The culmination of which after about nine months is a complex society of interdependent specialized cells. At birth a human is about 10^{14} (one hundred trillion) cells. This period of nine months during which the individual morphs from a few hundred fetal stem cells to trillions of specialized cells is quite complex. It involves not only programmed cell birth (which we call proliferation) and the specialization we call differentiation, but also programmed cell death. The first step is for cells to segregate by migration into three groups. Each of these groups has the potential of becoming several different kinds of tissues. But the members of each separate group have lost the ability to become all kinds of tissue. The cells of each group then begin to form into sub-groups with the potential to become an even more limited number of tissues. At birth there remains in each tissue cells with a limited capacity to become only a few types of cells. For example, the stem cells in the bone marrow can become one of several types of blood cells, and the stem cells in skeletal muscle can become muscle, bone, or perhaps adipose tissue (fat). These remaining "adult" stem cells are there for the purpose of tissue repair. If you damage a muscle the stem cells of the tissue are allowed to proliferate and then at some point, like during development, they differentiate and become part of the working muscle. Cell division is one of the most highly restricted processes in the adult. When damage happens to a tissue the restrictions on cell proliferation are relaxed for repair. Cell division is a complicated process involving multiple stages with several checkpoints to examine the fidelity of the copying of the genetic material. Mistakes (mutations) that escape the surveillance process during stem cell proliferation are the origins of cancer. In this regard, tissue damage causes cancer.

Cells are specialized: The smallest living compartment in the body is still the individual cell. Each type of cell is specialized to carry out a limited number of functions that are important to the survival of the society as a whole. Each cell is a living unit because it carries out the fundamental processes of life. A cell acquires, from the common water environment of the body, energy storing molecules (nutrients) and couples the release of the energy in these molecules to energy requiring processes (such as making other molecules). Under selective conditions, most cells can also use this energy to reproduce themselves. However, the cell itself is a collection of small compartments (organelles). All life processes are carried out within the various compartments of the cell. If an individual cell is removed from the body and placed in an appropriate environment, it will continue to function and live. However, unlike the single cell organism, an individual human cell outside the body will only survive in an exquisitely controlled environment. The advantage to this is that the human organism as a whole does not depend on the survival of a single cell. In fact, the lifespan of many individual cells of the body such as skin cells is quite limited, and at times, some cells are even directed to die (commit suicide), a process known as apoptosis.

The water environment must be controlled: A single cell organism is subject to whatever changes occur in the water environment around them: if it gets too bad, they die. Complex societies of cells attempt to control aspects of their environment. Humans are a complex society of cells that must attempt to control virtually all aspects of the water environment around their cells. One of many aspects that must be controlled is temperature. Amphibians and reptiles ("cold-blooded animals") simply find a place to wait out the cold periods and become hypothermic and dormant during the cold times. This greatly reduces their demand for high energy molecules and they let nature warm them back up. Of course this leads to a lot of down time, depending on the climate where they live. In many respects, the level of control sought by humans and other warm-blooded animals is the ultimate in environmental control. Warm-blooded animals like mammals and birds are metabolically free: they can control their internal temperature. However, it is expensive to be warm-blooded. Even among warm-blooded animals some in northern climes hedge their bets by being able to become hypothermic and hibernate. However, animals that are capable of becoming routinely hypothermic and then warming themselves back up also have specialized cells called brown adipose tissue that, like white adipose tissue, contains alot of energy storing molecules (fat). Fat is pure stored energy. If subjected to starvation how long you live is a direct function of how much fat your body has stored up. However, brown adipose tissue also has the relatively unique capability of using this stored energy to produce heat to warm the hibernating, hypothermic animal back up. Other than briefly during the postnatal period, humans do not

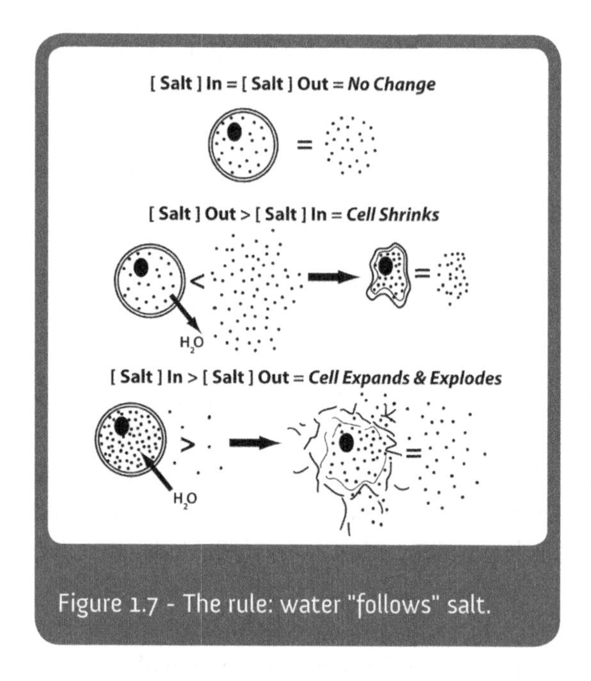

Figure 1.7 - The rule: water "follows" salt.

have brown fat. If a human becomes hypothermic below about 85 degrees F, they die unless someone comes along and takes measures to warm them up. However, they do accrue the benefits of hypothermia in that damage as a function of time is minimized by the cooler temperature. Controlled induced hypothermia is in fact used in medicine to reduce damage to the brain.

Besides temperature, virtually all aspects of the common water environment of the body, (the blood) must be controlled. The human body is a water based ecosystem in which virtually every aspect must be exquisitely controlled or the consequence is disease and death. At its most basic, blood is salt water, as is the environment inside cells. As discussed above, when the compositions of two water environments sitting side by side are different, the natural drive for disorder is a strong force to make them identical in all regards. Maintaining any difference between the salt content of the water inside cells and the salt content of the general water environment is very expensive in terms of organizational energy. To minimize this problem, the cells of the body attempt to exquisitely control the salt concentration of the general water environment.

Concentration is the amount of salt per volume of water. Therefore, two compartments sitting next to each other such as the inside of a small cell and the relatively vast general water environment of the body can be made the same by moving either water or salt from one compartment to the other. Practically, water moves far more easily through the oily barrier around cells than do salts. As a consequence, if the salt concentration of the general water environment goes up, the drive is to move water out of cells. The result would be that the cell would shrink. Conversely, if the salt concentration of the general water goes down, the drive would be for water to move into the cell. The result would be that the cell would swell (Fig 1.7).

Resisting the pressures created by even small differences in the salt concentrations inside and outside of a cell demands a huge input of organizational energy. If the society allows the salt concentration of the general water environment to change up or down the result is shrinking and swelling of cells. As an illustration, if you take a red blood cell and drop it in a glass of tap water, water will quickly flow into the cell and it will explode.

Cells require an energy source: Sugar and fat molecules are the most important sources of organizational energy for all of the cells of the body. Cells take in both sugar and fat from their environment and break them down into smaller molecules (catabolism). In this process, the relatively small amount of organizational energy in the bonds of these molecules is released and is used to produce other molecules (ATP) that contain very high-energy unstable bonds. It is the very high-energy, unstable bonds that are used to do the work of driving unfavorable processes (anabolism). That is why ATP is often referred to as the "energy currency" of the cell (Fig 1.8).

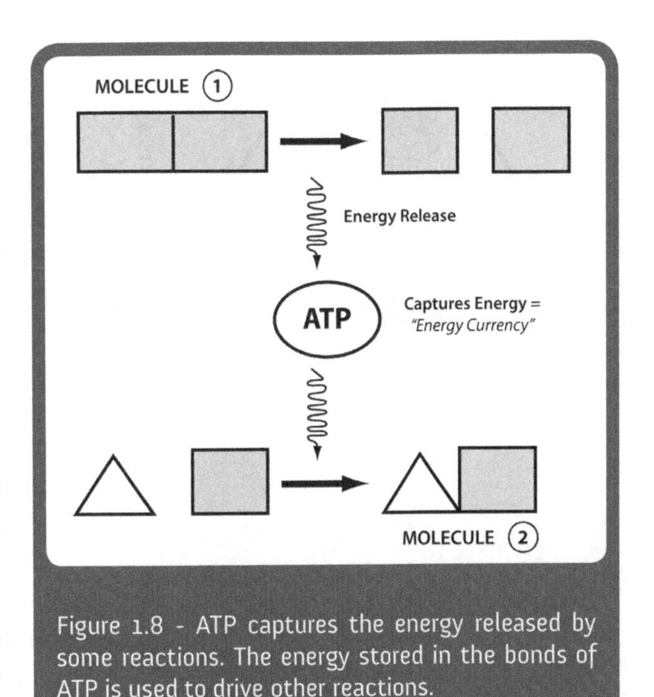

Figure 1.8 - ATP captures the energy released by some reactions. The energy stored in the bonds of ATP is used to drive other reactions.

There are two very important differences between sugar and fat. The first is that sugar is soluble in water and fat is not (remember the adage—"oil and water don't mix"). This is important because if you have a lot of sugar in a water environment you must have a lot of water to keep the sugar concentration just right. In contrast, because fat does not dissolve in water, it does not "play" in the concentration game. In a functional sense this is why if you eat a lot of sugar you become fat. The body simply converts the water-soluble sugar to fat for storage. The second difference between sugar and fat is that a little bit of the organizational energy in sugar can be extracted by a process that does not involve oxygen. Extraction of all the organizational energy in the bonds of fat and the remainder of the energy in sugar requires oxygen. The ability to get out a portion of the energy stored in sugar without the use of oxygen becomes important for activities such as exercise where a large amount of organizational energy is required in a short time. Many people talk about doing "aerobics." In aerobic exercise the demand for energy is balanced by the ability to bring in and deliver oxygen to the tissues. When a person does aerobic exercise his or her breathing rate generally increases to bring in more oxygen. When a person stops aerobic exercise, his or her breathing immediately returns to normal. In contrast, in anaerobic exercise the individual is exercising beyond his or her ability to bring in and deliver oxygen to the tissues. When a person stops anaerobic exercise, the person continues to breathe hard for a period of time while the liver pays back the "oxygen debt" run up by the musculature. An oxygen debt is caused by using the small amount of organizational energy in sugar that can be obtained without using oxygen. These partially used sugar molecules must be eliminated by the oxygen process. Maintaining a balance of sugar and fat in the general water environment is necessary so that all cells can obtain enough organizational energy to maintain themselves. However, there is a special caveat associated with sugar. Certain types of specialized cells, especially those of the brain, require sugar for moment-to-moment function.

If enough sugar is not available, then they will dysfunction and die. Like salt, too much sugar or fat in circulation is also detrimental to the system as a whole.

Figure 1.9 - Oxygen is a waste product released from plants as they synthesize molecules for usable energy. Carbon dioxide is a waste product released from animals as they break down molecules for energy, but is used by plants to synthesize molecules. It's a cycle!

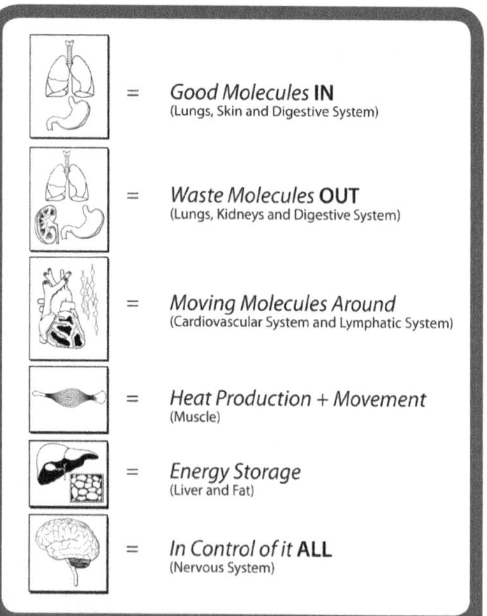

= *Good Molecules* **IN**
(Lungs, Skin and Digestive System)

= *Waste Molecules* **OUT**
(Lungs, Kidneys and Digestive System)

= *Moving Molecules Around*
(Cardiovascular System and Lymphatic System)

= *Heat Production + Movement*
(Muscle)

= *Energy Storage*
(Liver and Fat)

= *In Control of it* **ALL**
(Nervous System)

Figure 1.10 - The human body has a lot of work to do, and the brain controls it all.

The last major type of molecule that must be maintained just right in the general water environment are the two gasses, oxygen (O_2) and carbon dioxide (CO_2). As discussed above, plants obtain their organizational energy from the sun. They take up carbon dioxide and water, use the energy of the sun to make sugar, and then release oxygen as a waste product into the environment. Non-photosynthetic organisms, like humans, need to take oxygen into their general water environments. Individual cells take in this oxygen and use it to liberate the organizational energy captured by plants from sunlight (Fig 1.9).

Therefore, in non-photosynthetic organisms like humans, the resulting waste products are carbon dioxide and water. The water is no problem: it just adds water to the environment. However the carbon dioxide is a waste product that must be removed from the general water environment of the body. Humans remove carbon dioxide from their bodies in the process of breathing.

Humans are not static: Functioning of complex living organisms like humans is incredibly dynamic. Necessary molecules must be taken into the system and waste molecules must be removed. Taking molecules into the system involves specialized cells that form the lungs, the digestive system, and to a lesser extent the skin. Removing waste molecules involves specialized cells in the lungs, the kidney, and the digestive system. Distributing the molecules and heat of the general water environment involves specialized cells of the heart, blood vessels, and a sewer-like drainage system known as the lymphatics. Supporting and moving the system around the world involves specialized cells in muscle and bone. Heat production, so the system can maintain a constant temperature, involves the musculature. General storage of energy containing molecules involves adipose tissue and the liver. Production and detoxification of many molecules involves the liver.

Finally, "someone" must be in charge and coordinate the actions of the diverse collection of individuals that make up the society. This function is carried out by the central nervous system (Fig 1.10).

There are priorities in the human body: In the society of cells that make up the human body, all groups of specialized cells are not treated equally. There is a prioritization in both the distribution of resources and protection from changes in the environment. The highest priority group of cells is the brain and the spinal cord. Collectively, the brain and spinal cord are referred to as the central nervous system. The central nervous system both controls virtually every aspect of the body's internal function, and also its interaction with the external world around it. It is mainly the evolution of the central nervous system into the highly organized and complex structure that also makes humans different from other living organisms. The central nervous system is a highly organized collection of very specialized cells. Because of its organization and specialization, it has two very unique weaknesses. First, unlike liver, skin, and many other types of cells there is never a replacement of nerve cells. If a nerve cell dies for any reason, it and its function are gone forever. That is why if a person breaks his or her neck and kills a relatively few cells in the communication path to the rest of the body, much of the entire body may be paralyzed. In contrast, you can remove almost the entire liver and the cells will be replaced. The second unusual quality of the central nervous system is its dependence on the rest of the body to provide for its needs on a moment-to-moment basis. Unlike most types of cells, nerve cells store very little. If the delivery of molecules to the brain by the blood is disrupted for even a brief period of time, nerve cells begin to die. Because of its importance and weaknesses, the body has evolved such that the central nervous system can control and thus set the priorities of virtually every aspect of body function. However, because of this, the nervous system is also a very selfish collection of cells. The brain has the highest priority of all cells in the body and will protect itself "at all costs." This is because if the brain dies, then the body dies.

How does the central nervous system control the body? There are two aspects that are critical to control. The first aspect is the ability to collect information about what is happening all over the body. The second is the ability to send directions to all cells in the community. The central nervous system has two mechanisms for accomplishing both tasks. The first is direct contact with cells all over the body. Nerve cells send out processes like tentacles that distribute throughout the body. Individual nerve cells provide the brain with direct contact with other cells all over the body. Nerve cells are extremely unique in structure. The part of the cell distributed throughout the body is like a microscopic thread. It is so small in diameter that it cannot be seen by the naked eye. However, it can be so long as to extend from the waist to the foot. Some nerve cells carry information from the nervous system, giving directions to other cells. Other nerve cells bring information into the nervous system.

The second method by which the central nervous system collects information and gives direction is by means of the circulating blood. Blood entering the brain carries information to the brain. This information is in many forms. The blood carries unique molecular messages from different hormone producing cells throughout the body. For example, after one eats, the nutrients entering the blood cause special cells in the pancreas to release a hormone called insulin. Insulin not only directs many cells to store the excess nutrients but also serves as a signal to the brain which helps reduce the hunger drive. In addition, the concentration of various molecules

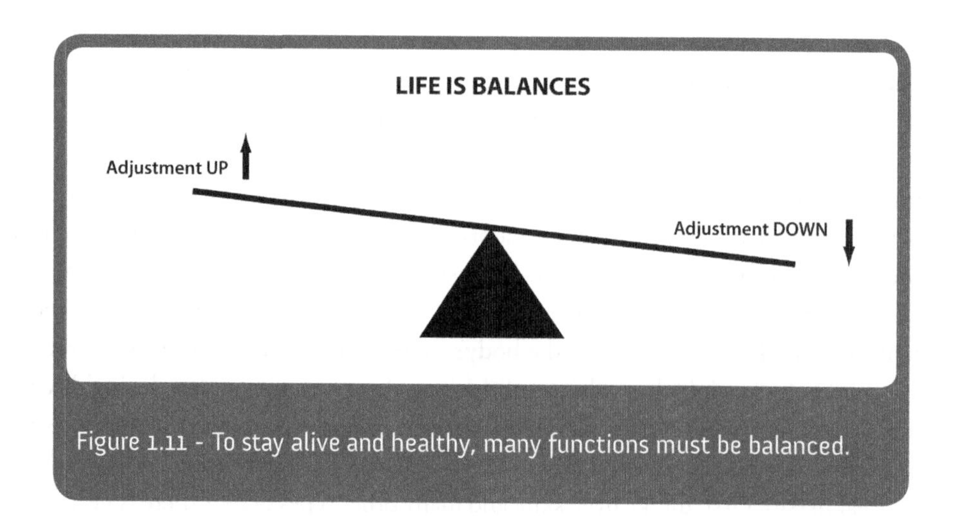

Figure 1.11 - To stay alive and healthy, many functions must be balanced.

represents information to the brain. For example, the salt concentration of the blood flowing through the brain provides information necessary for controlling the kidney. Even the temperature of the blood entering the brain provides information to the brain. Similarly, blood leaving the brain carries molecular messages to groups of cells throughout the body.

This information is in the form of unique molecular hormone signals produced by the pituitary gland under the direction of the brain. These hormone signals direct many types of cells all over the body. With its two methods of gaining information and distributing directions, the central nervous system is able to fine tune and coordinate the myriad of functions of the body in order to adjust to a wide range of changing circumstances.

It is all a question of balance: Life is like a tightrope walk. A complex biological system such as the human body functions on a balance of forces (Fig. 1.11). Plants capture the energy of the sun and animals eat the plants (and/or other animals that eat plants) and use the energy to maintain a balanced order of molecules against the natural force to disorder molecules. When any living individual, from a cell to the entire human body, cannot obtain sufficient energy to maintain molecular order it dies. However, the concept of balance extends from the simplest molecular level to the most complex functional process in the body. The entire body functions on a balance of processes. Skin cells die at a certain rate and they reproduce at a certain rate. Disrupt the balance and the surface barrier of the body is affected. There are mechanisms for lowering blood pressure which are balanced against mechanisms for raising blood pressure. Disrupt the balance and blood pressure is either too high or too low. There are mechanisms for generating heat to maintain the warmth of the body and there are mechanisms for removing heat from the body. Disrupt the balance and the result is either hyperthermia or hypothermia. There are mechanisms for increasing the signal activity in the brain and there are mechanisms for decreasing the signal activity in the brain. Disrupt the balance and the result is either seizure or coma. In all cases, the balances must be maintained within the context of a changing world. Damage to the skin can be repaired by a temporary imbalance such that more cells are made than die. When the damage is repaired the system returns to balance.

Blood pressure can be increased for running and decreased for sleeping.

Heat production and conservation by the body can be increased on a cold day, or decreased to allow more heat loss from the body on a hot day. Both occur simply to maintain optimal body temperature. Signal activity in the brain can be increased to focus attention or decreased for sleeping.

The extraordinary numbers of interconnected balances that represent a living individual are adjustable to changing circumstances, but only within limits. If circumstances (internal or external) change too much, then imbalances occur and the body malfunctions. This results in disease and potentially death.

chapter two

molecular behavior and the chemistry of water environments

Humans, like everything else on earth, are made up of matter: Humans are collections of molecules. Granted, they are highly organized collections of molecules—but there is nothing "magic" or "mysterious" about the molecules that make up the human body. However, there are some interesting aspects about the molecules that occur in living things. As we will see below, water is a very important molecule in living things. In addition, a relatively few elements are used to build the molecules that occur in the human body. Carbon (C) is the element that is present in the largest amount in most of our molecules (thus the molecules in living things are said to be "organic," which means carbon-based). Other important elements in our molecules include oxygen (O), nitrogen (N) and phosphorus (P). Another interesting and important thing about the molecules in living organisms is that relatively small molecules are used as "building blocks": they are joined together to form very large molecules called "macromolecules." Despite these interesting features, the "rules" by which the molecules in humans behave (how they are formed and how they interact) follow the rules of chemistry. Therefore, in order to understand how a human works, it is necessary to have a basic understanding of chemistry.

Making a living organism out of matter depends on water: Life on our plant is based on a liquid: water. Though intermediate forms exist, matter comes in three basic states: solids, liquids, and gases. The nature of life requires a fine balance between stability and mutability (the ability to change). Solids require too much organizational energy for change, and gases require too much organizational energy for stability. A liquid provides the best balance between stability and mutability. The characteristics of the liquid should also represent a balance in reactivity. The liquid should be sufficiently reactive that it promotes reactions between molecules, but should not be so reactive that it attacks most other molecule. And of course, it should be ubiquitous: the environment should contain large amounts of liquid. For life on earth, these characteristics describe water

(H_2O). Water is a liquid over a large range of ambient earth temperatures and provides a reasonable balance of reactivity.

About 70% of the human body is water. In this water are molecules, which are formed from different elements interacting with each other, as well as some free elements. Elements are the basic building blocks of matter. Elements such as carbon, hydrogen, oxygen and nitrogen are used to build small molecules. Small molecules such as sugars, amino acids, and nucleotides are used to build larger molecules. Large molecules are used to build structure. The structure is used to separate the body into water environments. There are large water environments such as the blood and there are microscopic water environments such as the nucleus of a cell. Each individual water environment contains its own collection of molecules and elements.

These molecules and elements have varying degrees of reactivity with the water and with each other in their environment. Each different element and molecule adds additional properties to the mix. This allows many different kinds of interactions to occur.

The structure of atoms and molecules: Elements are the basic building blocks of matter: you can't chemically change an element into something else. There are a limited number of different elements (approximately 100). These elements are also sometimes called atoms (the word "atom" actually refers to "one" of an element). Elements were formed ostensibly at the beginning of existence and continue to be formed today by nuclear fusion. Some examples of elements that you are probably familiar with include hydrogen, carbon, nitrogen, iron, sodium and helium. Each type of atom has its own unique structure but all have the same common form. The form of atoms is quite similar to a planet with satellites. There is a nucleus at the center, and orbiting around the nucleus are satellites called electrons. The difference between the atoms of different elements is the size of the nucleus and the number of orbiting electrons. The nucleus of each atom carries one or more positive electrical charges (protons) and each orbiting electron carries a single negative electrical charge. The number of orbiting electrons is equal to the number of positive charges in the nucleus so that the atom is electrically neutral.

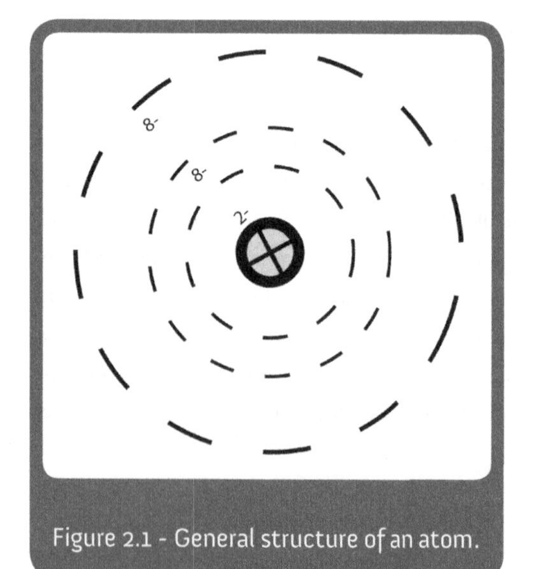

Figure 2.1 - General structure of an atom.

There are small atoms with few electrons and there are large atoms with many electrons. The electrons of an atom are organized around the nucleus in orbits. Each orbit of an atom has a capacity for only a defined, limited number of electrons (Fig. 2.1).

The first orbit, which is closest to the nucleus, can only hold two electrons. The next two orbits can contain up to eight electrons. (An atom can contain more electrons and therefore may have more orbits, but we will not be concerned with these larger elements right now). Small atoms therefore have

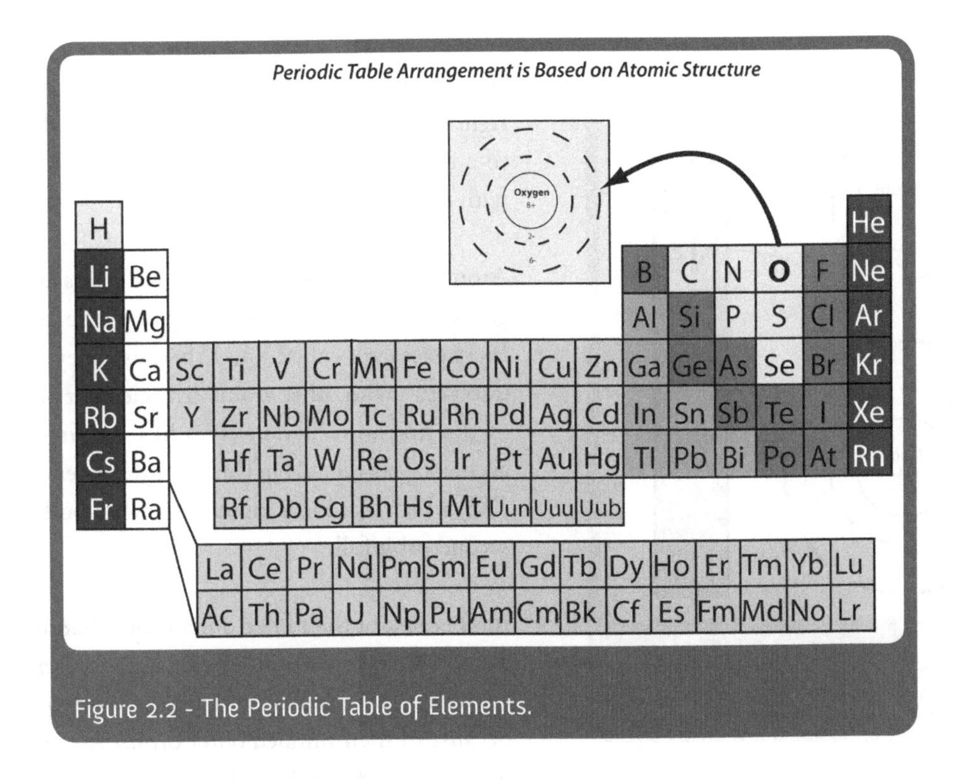

Figure 2.2 - The Periodic Table of Elements.

few orbits and few electrons, and large atoms have many orbits and many electrons. As the number of positive charges in the nucleus increases, electrons are added first to the inner orbits filling them up. However, depending on the number of positive charges in the nucleus, the outermost orbit of an atom may or may not be completely filled. How an element reacts with other elements to form molecules depends on the number of electrons present in its outermost orbit.

The relationship between the structure of atoms and the number of electrons in the orbits can be seen by looking at what is known as the periodic table (Fig. 2.2). In the periodic table, each different element is abbreviated as a letter or letters: for example, C stands for carbon, N for nitrogen, H for hydrogen.

The elements are arranged in the table based on the number of electrons they contain. The first row contains only two elements, hydrogen (H) and helium (He). This is because the innermost orbit can only contain either one or two electrons. H has one electron and is placed at the far left of the first row. He has two electrons and is placed on the far right. The second row of the table presents elements that have one or more electrons in the second orbit (remember, the second orbit can contain up to eight electrons). They are arranged from left to right in order of increasing numbers of electrons. Lithium (Li) contains a total of three electrons: two in the first orbit and one in the second. Since its second outermost orbit only contains one electron, it is placed at the far left, under H. Conversely, neon (Ne) contains a total of 10 electrons: of course two are in the innermost orbit, which always fills up first, and eight in the second orbit. Since the second orbit is filled, it is located on the far right of the table, under He. Therefore, if you look at the location of carbon (C), you will see that it has a total of six electrons, two in the first orbit and the remaining four in the second orbit.

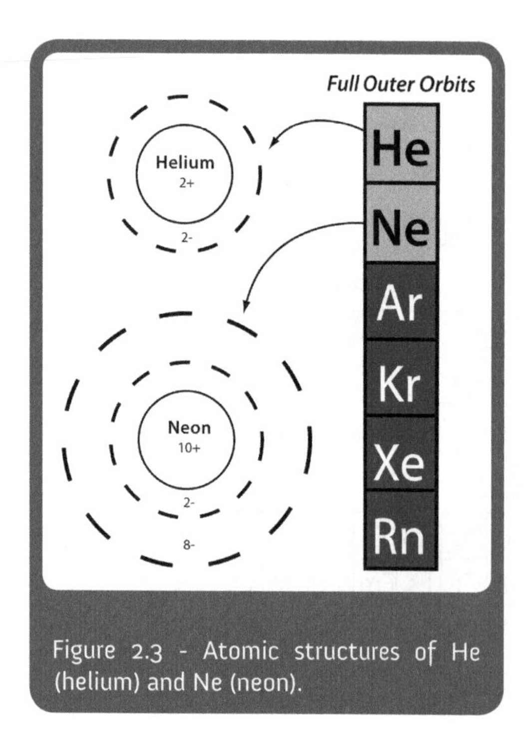

Figure 2.3 - Atomic structures of He (helium) and Ne (neon).

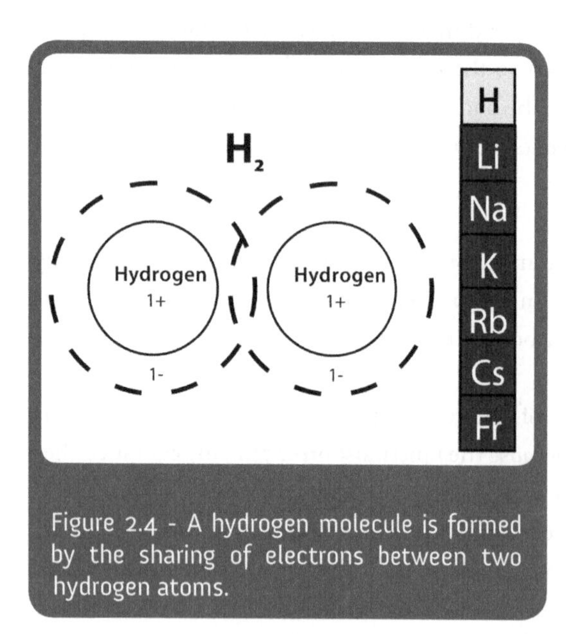

Figure 2.4 - A hydrogen molecule is formed by the sharing of electrons between two hydrogen atoms.

Actually there are only a few atoms that contain just the right number of positive charges in the nucleus to add up to a situation where their outermost orbit is fully filled with electrons.

Atoms with completely filled outer orbits are called inert gases and appear at the far right of the periodic table. Inert gases are called "inert" because they are singularly unreactive with other atoms. The element on the far right of the first row, helium (He) fills the first orbit and is an inert gas (Fig. 2.3). The second row ends with Neon (Ne) which like He is an inert gas with both the first and second orbit full.

It is because of their filled outer orbits that these elements do not interact with other elements to form molecules. However, most elements are reactive with other elements because of their unfilled outer orbits. Atoms "want" their outermost orbit filled. They will interact with other atoms, sharing electrons in order to achieve filled orbits. It is the interaction between the unfilled outside orbits that is the basis for constructing molecules out of atoms.

Some simple molecules: We have mentioned both hydrogen (H) and oxygen (O) as examples of two different elements. You are probably familiar with the word oxygen and hydrogen, and know that they are both gasses that are part of the air we breathe. However, the hydrogen and oxygen gasses contained in air are not free atoms: hydrogen gas is a molecule formed from two atoms of hydrogen. Likewise oxygen gas is a molecule made up of two oxygen atoms. "Free" hydrogen or oxygen atoms don't exist naturally, because both types of atoms have unfilled atomic orbits. Atoms will naturally interact with each other to form molecules because they "want" their outer orbits to be full.

Hydrogen is the smallest type of atom. In its small nucleus is one positive charge (+). Therefore, orbiting around the nucleus is one negatively charged electron (-). The combination of one positive charge and one negative charge is an electrically balanced atom. However, the hydrogen atom is reactive with other atoms

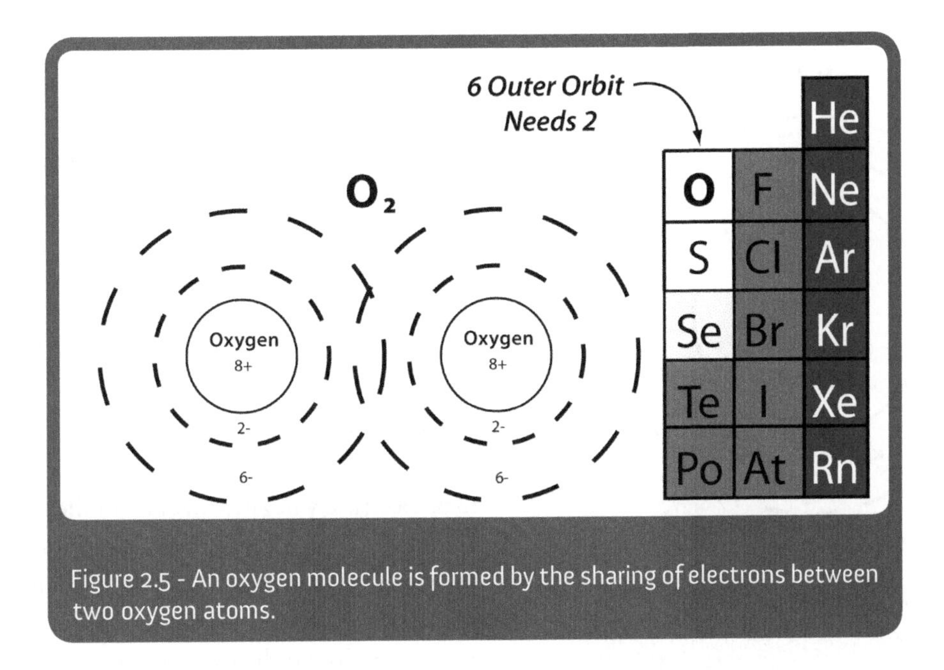

Figure 2.5 - An oxygen molecule is formed by the sharing of electrons between two oxygen atoms.

because it only has one electron in its first (and only) orbit. Remember, there is room for two electrons in this first orbit. When there are two electrons in this first basic orbit of an atom, that orbit is filled. A great deal of stability is gained by filling this orbit. In fact, the desire to fill this orbit is so strong that hydrogen is most stable when two of the hydrogen atoms get together to form a molecule (H_2) and share with each other their single electrons (Fig. 2.4). As a result of this sharing, half of the time each of the hydrogen nuclei has two electrons in its orbit.

Oxygen is a much larger atom than hydrogen. It contains eight positive charges in its much larger nucleus. These positive charges are balanced by eight electrons orbiting around it. Two of these electrons are filling up the first basic orbit. Therefore, the first orbit of oxygen is filled. The remaining six electrons of the oxygen atom are in a second orbit. Unlike the first orbit of atoms (which is filled when it contains only two electrons), the second orbit of an atom is only filled when it contains eight electrons. Because oxygen atoms only have six electrons in their second orbit, they are also reactive with other atoms. Like hydrogen, oxygen atoms form a molecule of O_2 based on the sharing of electrons. However, in this case, each atom of oxygen shares two electrons with the other oxygen atom (Fig. 2.5).

Water, another simple molecule: Water is a molecule composed of two hydrogen atoms and one oxygen atom. Although water is another simple molecule, water is quite different than molecules of either hydrogen or oxygen, because of the interaction of the atoms that make it up. The "marriage" between two hydrogen atoms and one oxygen atom is not an extremely strong relationship.

The reason for this lack of strength is that there is a large imbalance in size between the small nuclei of the two hydrogens and the larger nucleus of the one oxygen. What happens when you build a molecule out of two hydrogen atom and one oxygen atom? The answer is that like H_2 and O_2, the atoms of such a molecule share electrons. However, in the case of one oxygen atom and two hydrogens, it is not an equal sharing. The single electron of each of the hydrogen atoms spend far more time filling up the outer orbit of the oxygen atom than do electrons of the oxygen atom spend filling up the basic orbits of the two hydrogen atoms (Fig. 2.6).

For this reason, water is an internally electrically imbalanced molecule. The oxygen part of the molecule tends to be electrically negative because of the two extra electrons that spend most of their time around it. The two hydrogen parts of the molecule tend to be electrically positive because their electrons are spending much of their time around the oxygen. Thus, a water molecule has a negative and two positive poles.

Because of the electrical polarity, water molecules tend to associate with each other. Water molecules form a lattice-like structure. Water is a structured liquid. The negative oxygen pole of one water molecule is attracted to the positive hydrogen poles of other water molecules. The result is that the molecules of water are not entirely randomly moving in solution, but rather they form a constantly changing lattice of interacting water molecules (Fig. 2.7).

As one might guess, with all of these water molecules pulling on each other, every once in awhile a water molecule gets pulled apart. Water molecules get pulled apart in a very interesting way. The nucleus of one of the hydrogen atoms gets pulled away from the rest of the water molecule without its electron. The result is a positively charged hydrogen nucleus (H^+) free in solution, and a negatively charged oxygen-hydrogen pair (OH^-) (Fig. 2.8).

The free hydrogen nucleus is positively charged because the negative electron which balances the

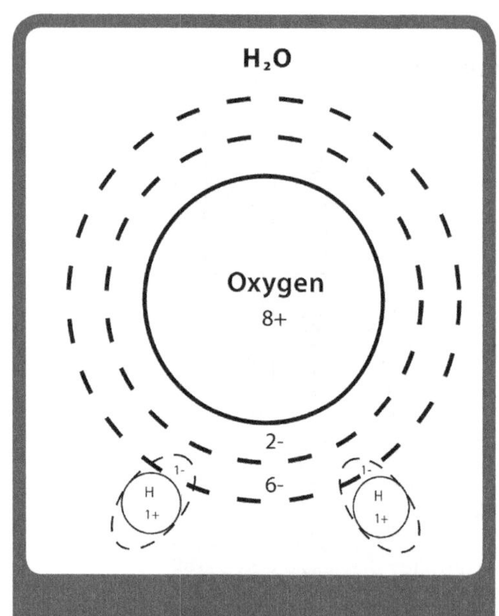

Figure 2.6 - A water molecule is formed by the sharing of electrons between one oxygen atom and two hydrogen atoms.

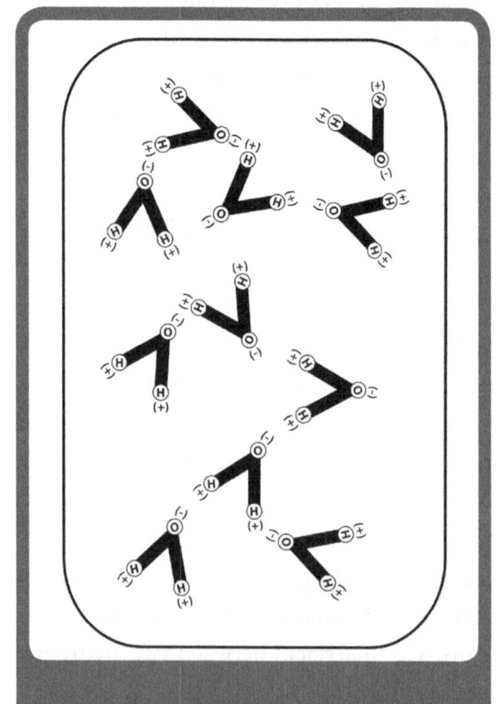

Figure 2.7 - Water is a structured liquid because of an unequal sharing of electrons between oxygen and hydrogen atoms.

nucleus is gone. Similarly, the oxygen-hydrogen pair is negatively charged because it contains one extra negative electron. The unit has nine positive charges in its two nuclei but 10 negative electrons in its joined orbits. The free hydrogen nucleus is called a hydrogen ion and oxygen-hydrogen pair is called a hydroxyl ion. Ions have charges.

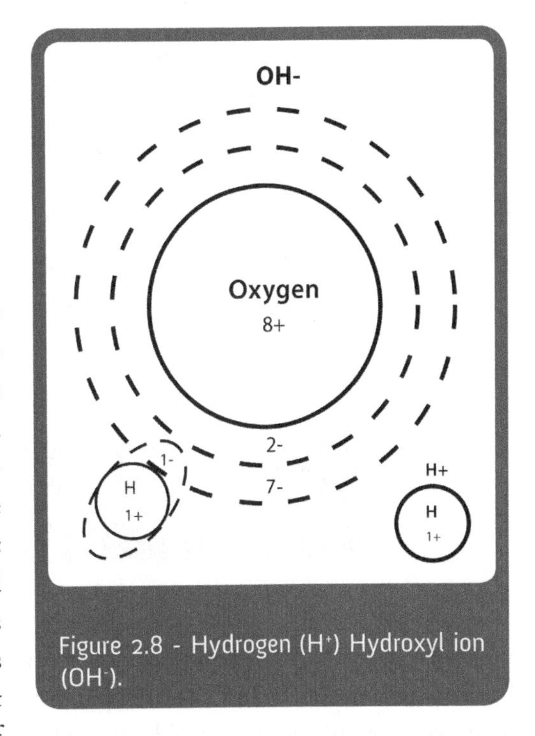

Figure 2.8 - Hydrogen (H^+) Hydroxyl ion (OH^-).

In a pure water situation, about one ten millionth of a mole of water molecules in every liter are pulled apart at any one time. In other words, every liter of water contains about one ten millionth of a mole of hydrogen ions. A mole is simply a number (2.22×10^{23}). Therefore, the hydrogen ion concentration is one ten millionth of a mole per liter. This could also be written in decimals as 0.0000001 moles/liter, or in scientific notation as 10^{-7} moles per liter. Because of their charge and small size, hydrogen ions are very reactive. Hydrogen ions have a tendency to interact with many other molecules. As a result, the hydrogen ion concentration is a very important characteristic of a water environment. It is a measurement of the reactivity of that water environment. In fact, the hydrogen ion concentration is used so often in describing water solutions that a special shorthand notation has been developed for it. This short-hand notation is called pH. pH is the negative log base ten ($-\log_{10}$) of the hydrogen ion concentration. For example, in pure water, which has a hydrogen ion concentration of 0.0000001 moles per liter, or 10^{-7} moles per liter, the pH is 7. If the hydrogen ion concentration of a water solution were 10^{-5} moles per liter then the pH would be 5. A water solution is considered acidic if it has a hydrogen ion concentration that is greater than pure water, and therefore a pH less than 7. Conversely, a solution is considered basic if it has a hydrogen ion concentration that is less than pure water, a pH greater than 7.

Differences in the pH of water environments are very important to biological function. Sometimes these differences in hydrogen ion concentration or pH are very large. For example, the inactivation of the digestive enzymes when they leave the acidic environment of the stomach and enter the small intestine involves a many thousandfold change in the hydrogen ion concentration of the water environment traveling through the digestive tract. On the other hand, sometimes important differences in hydrogen ion concentrations are very small. For example, the process by which oxygen is carried in the circulatory system is regulated by a change in the hydrogen ion concentration of blood of only a few hundred millionths of a mole. The blood in the veins has a small but very important difference in pH from blood in the arteries. Venous blood is slightly more acidic than arterial blood.

Acids: Water molecules are not the only source of hydrogen ions. A particularly important source of hydrogen ions is acids. Acids are molecules which are made up of one or more hydrogen atoms, and another atom or

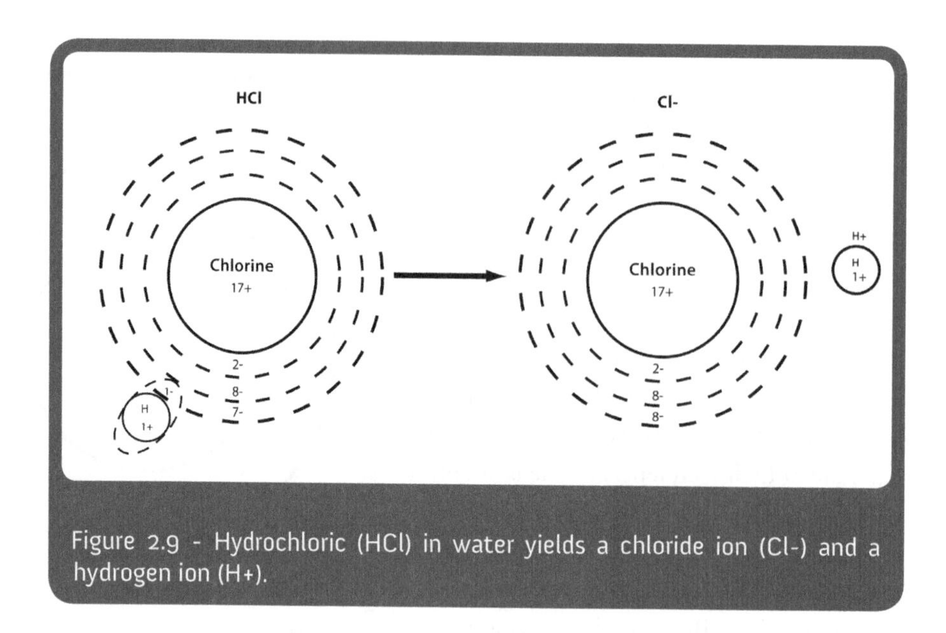

Figure 2.9 - Hydrochloric (HCl) in water yields a chloride ion (Cl-) and a hydrogen ion (H+).

molecular unit which has a great attraction for electrons. There are many types of acids, but they all share one behavior in common. When an acid is put in a water environment it has a great tendency to come apart. When it comes apart, like water, it yields the hydrogen nucleus without its electron—a hydrogen ion (H^+)—and a negatively charged ion which has kept the electron that belonged to the hydrogen atom. The difference between acids and water is that acids have a far greater tendency to come apart and liberate a hydrogen ion. Their ability to hold onto their hydrogen nucleus is far less, and therefore they are a much better source of hydrogen ions. For example, HCl (hydrochloric acid) is such a molecule. It is composed of one hydrogen atom (H), and one chlorine atom (Cl). Chlorine is an atom which is larger than oxygen and contains seven electrons in its outer orbit. Since the outer orbit is full when it contains eight electrons, chlorine is only looking for one more electron. Because of its size and because it is only looking for one more electron it is far more capable of taking an electron away from hydrogen than is oxygen. Because of this, a very large percentage of the HCl molecules come apart when they are put into water to yield electrically charged H^+ and Cl^- ions (Fig. 2.9).

HCl is, in fact, the source of the very large hydrogen ion concentration which gives the contents of the stomach such a low pH. Digestive enzymes of the stomach are activated by the high hydrogen ion concentrations. Cells of the lining of the stomach produce and release HCl into the contents of the stomach. The HCl comes apart, yielding large amounts of H^+ ions which in turn activate the digestive enzymes of the stomach. Another example of the biological role of acids involves respiration. When blood from the lungs reaches the tissues, it is carrying oxygen bound to a large molecule in the red blood cell known as hemoglobin. The oxygen must be released from the hemoglobin so that the cells can use it. The ability of hemoglobin to bind oxygen is highly dependent on the pH of the blood. If you raise the hydrogen ion concentration of the blood just a little bit, then the ability of hemoglobin to bind oxygen is greatly reduced. The change in the H^+ concentration involves the carbon dioxide (CO_2) that leaves the tissue cells and enters the blood. It then enters the red blood cell where there is an enzyme that makes the molecule carbonic acid (H_2CO_3) out of carbon dioxide (CO_2) and

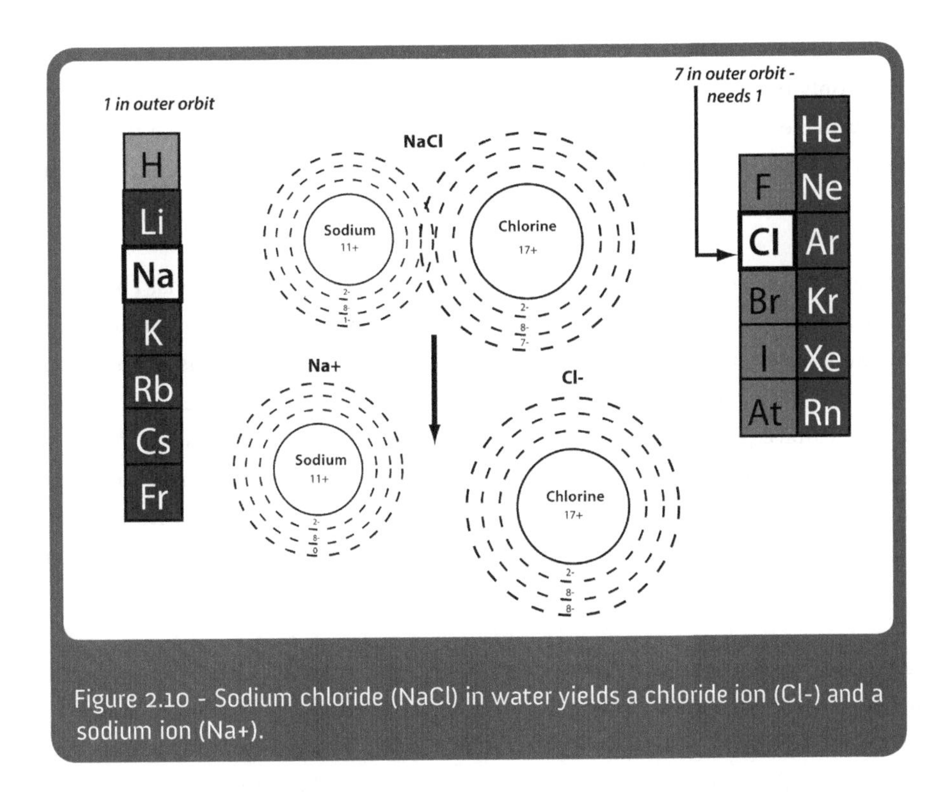

Figure 2.10 - Sodium chloride (NaCl) in water yields a chloride ion (Cl-) and a sodium ion (Na+).

water (H_2O). Some of the carbonic acid molecules come apart releasing H^+ and HCO^{3-}. This results in a very small increase in the concentration of hydrogen ions in the water environment of the hemoglobin molecule.

The small increase in the hydrogen ion concentration causes oxygen to come off of the hemoglobin molecule. The oxygen can then leave the blood and enter the tissue cells. These are just two of the many examples of the regulatory role of hydrogen ions.

Salts add more ions to the mix: Hydrogen ions are not the only ions which are important to biological function. Many other positive and negative ions exist and play important roles in life. A primary source of these ions is salts. A salt is a molecule very much like an acid. It is made up of an atom like hydrogen which easily loses an electron in water, and an atom such as chlorine which easily takes an electron in water. Sodium chloride (NaCl)—table salt—is such a molecule. Sodium (Na), like hydrogen (H), has only one electron in its outer orbit. The difference between hydrogen and sodium is that hydrogen has only one orbit with one electron. In contrast, sodium has an inner orbit with two electrons, eight electrons in a second orbit, and a third outer orbit with only one electron. Like hydrogen it has a very difficult time holding onto this one electron. Chlorine, as you will recall, has seven electrons in its outer orbit and is only looking for one more. NaCl is a very stable molecule until it enters the water environment of charge. When NaCl is put into the charged environment of water it comes apart. The sodium leaves without its electron yielding a sodium ion (Na^+), and the chlorine keeps the electron yielding a chloride ion (Cl^-) (Fig. 2.10).

Sodium chloride is only one example of a salt. There are many salts, all of which ionize in water. There are many other atoms which exist and function as ions in the body. They all have the same characteristic: they lose or gain electrons in water. For example calcium, iron, potassium, magnesium and iodine all exist and function as ions.

You probably noticed that these atoms are called minerals on the labels of many foods. There are daily requirements for these ions. They are all nutrients which are important for health. The important characteristic that they all have in common is that they are electrically charged. They all are either missing one or more electrons and are positively charged, or have one or more extra electrons and are negatively charged. Why is charge important? Some ions have positive charges (+) and some have negative charges (-). Water molecules have a negative pole around the oxygen and positive poles around the hydrogens. There is an attraction between positive ions and the negative pole of water. There is an attraction between negative ions and the positive poles of water. There are attractions between positive ions and negative ions. All of these attractions promote reactions between molecules.

Organic molecules—molecules built of carbon: Although water and ions are the medium of life, it is organic molecules that are the basis of life. Organic molecules are relatively stable molecular units constructed of carbon (C), with additional hydrogen (H), oxygen (O) and sometimes nitrogen (N) atoms. In addition, some important organic molecules contain other atoms such as iron (Fe), phosphorus (P), iodine (I) or sulfur (S). Carbon atoms have six electrons, two in their inner orbit and four in their second orbit.

As a result, they will share four electrons with other elements. For the most part, the sharing of electrons by carbon and another atom (for example with oxygen, hydrogen, or another carbon) is not an equal sharing, but is not as extraordinarily unequal as a salt. Therefore they do not tend to "come apart" in water (we call this type of sharing a "covalent bond," as opposed to an "ionic bond" such as forms between sodium and chlorine). Organic molecules are stable in a water environment. Organic molecules can be grouped into several general categories based on their structure and behavior.

Each category has its own unique role to play in life. Fig. 2.11 provides some examples of "small" organic molecules (although larger than water, they are still relatively small). In living organisms, these small organic molecules are often linked together into much larger molecules called "macromolecules." Sugars are small organic molecules that often form a ring structure. Glucose is an example of a sugar molecule. Many individual sugar molecules linked together form a type of macromolecule called a carbohydrate. Another important class of small organic molecules is called amino acids. Amino acids all contain nitrogen (N) as well as C, H, and O. Individual amino acids are joined together to form a type of macromolecule called a protein. Likewise, smaller lipid molecules are joined together to form larger lipids. A slightly more complicated type of small molecule in living cells are called nucleotides: these smaller units are linked together into longer chains called nucleic acids (DNA and RNA).

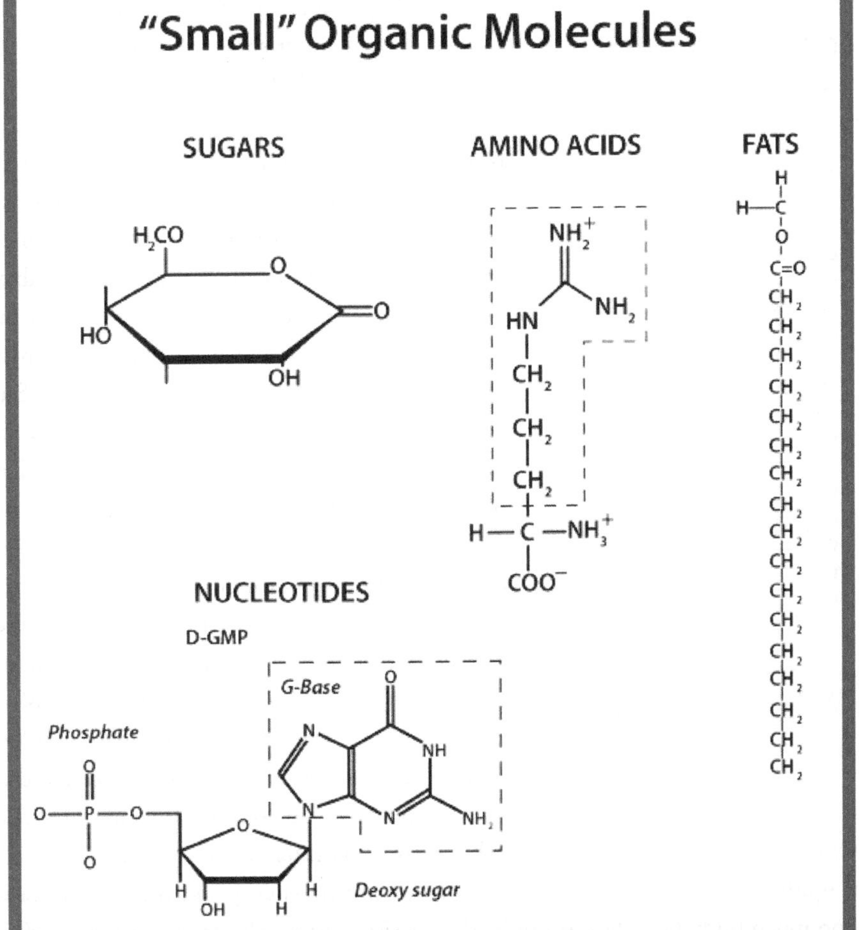

"Small" Organic Molecules

Figure 2.11

As we said, organic molecules are really the basis of life. Everything that happens in a living thing happens because of the molecules that are present. The bonds of organic molecules provide the basic energy of life. Organic molecules provide the genetic information on which an organism is assembled and run. Organic molecules also provide the compartmental structure of life.

Lipids: the antithesis of water: Fats and oils are types of lipids. The polar charged environment of water and ions contains many molecules. However, molecules that exist in the water environment are based on unequal relationships between atoms which are much different in character. Like water, the relationship between unequal atoms results in charge or polarity.

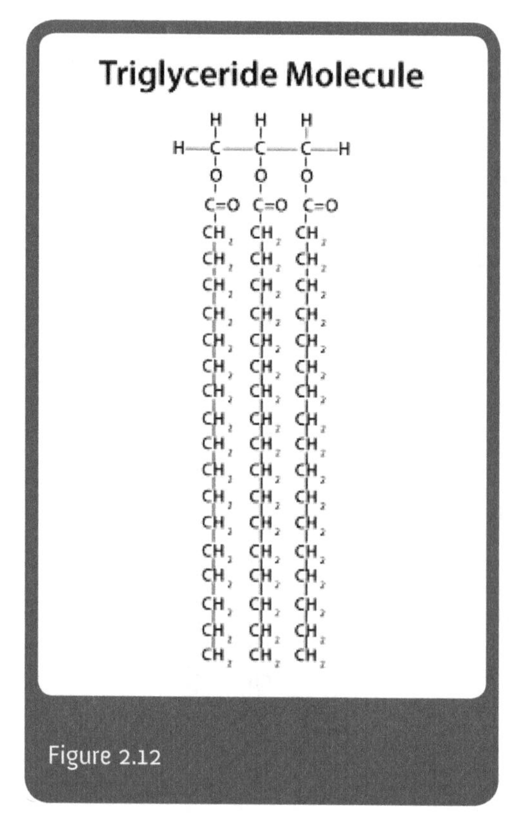

Figure 2.12

A very important atom for building lipids is carbon. Remember, the outer shell of carbon contains four electrons, so it is exactly half full. Because the outer shell is exactly half full carbon generally participates in near equal sharing relationships with other atoms. When atoms form bonds by equal sharing of electrons, this bond is much stronger and less likely to come apart in the water environment. In contrast to molecules like water, lipids are built on stable relationships between equal or near equal atoms. Because of this no part of the molecule is able to pull electrons away from any other part, and the lipid is non-polar and non-charged. Such a molecule has no basis for a relationship with either water or charged molecules and ions. For example, if you build a molecule out of a string of carbon atoms, you create a molecule constructed of equals. No part of the molecule is much stronger than any other part and the molecule does not have the type of unequal distribution of electrons that occurs in water molecules. Such a "balanced" molecule also does not have the polar character that comes from the unequal distribution of electrons (Fig. 2.12). An important consequence of this difference between non-polar lipid environments and the charged water environments is that they do not mix. Oil and water don't mix. Lipids serve many functions in the body. Body fat, which we are all familiar with, is made up of cells that store a large amount of lipids. Lipids are both an efficient way of storing energy and provide insulation that keeps us from losing too much body heat. For example, the fat tissue that collects around the surface of the body provides insulation as well as a source of stored organizational energy in the form of carbon bonds. This is a very efficient way of storing energy because it does not involve water. The fact that water is not involved is important for two reasons. First, the absence of water means that the stored energy is lightweight. Sugar is the other major type of molecule that stores organizational energy. However, sugar does interact with water. If the same amount of energy was stored as sugar, it would require six volumes of water which would greatly increase the weight of stored energy. Practically, this would greatly limit the amount of stored energy an animal could carry around. Second, lipids are very poor in conducting heat while water is very efficient in conducting heat. Therefore, the stored fat without the water serves as insulation for the body.

Lipids are also involved in constructing barriers (called membranes) between water compartments. These membranes both separate and define a cell (the cell membrane), as well as separate and define other compartments within a cell (such as the nucleus of a cell). Their role in the construction of water compartments is to provide a barrier to charged ions and polar molecules. Charged ions and polar molecules cannot move across the environment of "no charge." Lipids are a major part of the cell membrane, which encloses and defines the water compartment we call a cell. Within a cell are other smaller water compartments also formed by

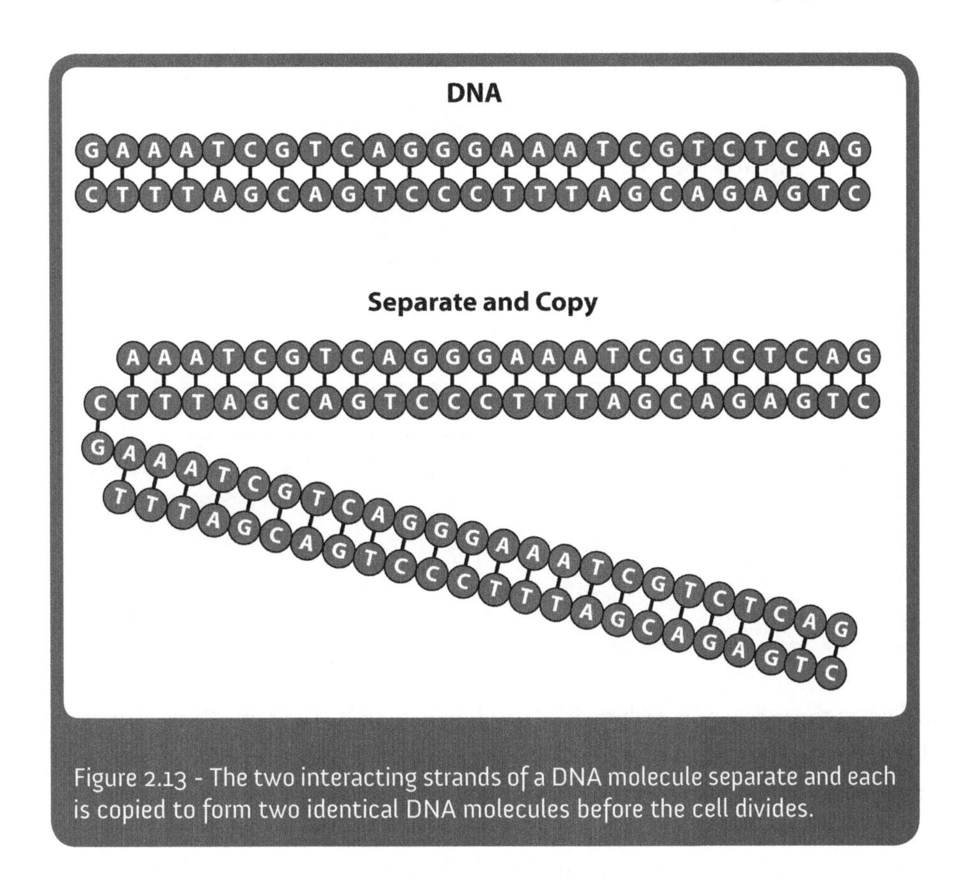

Figure 2.13 - The two interacting strands of a DNA molecule separate and each is copied to form two identical DNA molecules before the cell divides.

lipid membranes. One of the most important water compartments in a living human cell is the nucleus. The nucleus of a cell contains the genetic material, the directions for life.

Genes, the directions for making and running a living organism: An entire individual forms, develops, grows, lives and even dies based on the information contained in the first cell that is produced at conception. This information is contained in a kind of nucleic acid called deoxyribonucleic acid, or DNA for short. All information necessary for directing these processes is written in the language of deoxyribonucleotides within the DNA molecule. (Deoxyribonucleotides are 1 of 2 different kinds of nucleotides: the other type is called ribonucleotides) Nucleotides.are small modified sugar units constructed of five different atoms: carbon (C), hydrogen (H), oxygen (O), nitrogen (N), and phosphate (P). There are four different deoxyribonucleotides (which we abbreviate as A, G, C, and T). These four deoxyribonucleotides serve as an "alphabet." The four letters of the deoxyribonucleotide alphabet are linked together in patterned sequences to construct deoxyribonucleic acids (DNA) (Fig. 2.13).

The DNA in a cell, its genetic information, is simply patterned strings of millions of deoxyribonucleotides bonded together. A single, large DNA molecule is called a chromosome. A chromosome contains many individual genes, or "words" spelled out by the deoxyribonucleotide alphabet. The information is contained in the sequence of the four letters of the deoxyribonucleotide alphabet. When the first cell that started the individual divided to produce two cells, all of the DNA sequences were copied, and each new cell got a copy. The same

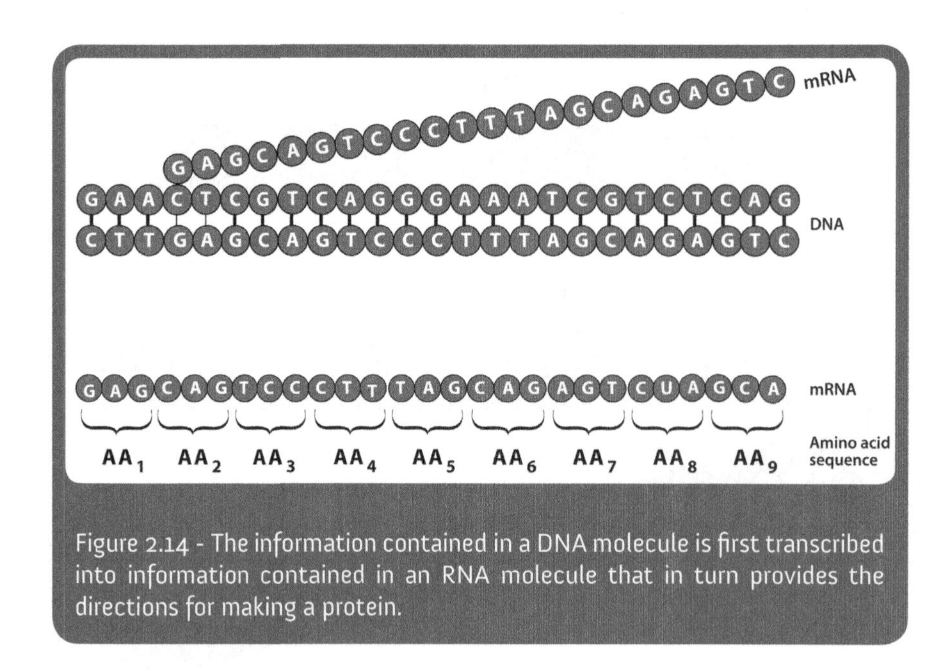

Figure 2.14 - The information contained in a DNA molecule is first transcribed into information contained in an RNA molecule that in turn provides the directions for making a protein.

thing happened when those two cells each divided yielding four cells. The same thing also occurs in the adult when new cells are formed by cell division. Each new cell, whether it be a skin cell or a liver cell, is provided with a copy of the entire genetic information of the individual.

There are two basic types of information contained in the large DNA sequences referred to as chromosomes. One type of sequence describes how to make a particular protein. Proteins are large molecules constructed of a different molecular alphabet, the alphabet of amino acids. Each three nucleotides in the DNA sequence codes for one amino acid in the sequence of a protein. The second type of information coded for in the DNA sequences is necessary for controlling when a particular protein or group of proteins is made. For example, both a liver cell and a skin cell of an individual contain the exact same DNA sequences. However, because of control, the liver cell is making those proteins necessary for liver cell function and the skin cell is making those proteins necessary for skin cell function. No type of cell makes (expresses) all of the proteins coded in the genetic material.

Reading the genes: A gene only express its influence on the functioning of a cell when a protein is made from it. In order to make a protein, the DNA sequence coding for the amino acid sequence of the protein (a gene) must be first copied into mRNA. This is because the DNA molecule is very large and cannot leave the nucleus of the cell, and each chromosome contains many genes. So the sequence in a particular gene (or part of a chromosome) is copied and leaves the nucleus to the site in the cell where proteins are made. The copying process involves another molecular alphabet, which is very similar to the deoxyribonucleotide alphabet. This second alphabet is constructed of ribonucleotides. A major difference between deoxy-ribonucleotides and ribonucleotides is that the sugar part of the molecule is a little bit different. Like the DNA alphabet, the alphabet of ribonucleotides contains four molecular letters. For the purposes of copying a DNA sequence, each ribonucleotide has a corresponding deoxyribonucleotide to which it pairs. The first step in making a protein is to construct a ribonucleotide sequence that corresponds to the DNA sequence that describes the protein (Fig 2.14).

Such sequences of bonded ribonucleotides are called ribonucleic acid (RNA). An RNA sequence that codes for a particular protein is called a messenger RNA (mRNA). The mRNA sequences are messages sent out of the nucleus to other parts of the cell directing the construction of the proteins coded in the DNA sequence. The entire process coded in the DNA simply describes when and how to make proteins. The directions for making the proteins of the body are all coded for in the DNA, with three deoxyribonucleotides (and therefore three corresponding ribonucleotides) representing one amino acid.

Proteins, the capacity to do things: Proteins are sequences of amino acids bonded together (Fig. 2.15).

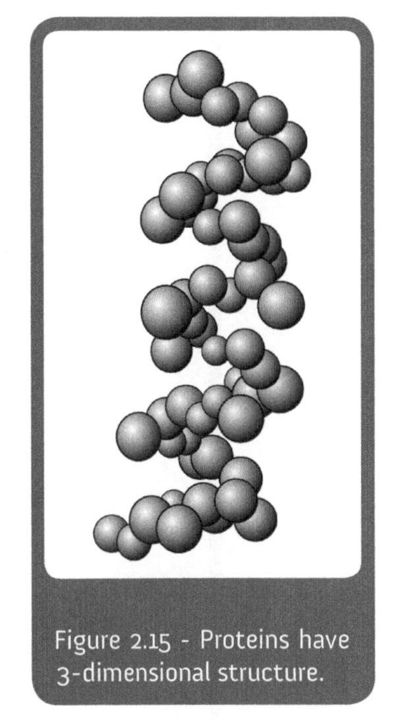

Figure 2.15 - Proteins have 3-dimensional structure.

For example, a DNA sequence 300 nucleotides long would code for a protein sequence 100 amino acids long.

Amino acids, which are constructed of the atoms carbon (C), hydrogen (H), oxygen (O), nitrogen (N) and sometimes sulfur (S), form the molecular alphabet from which proteins are built. The alphabet of amino acids consists of 20 basic molecular characters: there are 20 different amino acids. Each amino acid is represented in the DNA by a specific combination of three deoxyribonucleotides. When a DNA sequence is expressed by a cell the result is a protein. The presence of that protein allows the cell to carry out some function. For example, all enzymes are proteins. Enzymes are molecules that catalyze, or cause a reaction to go at a rate fast enough to be useful. There are many different enzymes. Usually a specific enzyme will cause one specific reaction to occur.

There are a number of different enzymes involved in making proteins. There are a number of different enzymes involved in making DNA and RNA. Different enzymes control the release of energy from sugar bonds. Specific enzymes coded for in the DNA cause the myriad of life's reactions to occur. Besides functioning enzymes, specific proteins carry out many other functions. Hemoglobin is the protein that carries oxygen in the blood. Insulin and many other hormones, or chemical messengers, are proteins. Muscle strength is a construction of proteins. Antibodies that protect the body from disease are proteins. There are even proteins in the surface membrane of every cell that control the entry and exit of charged molecules, ions and water. All that every cell is and does is a function of the particular proteins that it makes.

Potential energy: The last factor necessary for life is energy. Life runs on potential energy. Potential energy is stored energy. The "potential" is to release the energy. There are many types of potential energy. For example, when one uses energy to pull the chain and raise the weights of a cuckoo clock some of the energy used to

Figure 2.16 - The energy flow of life: Plants take in carbon dioxide and water, and use energy from the sun to make carbohydrates. Animals take in carbohydrates and oxygen and release carbon dioxide and water when those carbohydrates are broken down to obtain energy. It's a cycle!

raise the weights is stored as potential energy. Because of gravity the weights are pulled down towards the floor. The mechanism of the clock couples the release of the stored potential energy as the weights return to the floor with the moving of the hands of the clock. When the weights have returned to the floor, all of the stored potential energy has been used and the clock stops. Potential can also be quite ephemeral. If one puts a single drop of food coloring into a glass of water, the concentrated drop of food coloring in the glass of water contains potential energy. The potential is released as the food coloring distributes equally throughout the water. The potential energy stored in the unequal distributions of molecules and ions is a major driving force of life. Potential energy is also stored in the bonds of molecules. Plants use the radiation energy from the sun to build sugar molecules in a process called photosynthesis (Fig. 2.16). Sugars are relatively small molecules constructed of C, H and O. Some of the energy from the sun is stored in the bonds of the sugar molecule. This energy can be used to build the molecules of the trunk of the tree. Energy is now stored in the molecules of the wood. When the wood is burned the energy is released as heat and light. Life runs on the same type of energy. Therefore, the source of biological organizational energy is the sun. Plants take-up carbon dioxide (CO_2) and water (H_20) and use the energy in sunlight to make molecules of sugar.

(energy in)

$$CO_2 + H_2O \rightarrow (C.H_2O) + O_2$$

Plants, through the process of photosynthesis, can use the radiation energy of sunlight to forge bonds between atoms. The energy is stored as potential energy in the bonds. Such potential energy remains in the bonds until oxygen is used to release the energy.

(energy out)

$$(C.H2O) + O2 \rightarrow CO2 + H2O$$

Release of this stored energy in the presence of oxygen can be explosive as it is with fire, or it can be a highly controlled exchange of energy among bonds between atoms as it is in living things

All of life runs on energy from the sun originally stored in the bonds of sugar molecules. Each molecule of sugar stores a small amount of energy. However, the amount of energy in a single sugar molecule is not really enough to do anything. It is not enough to make a bond between two amino acids in a protein. It is not enough to move a muscle. In order to use the energy in a sugar molecule it must first be condensed. Cells take the little bit of energy contained in the bonds of sugar molecules and synthesize molecules with high energy bonds. They build high energy bonds onto nucleotides. The most well-known of these molecules is called adenosine triphosphate (ATP). There is a great deal of energy in the bonds between the last two phosphates (P) of an ATP molecule. In these bonds there is enough energy to run life.

Sugars are also the basic molecular building blocks of carbohydrates. These basic building blocks of sugar can be bonded together to make large macromolecular polymers (Fig. 2.17). These polymers are branching strings of sugar molecules bonded together.

For example, sugar is stored for the body's use in the liver as a large macromolecule called glycogen. When there is an excess of sugar in the blood following eating, hormone signals direct the cells of the liver to take up sugar and construct large glycogen sugar polymers. In this way the excess sugar is stored in the liver until such a time when it is needed by the body. When the sugar is needed other hormone signals direct the cells of the liver to break down the glycogen and release the resulting sugar back into the blood.

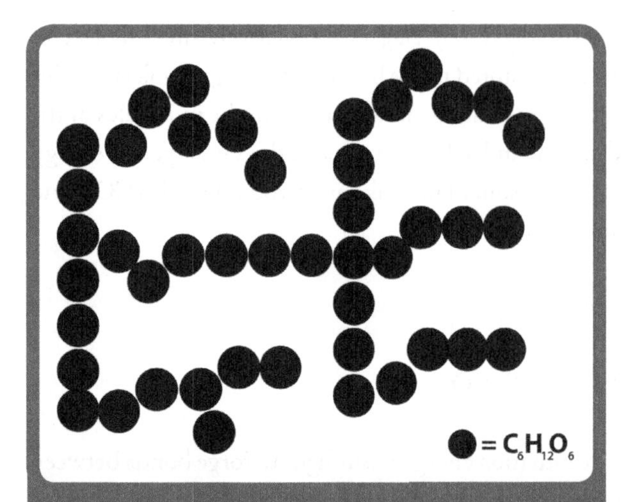

Figure 2.17 - Glycogen is a sugar polymer; it is formed by linking together many small sugar molecules.

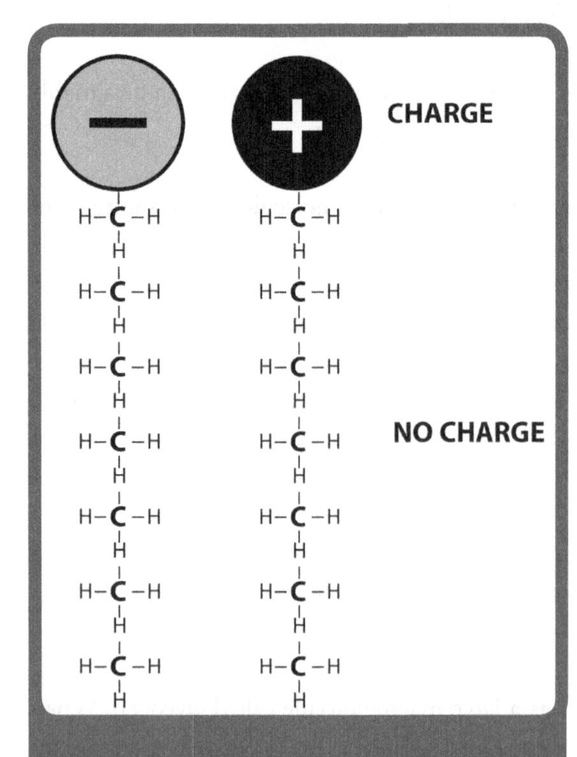

Figure 2.18 - Phospholipids have a charged region and an uncharged region.

Life exists as compartments of water: Compartmentalization is necessary in order to separate the living water environment from the nonliving water environment.

It is also necessary for building a big living organism like a human. The necessity derives from a surface-to-volume problem. Adaptability (change) requires that there be an exchange of molecules and ions with the environment. The size of any compartmentalized volume will limit the rate of exchange with the environment around it. This occurs because of simple geometry. For example, consider a living organism in the shape of a sphere. As the radius of the sphere increases its volume increases as the cube of the radius ($V = 4/3\ \pi r^3$), but the surface increases only as the square of the radius ($A = 4/3\ \pi r^2$). The result is that as the sphere gets larger and larger, the amount of surface relative to the amount of volume gets smaller and smaller. Since exchange with the environment occurs at the surface, as the sphere gets larger exchange with the outside world cannot occur fast enough to take care of the center of the compartment. The center would therefore become a "stagnant pool" if the size is too large. The answer to this problem is to build small compartments surrounded by water barriers. The barrier must be built out of molecules that do not readily mix with the water in order to provide a degree of stability to the barrier. Such charge barriers, called membranes, are constructed of special hybrid molecules called phospholipids (Fig. 2.18).

As the name implies this is a molecule with two parts. This molecule is made by bonding a molecule with charged phosphates to a lipid molecule. When a group of such molecules are put into water they have a problem: part of the molecule wants to be with the charged water environment and part of it does not. The solution is for them to stay by themselves.

They form a structure in which they put their charged parts together facing out towards the water and their uncharged parts together creating a small oily non-polar environment encapsulated in charge. This is the molecular behavior on which biological barriers to charge (membranes) are made. Proteins are then used to construct channels in these charged phospholipid barriers (Fig. 2.19). Given these relatively stable charged barriers containing protein channels, organizational energy can be used to move molecules and ions between compartments, creating compartments with different compositions. Compartments with different compositions store potential energy. A battery is such a device (Fig. 2.20).

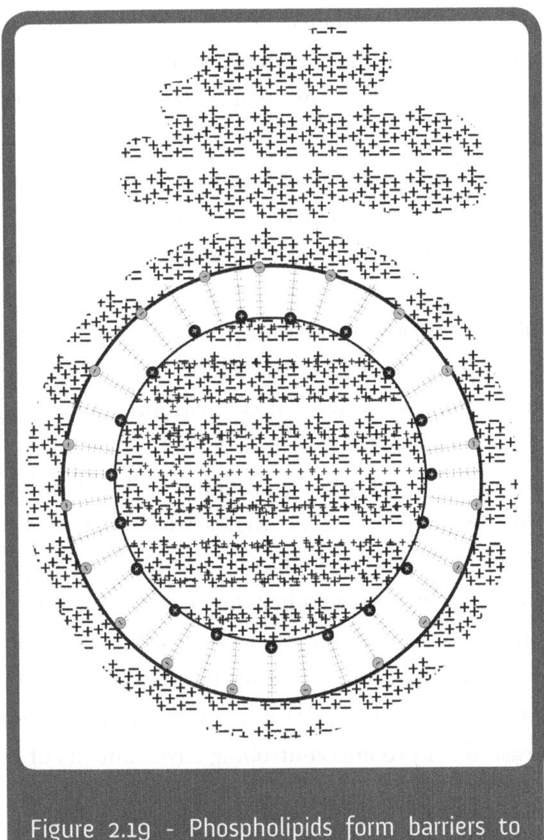

Figure 2.19 - Phospholipids form barriers to charge.

The potential is for compartments side-by-side with a membrane in between to become the same. One very important potential across a membrane is an electrical potential.

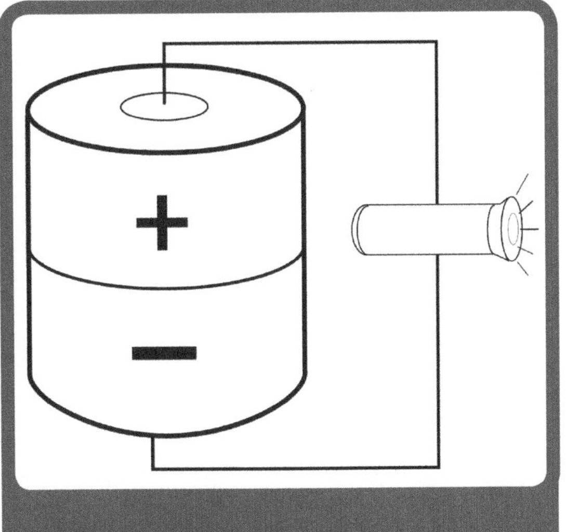

Figure 2.20 - An electrical potential can be used to do work.

Electrical potentials between water environments across membranes play a fundamental role in biological function. For example, such electrical potentials are the basis of the flow of information within the brain and from the brain to the rest of the body. An electrical potential is simply an unequal distribution of charges. If you take a solution containing many positive and negative ions and move all of the positive ions to one place and all of the negative ions to another, there is an electrical potential between the two places. We call the potential "electrical" because it is based on charges. We call it a "potential" because there is the potential for all of the ions to flow back together and be evenly distributed in the solution. Separating and concentrating charges like this also represents potential energy. One must use energy to move the charges apart. This energy is used against the natural force for everything to be equally distributed. What we have done is to use energy to structure the environment. The energy is potential as long as the environment remains structured. This potential is released when the ions flow back together and the environment is no longer structured. The release of this energy as the ions flow back together can be used to do work. It can also be measured and described in terms of a voltage. Voltage is simply a unit of electrical pressure. Living organisms use the energy stored in the structure of molecules to create and use such electrical pressure to carry out many functions. The presence of the membrane, the charge barrier, in between the water environment provides some stability. The protein channels in the charge barriers allow for change.

Life—using and controlling proteins: Without proteins, a cell does not have the capability to do anything. It is proteins that carry out all life functions. One of the most important characteristics of proteins is that they are very sensitive to their environment. Proteins are large molecules which contain many charges as well as areas with no charges based on the type and order of the amino acid units that makes it up. These long sequences of stably bonded amino acids coil and roll up and in doing so form three-dimensional structures. How they coil and roll up, that is, what shape they form, also depends upon their environment. The ions in an environment play an important role in controlling the shape of proteins. When they coil and roll up one way they are active, and in other ways they are inactive or their activity is modified. This is why the high hydrogen ion environment of the stomach activates the digestive enzymes of the stomach and why the low hydrogen ion environment of the small intestine inactivates them. This is why small changes in the hydrogen ion concentration of the blood greatly influence the ability of hemoglobin to carry oxygen. The entire process of life is really a cascade of proteins controlling environments of other proteins which in turn control other environments.

chapter three

similarities and differences between cells living in the society

Cells—the fundamental units of life: As we discussed in earlier chapters, the body is made up of individual cells living and working together in the community of the body. So what exactly is a cell? An individual cell is very small—it can't be seen without a microscope. A cell is considered as the smallest unit of life, because it is capable of doing the things of life. It interacts with its environment. It takes up useful molecules and gets rid of waste molecules. It breaks down larger molecules to produce usable energy in the form of ATP molecules. It can use the energy stored in the bonds of ATP to synthesize other large molecules. However, because life requires change and adaptation and because big molecules wear out, big molecules are constantly being broken down into small molecules and the pieces used to make big molecules again, a process called "turnover" (Fig. 3-1). Under defined circumstances a cell can also reproduce, or make new copies of itself. All of these things are what defines "living"—and if maintained in the right environmental conditions, human cells can continue living outside the human body.

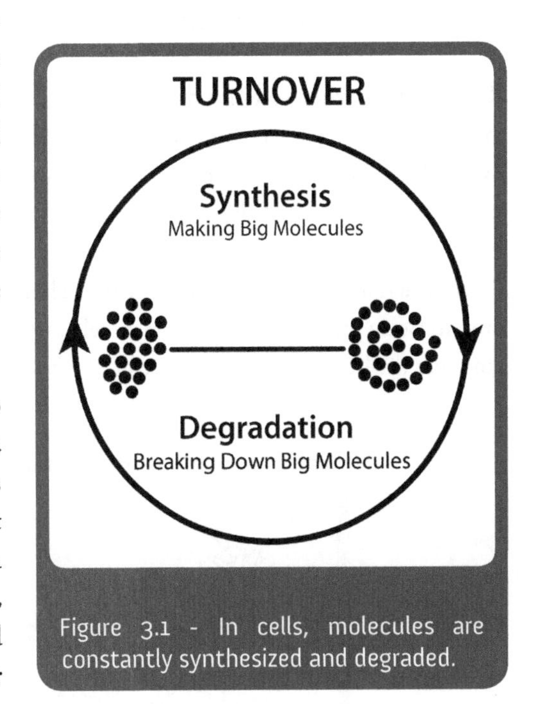

Figure 3.1 - In cells, molecules are constantly synthesized and degraded.

As we discussed in the last chapter, living things are no different from everything else: they are made up of matter. However, a cell is not just a little bag of molecules—cells are highly organized structures. So let's look briefly at what a "typical" cell looks like (Fig 3.2). The outer boundary of a cell is called the cell membrane. As discussed in Chapter 2, this membrane acts as a barrier that both defines the cell and separates the water environment inside the cell from the water

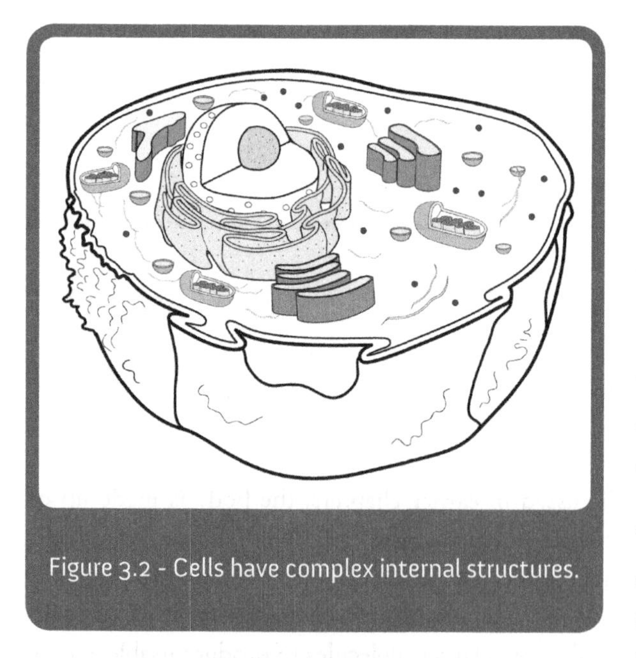

Figure 3.2 - Cells have complex internal structures.

environment outside the cell. The cell membrane is made up of phospholipids molecules with proteins dispersed throughout it. The membrane of the cell is actually a "double layer," or lipid bilayer. Inside the cell is another separate compartment, called the cell nucleus. The nucleus is separated from the rest of the cell by a lipid membrane of its own called the nuclear membrane.

The nucleus contains the genetic material of the cell—the chromosomes (DNA molecules and associated proteins) reside in the cell's nucleus.

There are also many other structures within the cell that can be seen if one looks with very powerful microscopes. These internal structures are called organelles ("little organs"), and different processes within the cell occur in these different structures. For example, mitochondria are organelles where most of the reactions that break down molecules to obtain high energy molecules (eg. ATP) takes place.

Ribosomes are the protein synthesis machines of the cell. There are many other structures involved in processing the proteins that the cell makes. There are other structures involved in breaking down and removing waste molecules. The watery environment within the cell membrane and outside all of these internal structures is called the cytoplasm of the cell. Many reactions occur in the cytoplasm as well.

There are many different types of cells: There are many (about 200) different kinds of cells that make up the human body. (In addition to human cells, there are "outlaws" such as bacteria and fungi that invade and live amongst the human cells in the society.) However, none of these cells really look like the "typical" cell shown in

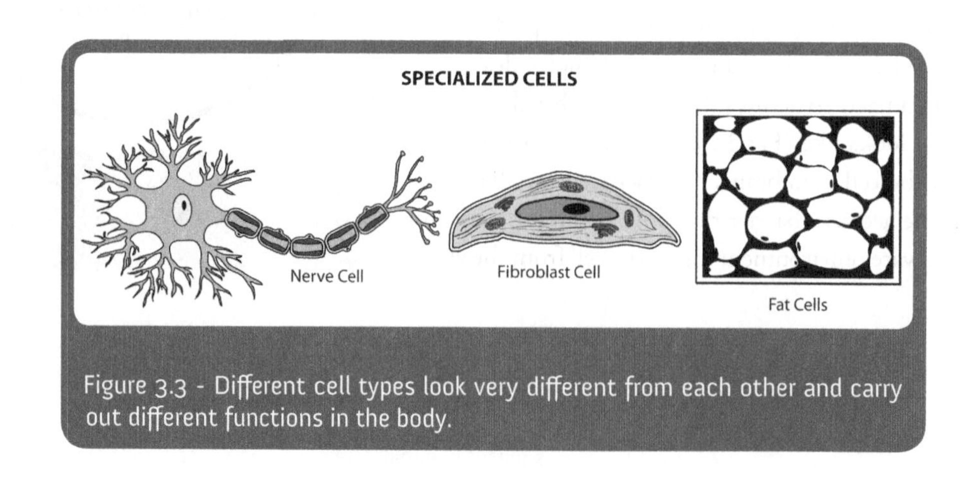

Figure 3.3 - Different cell types look very different from each other and carry out different functions in the body.

Fig 3.2. Cells have different shapes, different amounts or types of internal organelles, and do different things in the human body. We say that cells become "differentiated" or "specialized": they become different from each other and capable of carrying out special tasks within the society of cells that is the human body. Some examples of how different some cells of the body look from our "typical" cell can be seen in Fig. 3.3.

All of the different types of cells (whether human or foreign "invaders") are different from each other based on the proteins that they make. Remember, different cells of an individual's body all have the same DNA molecules, which contain the instructions for making all of the proteins in any cell type. However, in a particular type of cell, only some of these instructions are actually "read" and turned into proteins. It is the cell specific proteins (the proteins made by that cell type) that allow a cell to "be what it is" and do what it does in the body.

Of course, these different kinds of cells are not just free floating in a water environment. Various groups of specialized cells are organized together into higher level structures (tissues, organs, and organ systems). There are tissues and organs designed to take in and process food, to take in oxygen and excrete carbon dioxide, to process other waste products from the body, to protect the body from invaders, to move materials around the body, to reproduce new individuals, to control both the internal environment and interactions with the external world. Although these different types of cells are specialized (different from each other) there are many things they have in common. However, it is the differences between cell types—whether human or foreign invaders—that is very important to both health and treating disease. The differences provide the means for selectively modifying the behavior of particular cells.

Energy: A common function of all cells: Although there are many different cell types in the body specialized for carrying out specific jobs in the body, there are many common functions that all cells carry out. Some of the most common functions shared by cells are those associated with the production of useful high energy molecules (ATP). Energy production, like all functions of the cell, involves proteins. Proteins are the tools for accomplishing the task. Molecules such as sugar have potential energy stored in the bonds between their atoms. However the amount of energy in the sugar bonds is not sufficient to drive chemical reactions. Molecules such as sugar enter the environment of a cell and are taken into the cell through protein channels in the cell membrane. Some of the proteins of the cell then act on the bonds of the sugar to release the energy and condense this energy into very high energy bonds such as those in ATP.

The production of ATP from a molecule such as sugar can be accomplished in two basic ways. One way does not involve the use of oxygen while the other way does. However, only a small fraction of the total energy in a sugar molecule can be obtained by the cell without using oxygen. When oxygen is used, all of the energy is extracted and the molecule is reduced to the waste products of carbon dioxide (CO_2) and water (H_2O).

Although all cells of the body can concentrate energy from molecules with or without oxygen, there are great differences between cells in their capacity for and in their dependence on the oxygen requiring part of the process. For example, some muscle cells have a very large capacity to use the non-oxygen requiring process and a relatively low dependence on the oxygen requiring part of the process. Muscles which use metabolic energy both to do mechanical work and to generate heat have very large capacities to produce high energy molecules. If a muscle needs energy but is unable to obtain enough oxygen (such as often occurs when one is running) it simply takes all of the energy it can out of a molecule without using oxygen. It then releases back into its environment and into the blood the unused part of the molecule. This unused part of the sugar molecule is called lactic acid. Lactic acid is then taken up by the liver for reprocessing. All of these partially used molecules simply represent an "oxygen debt" that must be paid back at a later time. When sufficient oxygen becomes available at a later time, the liver uses the lactic acid either to make more ATP or to build sugar. It is in fact the buildup of lactic acid in the muscle and its environment that cause muscle aches after one has exercised.

In contrast to muscle, the brain is highly dependent on the constant availability of oxygen in its environment. If sufficient oxygen is not available even for a very short period of time, brain cells begin to die. Just as the brain is special in its control of its environment, it is also special in its critical dependence on the availability of oxygen. As a consequence, if a drug influences any part of the process by which cells produce energy, it will influence all cells to some extent but brain cells more than others.

Cellular Reproduction: For the most part, individual cells within our body don't live as long as we do. Cells wear out and die. In addition, sometimes our cells are damaged and "die prematurely." Because of this, cells must constantly be replaced in our body. This cell replacement is the job carried out by stem cells. Within all tissues are stem cells. These stem cells are not like the embryonic stem cells that have the capability to become any type of cell. Stem cells in adult tissues are there to provide replacement cells for the tissue. They are capable of dividing to reproduce more stem cells like themselves and differentiating to become mature replacement cells of the tissue.

Like all other functions, it is protein enzymes that carry out the process of cellular reproduction. Most fundamental to the process of cell division is copying the genetic information, the DNA, of the cell. There are proteins whose entire function is to participate in making copies of those long strings of linked deoxyribonucleotides (DNA) that are the genetic information of the cell. Without these proteins, the DNA molecules cannot be copied. Once this information is copied, the cell can pull apart into two new individual cells, each with a copy of the genetic information. How often the stem cells of a tissue divide is one of the greatest differences that exist between populations of cells. Some cells like the stem cells of the skin and the lining of the digestive tract divide very often. The entire population of mature differentiated cells is replaced every few days. Other stem cell populations like those of the liver and skeletal muscle are generally quiescent and only enter the proliferative (dividing) stage in response to injury that requires repair. At the extreme are stem cells of the central nervous system. The central nervous system is composed of two fundamental types of cells, nerve cells and helper cells called glial cells. Both nerve cells and glial cells arise from the same stem cell population. However, in the adult, the stem cells of the nervous system can only differentiate into glial cells. It is the inability of stem cells of the nervous system to become nerve cells that makes the repair of a damaged nervous system so difficult. If most of the liver were removed, the remaining stem cells would divide, differentiate and replace the liver in about one month. In contrast, if some of the nerve cells of the brain are lost there is no ability to replace them.

The most the brain can do in terms of repair is to attempt to redistribute the function of the lost nerve cells among those nerve cells that are left. That is why there are many people who are paraplegics for want of a few cells to transfer information across a damaged part of the spinal cord. The cells of the immune system are also unusual. They arise from bone marrow stem cells that proliferate. They mature and differentiate, either in the bone marrow or in the thymus. However, these mature, differentiated stem cells do proliferate in response to immune challenge (we will talk about this more in a later chapter).

Because stem cells are the cells in tissues that proliferate, they are also the cells that give rise to cancer. Cancer is simply a stem cell that entered the proliferative stage and because of some "mistake" that cell and it's daughter cells continue to proliferate and do not differentiate. Many purposeful effects of drugs as well as their side effects discriminate among groups of cells because of their rate of cell division. For example, because cancer cells are dividing at a very rapid rate, many cancer drugs are designed to interfere with cellular functions associated with cell division. As a consequence, the side effects of these drugs are seen in groups of cells such as the skin and the lining of the digestive system which are also rapidly dividing as part of their normal function. These drugs also make the person immunodeficient because they prevent cells of the immune system from dividing.

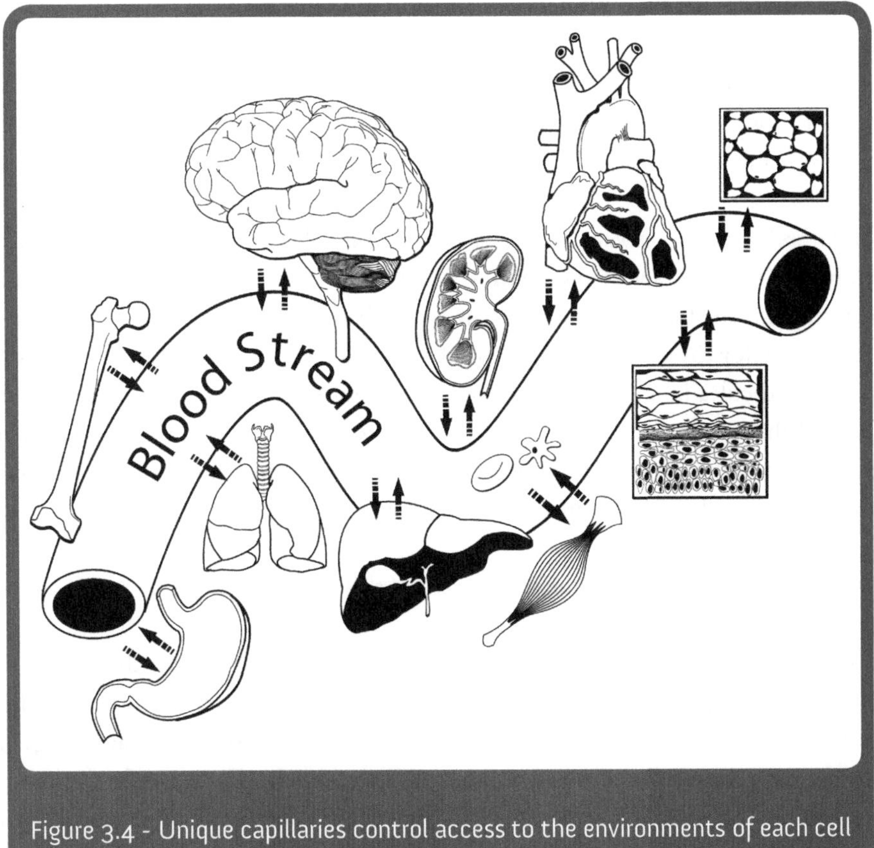

Figure 3.4 - Unique capillaries control access to the environments of each cell type.

Protein synthesis—A third commonality: Every task that a cell performs requires a protein or proteins to carry it out. All cells make proteins, and those proteins in turn determine what functions the cell can carry out. Like all other functions in the cell, the making of proteins is carried out by proteins. There are proteins that make the RNA message from the DNA that describes the amino acid sequence of a protein. There are also proteins that make the amino acid sequence that is the protein from the message. Some drugs such as certain antibiotics, interfere with the processes involved in making proteins and will therefore affect virtually every cell of the body that they come in contact with. The only variation in effect will be how active the process is in the cell at the time and how important the protein is to that cell. The major difference between one cell type and another is the proteins that they make.

Blood—the common environment of the body: The blood is the common water environment of the body. Most substances travel throughout the body in the blood. However, very few cells actually live in the blood. Most cells live in special water environments that are separated by barriers from the blood. These barriers are the capillary beds between the cells' water environment (interstitial spaces) and the blood. Just as the membrane of the cell forms a barrier that allows the inside of the cell to be a unique water compartment, the capillaries are barriers which allow each group of cells to live in their own special water environment. These barriers, which are constructed of cells rather than just membranes, select those molecules from the blood which can enter the environment of a particular group of cells (Fig 3.4).

There is tremendous variation in the selectivity of the capillary barriers between the blood and different groups of cells. The most selective one in the body is that of the brain. The capillary barrier protecting the brain selects against many molecules.

This allows for control of which molecules have access to the cells of the brain. In addition to this barrier, there is another unique feature associated with the control of the water environment of the brain. This unique feature is associated with the function of a group of cells called the choroid plexus. The choroid plexus acts like a kidney exclusively for the water environment of the brain. The choroid plexus controls the movement of water and small molecules in and out of the water environment around the cells of the brain. Together, the capillary beds of the brain and the choroid plexus are often referred to as the "blood-brain barrier." The blood-brain barrier also excludes many drugs. For example, penicillin and the newer antihistamines that don't make you drowsy are normally excluded from the water environment of the brain.

Therefore, these drugs cannot directly influence the cells in this region of the body.

The extreme opposite of the highly selected water environment of the brain is the rather non-selective environments of the liver and kidney. Both the liver and the kidney are directly involved with controlling the quality of the general water environment of the body. The liver has very extensive molecular storage and processing capabilities. It expresses proteins that allows it to interconvert molecules. For example, the liver can synthesize sugar from amino acids. The liver also stores molecules such as fat soluble vitamins and sugar for use by the body in general. The liver also constructs many molecules necessary for the composition of the blood. Finally, the liver detoxifies the blood by removing damaged or toxic molecules, and is important in removing drugs from the body. Because of these various functions, the quality and composition of the blood is extremely dependant on a normal functioning liver.

Similarly, the balance of the composition of the blood is dependant on the function of the kidney. As discussed previously, differences in the composition of water compartments provide a driving force for the movements of molecules and ions. What is important is the concentration (amount of the molecule or ion relative to the amount of water). A major function of the kidney is to maintain the proper concentration of small molecules and ions in the blood. Because this function is accomplished by controlled movement of water as well as ions and small molecules, kidney function is also very important for maintenance of correct blood pressure. The kidney accomplishes this function by filtering the blood, removing small molecules, ions and water, then returning just the right amount of each in order to create the correct balance in the water environment for the blood. The kidney is also the major organ responsible for removing toxic substances such as drugs from the body. Because of their involvement in controlling the quality of the blood, the cells of the liver and the kidney are exposed to and often concentrate virtually all drugs and toxic substances that enter the body. Between the extremes of control represented by the brain and the liver, there are the capillary barriers that control the environments of all of the other groups of cells in the body. These barriers will determine if and to what degree a substance will be able to interact with a particular group of cells.

Communications between cells: The proteins of the cell not only determine what that cell can do, but they also regulate the relationship between that cell and the other cells of the body. All cells are surrounded by membranes. Membranes limit the movement of charged molecules into and out of a cell. Molecules like amino acids and ions (like calcium) must enter or exit the cell through special channels in this barrier. As a result, the inside of the cell is a compartment which contains a unique water environment—the cytoplasm of the cell. Because the properties of the membrane control what molecules move in and out of the cell, the composition of the cytoplasm of a cell is to a great extent a function of the membrane.

Different types of cells have membranes that contain a different compliment of proteins and therefore have different functional properties. We said that these membranes, these charge barriers, were made up of phospholipids.

However, they also contain proteins which in a way "float" around the membrane with part of the molecule in the oily center of the membrane, part in the charged water environment inside and part in the charged water environment outside the cell. One can make such a protein simply by constructing an amino acid sequence, part of which has charge and part of which has no charge. Such proteins float, or are sometimes anchored in the membrane, with their uncharged area in the oily environment created by the uncharged tails of the phospholipid and with their charged area hanging out in the water on both sides. It is these proteins that make the membranes of each cell type relatively unique. That is not to say that the membranes of different cell types don't contain some of the same proteins. They do, but they also contain some that can be quite unique to that cell type.

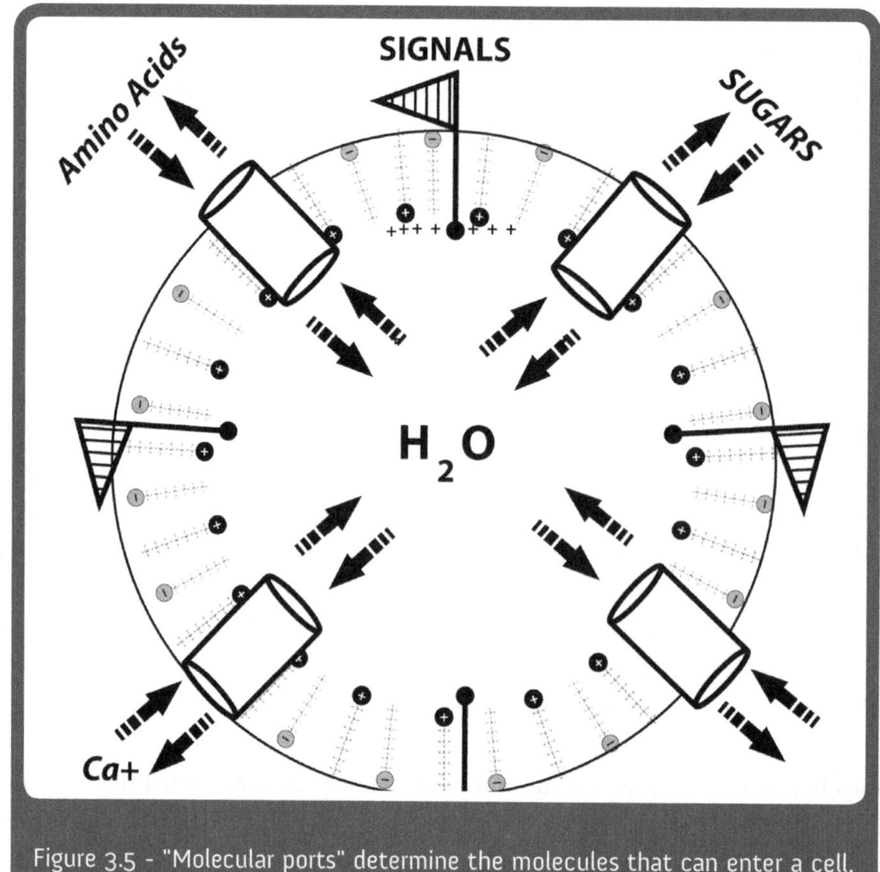

Figure 3.5 - "Molecular ports" determine the molecules that can enter a cell, and "information ports" determine the signals that a cell can respond to.

The proteins of the membrane of a cell, like all proteins, have capacities to do things. The proteins of the membrane determine what that membrane can do. Therefore, they determine the relationship of that cell with the cells around it. They also serve as "molecular ports" called channels. These molecular ports are channels that determine what charged molecules can enter or leave the cell through the charge barrier. Therefore, they control the quality of the internal water environment of the cell. Other membrane proteins are the "information ports" called receptors. They determine what environmental signals the cell can respond to (Fig. 3.5). As discussed in the first chapter, the complex society of cells that is an animal can only function because groups of cells communicate with each other. Groups of cells communicate with each other in patterns of chemical messages. The chemical message provides information in the form of the amount of the message.

There are many different chemical messages. The amount of each message changes in some time-varying pattern. As a result, each cell is constantly being bombarded by patterns of messages giving it directions. In this way, the cell is spoken to in many different voices. The particular message pattern must reach the cell in order to have a direct effect.

There are several types of chemical messengers. Some are called hormones, some are called neurotransmitters, and others are called growth factors. What they all have in common is that they are molecules which are produced by one cell for the purpose of imparting information to other cells. Each chemical message has specific commands for its target cells. Some messages may tell certain cells to grow faster. Some may tell a cell to take up sugar from its environment, or they may simply tell a cell to release another molecule which is a message to another group of cells. Those protein "information ports" floating in the membrane are the receivers for the information carried by the various chemical messengers. Each specific chemical message has it own unique information port (receptor) which fits it like a lock and key (Fig. 3.6).

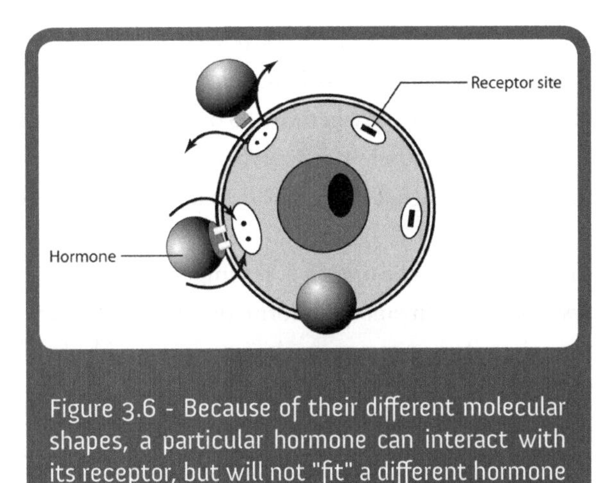

Figure 3.6 - Because of their different molecular shapes, a particular hormone can interact with its receptor, but will not "fit" a different hormone receptor.

The chemical messenger binds to the receptor protein on the surface of the cell and starts a cascade of reactions which influence some functions going on in the cell. If the cell has the appropriate receptor for the particular chemical message then it can respond to the signal. If it doesn't have the receptor then it is blind to that signal. For example, the hormone glucagon is a chemical message molecule released by a group of cells in the pancreas called "alpha islet cells." This chemical message molecule binds to special information ports (glucagon receptors) on liver cells and directs liver cells to release sugar into the blood. Many other cells which may contain plenty of stored sugar do not respond to this chemical message because they do not make the protein that is the glucagon receptor. In some cases, the same chemical message is directed to many different groups of cells at the same time. For example, adrenalin is a chemical message molecule which is directed at many different groups of cells. Such messages are designed to tell many different groups of cells to do many different things, all of which add up to some coordinated action of the body. Adrenalin tells the liver to release sugar so that more is available in the blood for other cells to use. It also tells some skeletal muscles to relax and others to contract more strongly. It tells some smooth muscles to open up the blood flow to the heart and brain while telling others to reduce blood flow to the digestive system.

It even tells some cells of the brain to pay more attention to what is going on in the world outside of the body.

Another characteristic of the receiving mechanism for chemical messengers is that the cell can adjust how sensitive it is to the message. Such adjustments are often in response to the amount of the message (the strength of the signal). When the strength of the signal increases, the sensitivity of the receiving mechanism decreases. Conversely, when the strength of the signal decreases, then the sensitivity of the receiving mechanism increases. This process of adjustment is very similar to what happens in virtually all "signal response" type sensory mechanisms, for example the eyes. When you go from a dark room outside into the bright light, you can't see very well because the strength of the light signal is too strong for the sensitivity of the visual system. After a period of time, sensitivity of the visual system decreases and the information in the strong signal can be obtained. Conversely, when you go back inside the dark room, there is a period of time during which the visual system cannot gather information. It is not sufficiently sensitive to evaluate the information contained in a low intensity light signal. Again the eyes must rebalance sensitivity with the strength of the signal.

The quality of adjustment is an important feature in the behavior of receptor systems. In effect, receptor systems have a "memory." Information received at one time affects the processing of information received at a later time. This property can be important with the use of some drugs. Some drugs are designed to act like a particular chemical messenger.

If these drugs are taken, they can affect the ability of the receptor system to process information. When the drug is no longer taken, there are withdrawal effects caused by the fact that the receptor systems are left for a period of time with an inappropriate sensitivity.

The body is a good environment for many other forms of life: The cells of the body itself are not the only living units in the body. Living in the community of body cells are other forms of life. These are viruses, bacteria, fungi and parasites. For the most part, parasites that infect humans are types of "worms" and protozoa that live part of their life cycle in humans. They can invade both the digestive system and other tissues. They are actually animals that live part of their life within other animals. Hookworms are an example of a roundworm infection while malaria is an example of a protozoan parasite. Both types of organisms are complex cells or groups of cells that function quite similarly to human cells. For this reason, targeting a parasite must be very specific and generally quite locational. Like parasites, pathogenic fungi are complex organisms. Fungal infections range from localized surface infections such as vaginal yeast infections and athlete's foot to infections that can be disseminated throughout the body such as candidiasis. Two unique features of fungi that represent differences from human cells are the structure of their surface barrier, and their means of obtaining nutrients. The surface barrier of fungi consists of both a cell membrane (like all cells) and a cell wall like plants. With respect to nutrient acquisition, fungi digest their food outside of the cell. They secrete digestive enzymes into their environment and then they absorb into the cell the results of the molecular digestion. Treating fungal infections can be "expensive" in terms of side effects because of the similarities between fungal and human cells. In addition, such treatments often kill useful fungi that live on the surface of the body and help protect us from invasion by other infective microorganisms.

In some respects, viruses are a problem that we share with parasites, fungi and bacteria. Viruses infect all of us. There are viruses that infect animals such as humans and there are viruses that infect bacteria such as Escherichia coli. In fact, viruses are generally quite specific with respect to who they infect. That is probably because a virus by itself is not quite living. A virus by itself is not a cell, cannot reproduce, and cannot make proteins. A virus is simply a bit of genetic information coated with a few proteins that has "learned" to make its own way in the world. Because it can't make proteins it cannot do anything by itself. It cannot produce high energy molecules such as ATP. It cannot make enzymes to form a membrane, and most important of all, it cannot reproduce. A virus cannot make copies of its genetic material. What a virus can do is bind and inject the genetic information that it contains into another form of life. This little bit of information that it injects takes control of the cell it infects. This cell has all of the capabilities of life, and now the viral genetic information has access to all of the capabilities of life. Viruses live as bits of information in other living forms of life. When the virus takes over a cell, they use the DNA or RNA reproducing enzymes of that cell to selectively reproduce the genetic information of the virus. They also take control of the protein synthetic enzymes of the cell and direct them to selectively produce the few proteins of the virus. These few proteins are the targeting molecules that coat the bit of genetic information of the virus.

Generally a virus eventually kills the cell which it has infected. However, before killing the host cell, the genetic information of the virus, which the host cell has been forced to reproduce, is repackaged in a coat of targeting proteins. The virus has reproduced itself. When the infected cell releases the viral packages of information, the few virus specific coded proteins target the virus information to a new cell. The virus binds to that new target cell, and the infection spreads. Because viruses live within our own cells, it is very difficult to kill them. At present our best defense is a good offense: kill the virus with the immune system before it can infect the cells of the body.

Bacteria are very different from viruses. Bacteria are living organisms with information of their own and the ability to make proteins from this information. Their information is in the form of DNA, similar to an animal

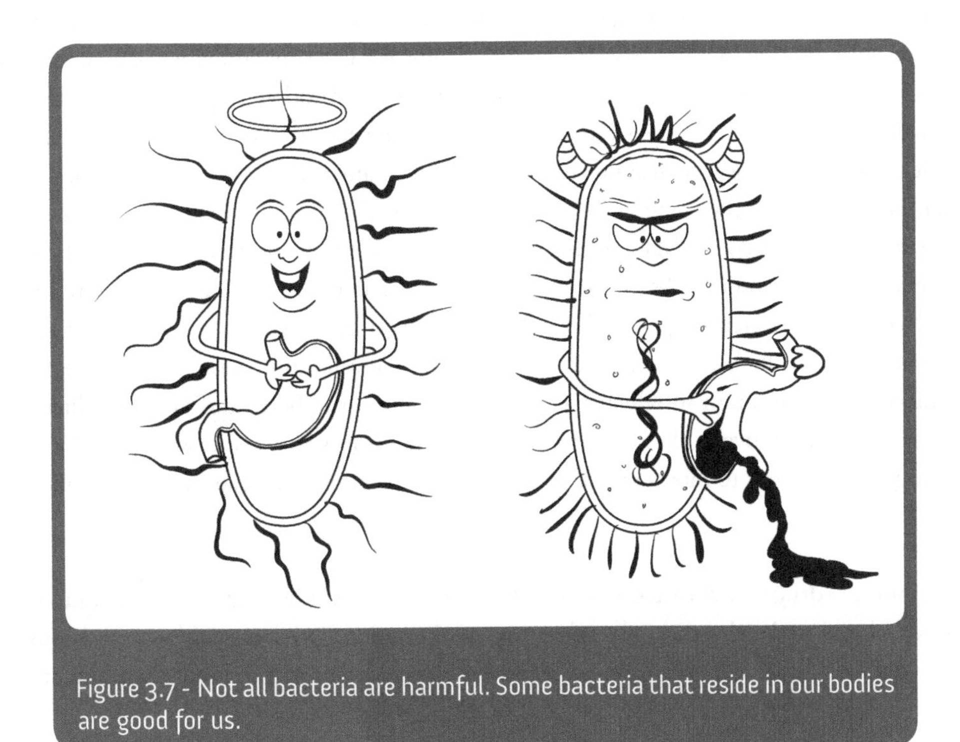

Figure 3.7 - Not all bacteria are harmful. Some bacteria that reside in our bodies are good for us.

cell. They make messages from this information in the form of RNA, similar to an animal cell. They also make protein from this information, similar to an animal cell.

Just like there are many different species of animals, there are many different strains (or types) of bacteria. Many bacteria are the causes of different diseases that a human contracts. For example, some forms of pneumonia are caused by bacteria. So is one form of meningitis. And an infection you might get after cutting yourself is due to bacteria. However, bacteria in general can get a "bad rep." Even though there are many types of bacteria that are harmful to our health, there are also "good" bacteria that live naturally in our body in a symbiotic (mutually beneficial) relationship (Fig. 3.7). These bacteria benefit us in different ways. For example, there are bacteria that live in our digestive systems and produce Vitamin K, a substance that we need to stay healthy. In addition, many non-disease causing bacteria live in various locations in our body and their presence prevents the invasion of other bacteria that may cause disease.

However, when we do get a bacterial infection, we often try to cure it with drugs called antibiotics. Because they work fundamentally in the same way as an animal cell, selective targeting of bacteria by a drug is difficult. However, there are some very important differences between animal cells and bacteria. Bacteria do not have a discrete nucleus. The genetic material of the bacteria is not collected into a special membrane surrounded environment inside the cell. In fact, unlike human cells, bacteria do not contain any internal membrane surrounded structures. For example, human cells produce high energy molecules such as ATP using internal membrane surrounded structures known as mitochondria. While bacteria produce very similar molecules in much the same way as a human cell they use their outer membrane for this purpose. In addition some bacteria, such as the ones that cause gangrene, produce all of their high energy molecules using the very limited metabolic process that does not require oxygen. There are also some differences between animal cells and bacteria in the cellular machinery that makes their proteins. Another important structural difference between animal cells and bacteria is their outside surface. Similar to an animal cell (and all cells), a bacteria has an outside membrane. However, around this outside membrane is another structure called the cell wall. This cell wall further protects the bacteria. Like everything else, the ability to make this cell wall is coded in the genetic material of the bacteria. Making this cell wall and the structures that constitute the cell wall is a unique difference between animal cells and bacteria. This is a difference that can be exploited in designing drugs that selectively kill bacteria. Specifically, a drug can target processes associated with the production and assembly of the cell wall. Another difference between animal cells and bacteria is the rate at which they divide. An infection is a situation in which bacteria have entered the body and are reproducing. Bacterial reproduction is very fast relative to the cells of the host animal. If it wasn't fast it would not be an infection! One way of targeting a drug is to interfere with the materials and processes associated with this rapid rate of reproduction.

Generally antibiotics (drugs used to treat bacterial infections) take advantage of the unique processes of bacteria. How to build molecules that take advantage of the unique processes of bacteria is quite difficult. In fact, it is so difficult that microorganisms so far have been much more successful than humans at making these drugs. Humans have simply exploited the molecules microorganisms have developed for fighting with each other.

However, even with the best of these drugs the effect is not entirely selective. Such drugs have some effect on cells of the human body as well. Quite commonly these "side effects" are seen in the rapidly reproducing cells of the body. In addition, there are side effects caused by these drugs killing some useful bacteria that live in and on the body. For example, living in the digestive system are bacteria that both aid in digestion and produce vitamin K that is used by the body. How does a drug discriminate between the useful bacteria and the infection? Often the answer is that it can't.

chapter four

a question of priorities: an organization chart of the body

Blood is the common water environment of the body: So far, we have developed the concept that the body is made up of many different groups of cells, further organized as tissues and organs. However each group of cells exists in its own unique water environment. The unique environment of each group is created by the capillary barriers (very small blood vessels) that exist between those cells and the blood. The blood is the common water environment of the entire body. The blood is the vehicle which carries nutrients, as well as information containing molecules to cells, carries away waste products from the cells, carries heat to and from the cells, and provides protection against foreign invaders that would do the body harm. Thus, the blood has a great deal of influence on how the various cells of the body function. In turn, the composition of blood is impacted by the combined effects of all of the different groups of cells of the body. However, access to the blood and the resources it carries is not equal among the various groups of different kinds of cells.

The body is a collection of different priorities: One could also view the body as being made up of a collection of many different priorities.

Priorities mean that all cells and their particular functions are not of equal importance all of the time. Priorities are necessary for two reasons. First, resources are limited. Second, the circumstances of the body change, and different functions are more important in different situations. An example of this can be found in a children's riddle (Fig. 4.1): you are walking through the woods and see a tiger—what do you do? (Answer: you hope the tiger doesn't see you!) But your body's "automatic" answer is to make the workings of your brain, heart, lungs and muscles more efficient than less important functions like reproduction or digestion, so that you can either escape from or fight the tiger.

Because all functions are not always equally important to the body, the body must sometimes promote some functions associated with some groups of cells over other functions associated with other groups of cells.

Figure 4.1 - Danger causes a useful stress response to occur in our bodies.

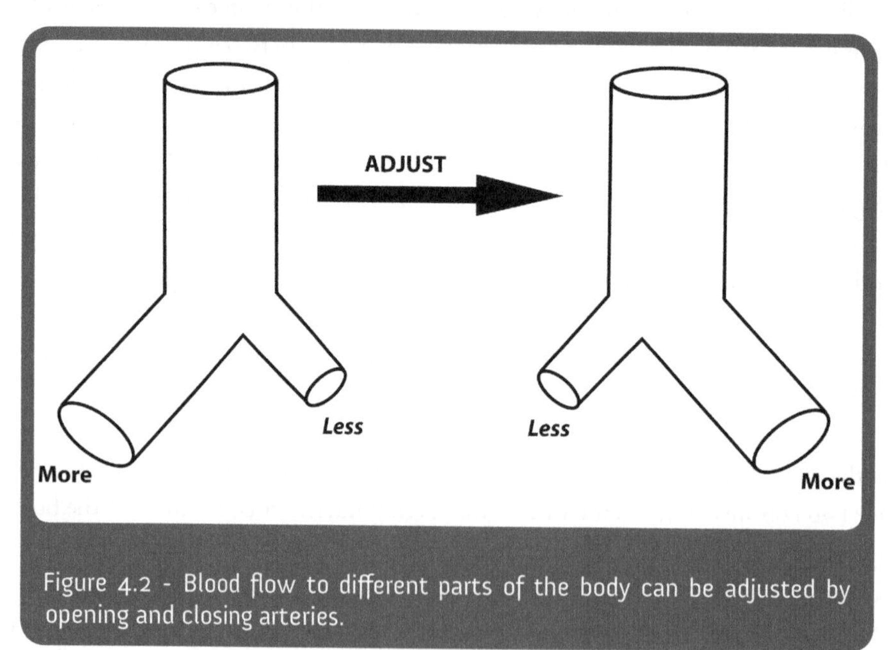

Figure 4.2 - Blood flow to different parts of the body can be adjusted by opening and closing arteries.

The major way that the body expresses these priorities is by controlling the relative access of the different groups of cells to the blood. In essence, the question of priorities is really a question of relative access to what is in the blood. In particular, there are priorities which determine the relative access of each group of cells to the nutrients, oxygen and heat that is carried by the blood. The blood carries all of these around the body and distributes them in a rather discriminating manner. The mechanism by which this discrimination takes place is the control of blood flow to the different tissues. Under different circumstances, more or less blood is allowed to go to different groups of cells in the body.

Blood is delivered from the heart to the various tissues of the body by arteries. Arteries are channels for blood which are surrounded by circular muscle (a kind of muscle called "smooth muscle." In general, this circular muscle is neither entirely relaxed (more open) nor entirely contracted (more closed). To send more blood to a tissue, the muscle of the artery is relaxed to some degree which increases the diameter of the path, opening it up to more blood flow. To send less blood to a particular tissue, the muscle of the artery contracts to some degree which reduces the diameter and less blood flows down that path (Fig. 4.2). The important thing to remember is that there is only so much blood available. Therefore, in order to send more blood to one tissue blood must be taken away from another tissue. Circulation is a "zero sum game." If all arterial paths in the body open at once, a condition known as shock, then the brain which is at the highest point is left without blood and the person becomes unconscious and may die if the situation is not rectified quickly.

Priorities can be influenced by circumstances and drugs: Many situations influence the priorities associated with the distribution of blood in the body. If you get angry or frightened, priorities change. If you eat dinner or sleep, priorities change. In addition, many drugs influence the process of controlled distribution of blood in the body. If you drink a cup of coffee or eat chocolate, you influence priorities. If you smoke you influence priorities. If you take birth control pills, alcohol, marijuana, cocaine, Librium, Valium, amphetamines, or even if you use a cream for muscle aches you are influencing the ability of the brain to control priorities. All of these situations and drugs, as well as many others, influence blood flow.

Priorities allow for change: Why doesn't the body simply make an ample blood supply available to all groups of cells all of the time? The answer to this question is that the body must maintain the capacity to adjust to changing circumstances. The circumstances of the body are always changing. There is more or less to eat. It is hot or cold outside. You stand up or sit down. Sometimes the body is in a dangerous situation. It is the controlled flow of blood around the body that allows the body the flexibility to deal with change. This flexibility is necessary in order to deal with, among other things, the effect of gravity, the change in environmental temperature and the dangers of the world.

Not all cells are equally sensitive to changes in blood flow: The factor that allows this system of priorities to exist is that all groups of cells and all functions being carried out by groups of cells are not equally sensitive to reduced blood flow. For example, the brain is very sensitive to reduced blood flow. If the brain were to treat itself like it often treats the skin and drastically reduce its blood flow, within seconds the person would be unconscious and within minutes there would be damage. The cells of the brain are highly specialized among themselves. The nerve cells of the brain generally are not replaced, and therefore damage is irreparable. In contrast, although there are variations on the theme, one skin cell is quite similar to another. Skin cells are replaced from stem cells quite rapidly and damage is relatively easily repaired. Similarly, all functions are not equally sensitive to reduced blood flow (Fig. 4.3).

If the body were to treat the function of pumping the blood, which is carried out by the heart, in the same way that it treats the function of digestion, it would be risking a heart attack. If the body reduces blood flow to the digestive system it simply risks a bit of indigestion. In order to understand the priority of a group of cells in the

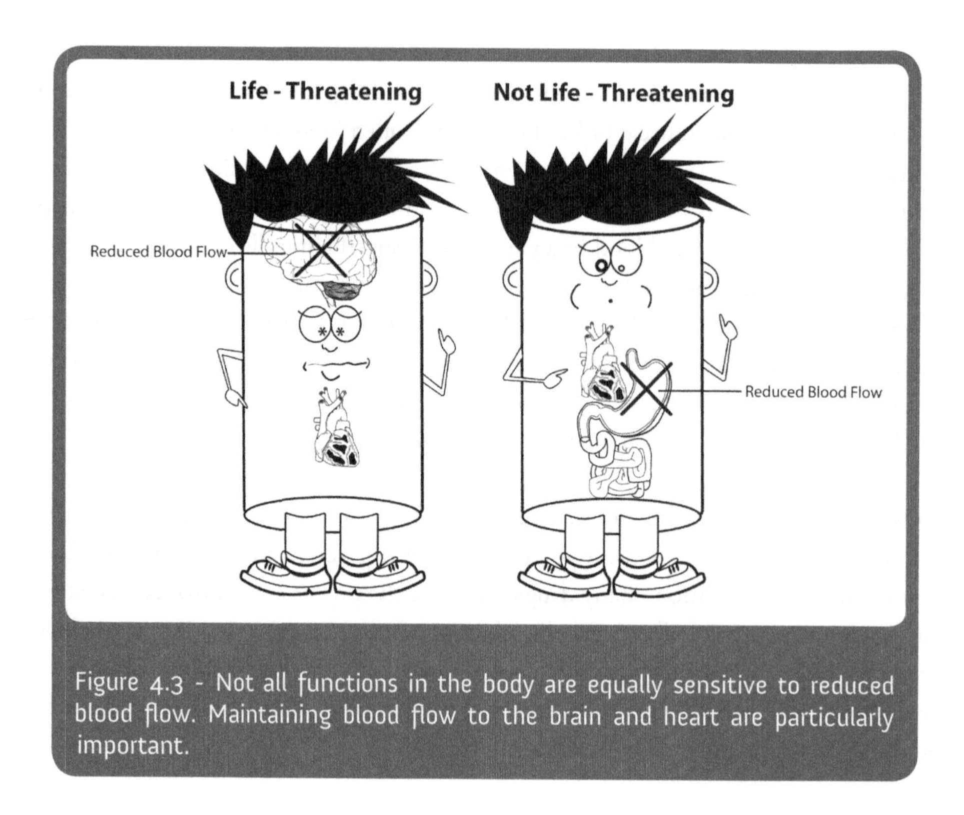

Figure 4.3 - Not all functions in the body are equally sensitive to reduced blood flow. Maintaining blood flow to the brain and heart are particularly important.

body, one simply must ask three questions. First, how necessary is the function of those cells to the immediate needs of the body? Second, how much damage is done to the group of cells when blood flow is reduced? Third, how easy is it to repair that group of cells?

The brain establishes priorities: At the center of all of the protective barriers and mechanisms of the body is the brain. The brain has always "known" what man has only figured out in relatively modern times: when it dies, the person dies. The brain and other parts of the nervous system are composed of many cells. All of these cells are structured and organized into a very elaborate but rather precisely patterned sequence of connections. Information flows into the brain through this precise sequence of connections. The brain both saves the information in the sequence of connections, and reacts to the information by sending messages out through the sequence of connections. Information concerning the outside world flows into the brain from the sense organs (the eyes, ears, etc.). The brain also receives information about virtually everything that is going on inside the body. Information about the body is gathered in two ways. First there are certain groups of cells in the brain itself whose function is to monitor, or pay attention to, certain aspects of the quality of the blood. For example, there are groups of cells which monitor the temperature of the blood. If the blood is too hot or too cold, then these cells cause information to be sent out from the brain telling the body to release or conserve heat. The second way that the brain gathers information is from "sense organs" inside the body itself. These "sense organs" pay attention to such things as the hydrogen ion concentration of the blood, the position of the right knee joint, and the movement of food in the digestive system. Like the information gathered by the eyes or the ears, this information is sent to the brain through the organized sequence of connected nerve cells. The brain uses all of the information flowing into it to respond to both the environment outside the body and to the environment inside the body (Fig. 4.4).

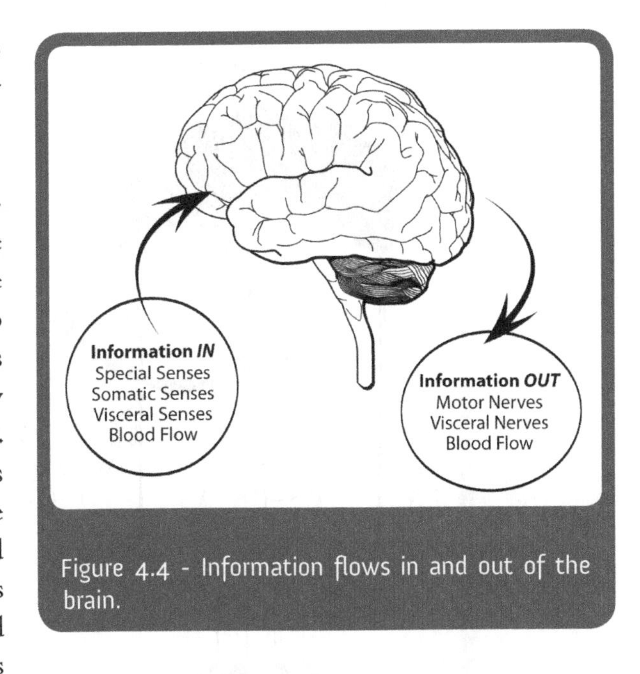

Figure 4.4 - Information flows in and out of the brain.

The brain has two ways of responding to information. The first way is through sequences of nerve connections. These sequences of connections are simply nerve cell talking to nerve cell talking to nerve cell and eventually telling a group of cells such as a muscle to do something. The second way that the brain can respond is through hormones. Hormones are made and released by special groups of cells in the body called endocrine glands. There are many different kinds of endocrine glands, and there many different kinds of hormones. Hormones are chemical messages that travel in the blood and deliver controlling information to different groups of cells in the body. Hormones impart information to cells through the receptors, the "information ports" we discussed in the previous chapter. Hormones are molecules like adrenalin, insulin, glucagon, androgens, estrogens, glucocorticoids, thyroid hormone and many others. In many cases, it is the brain (either directly or indirectly) that "tells" a gland to release a hormone. Using these two ways of responding (nerves and hormones) the brain, which lives at the center of all of the barriers and protective functions of the body, is able to control the environment inside the body and to deal with the environment outside of the body.

Barriers to the outside environment—skin, lungs, and digestive system: While it is the brain that is the group of cells which is most protected from the vagaries of the environments inside and outside the body, it is the skin, the lungs and the digestive system that are least protected. These three groups of cells form the barrier between the environment outside of the body and the environment inside the body.

Like the surface barrier of a cell, information and molecules move into and out of the body through these barriers. These barriers are also the groups of cells through which energy in the form of heat flows in and out of the body. Because they are the surface barriers of the body, they are also most often subjected to damage from the environment outside of the body.

The lungs: Perhaps the simplest of these three barriers is the lungs (Fig. 4.5). Gases, vapors, and heat move in and out of the body through the lungs. The lungs themselves are quite passively involved in this process. When the diaphragm and the muscles between the ribs contract, the chest cavity becomes larger and air is sucked into the lungs. When these muscles relax, the chest cavity becomes smaller and air is pushed out of the lungs (Fig. 4.6). A very important aspect of the expansion and contraction of the chest cavity is that it is carried out by voluntary, skeletal muscles. The importance of this is that each breath requires that signals be

Figure 4.5 - Lungs bring in oxygen and remove carbon dioxide.

sent from the brain to the muscles involved in expanding the chest. For this reason if a person "breaks his or her neck," severing the spinal cord at the level of the neck, he or she stops breathing. If he or she survives, the individual will be on a respirator for the rest of his or her life.

The air that moves into the lungs contains gases such as oxygen and nitrogen. It also contains vapors such as those of gasoline or ether anesthetics, and it contains the heat of the outside environment. The air that moves into the lungs contains one other thing as well: particulate garbage such as dust, pollen and other solids in the air. In the lungs there is a direct exchange of gases, vapors, and heat between the air and the blood. To understand this exchange it is important to know that gases such as oxygen, nitrogen and carbon dioxide dissolve in the water of the blood. Heat energy and molecules move from the air to the blood, and heat energy and molecules move from the blood to the air. This transfer is entirely passive. Everything moves from "more to less" that is, from a higher to a lower concentration. If the air contains a higher concentration of oxygen than the blood, then oxygen moves from the air to the blood. If the blood is hotter than the air, then heat moves from the blood to the air. The most active thing that the lungs do is to attempt to clean themselves. The lungs are continually carrying the particulate garbage that entered with the air back up and out of the passages.

Figure 4.6 - When the chest cavity gets bigger air is sucked into the lungs, and when it gets smaller air is pushed out.

The brain controls the function of the lungs by regulating how often it expands and contracts the chest cavity, and how large and small the cavity becomes each time it expands or contracts. Interestingly it is not the oxygen content of the blood that regulates breathing.

Rather the brain monitors the hydrogen ion concentration of the blood to determine when to send the directions for breathing. The brain also controls the function of the lungs by regulating the size of the air passages. The air passages can be opened up so that more air can be taken into the lungs. Because of the importance of the molecules which are exchanged between the air and the blood in the lungs, this group of cells is not subjected to big reductions in blood flow. What it is subjected to is large, rapid increases in blood flow associated with demands for more exchange such as occurs when one is running.

The other thing that is going on in the lungs is protection from invaders by the immune system. Invaders enter the body through its barriers. Living in the lungs are a population of immune cells known as alveolar macrophages (we will discuss this further in Chapter 10). These cells, whose name means "big eaters," scavenge the lungs for dead cellular debris and invading pathogens. Alveolar macrophages are the first line of defense against pathogens that attempt to enter the body through the lungs. Alveolar macrophages also play a major role in diseases such as asthma, "hay fever" and chronic obstructive pulmonary disease (COPD) which is generally caused by smoking.

SKIN: PROTECTION & HEAT EXCHANGE

Figure 4.7 - The skin serves both as a protective barrier against the outside of the body, and is the major place where heat is lost from the body.

Skin: A second outside barrier of the body is the skin (Fig. 4.7). The skin is often subjected to damage such as cuts and burns. Because of this, the skin has a very active stem cell population that is constantly replacing the surface. When damage occurs, proliferation of stem cells increases beyond the replacement rate so that repair can occur. The skin, because of its large surface area, is the major site of heat exchange between the body and the environment outside the body. If the blood of the body is too hot, then more blood is sent to the surface and heat is lost through the skin. If the blood of the body is too cold, then less blood is sent to the surface and less heat is lost from the body through the skin. Heat exchange through the skin is a rather simple process. Blood carries heat around the body. If the brain wishes to lose heat, then more blood is sent to the surface of the body and heat is lost. Conversely, if the brain wishes to conserve heat, then less blood is sent to the surface of the body (Fig. 4.8). The process is very similar to the way a house is heated using radiators.

Figure 4.8 - The amount of blood flow to the skin helps control body temperature.

When the radiator in a room is turned on, hot water or steam runs through it and heat is transferred by radiation from the water to the room. The water that returns to the basement is cooler because it has lost heat to the environment. If you turn down the radiator in a room then the flow of hot water or steam is reduced and there is less heat transfer to the room. If you go outside on a cold day, the brain reduces the blood flow to the surface of the body. In this way, the body conserves its heat. On a hot day, the brain increases the flow of blood to the surface, which allows for more heat transfer to the environment outside of the body. Sometimes the brain even dumps water over its surface in the form of sweat.

The sweat is hot water and therefore carries heat. The evaporation of this sweat removes that heat from the body.

Because of the importance of body temperature to the brain, much of the regulation of blood flow to the surface is associated with simple heat conservation or loss. However, increased blood flow also occurs as a result of damage. Because of the high potential for damage to the skin, a substantial natural immune surveillance and response capability is associated with the surface of the body. When the skin is damaged immune cells called dendritic cells release a chemical message called histamine that increases blood flow to the area and attracts other immune cells that become macrophages. Macrophages in the skin do the same thing they do in the lungs, clean up cellular debris and engulf (eat) pathogens. This is why when you cut yourself, the area around the cut will become red and puffy.

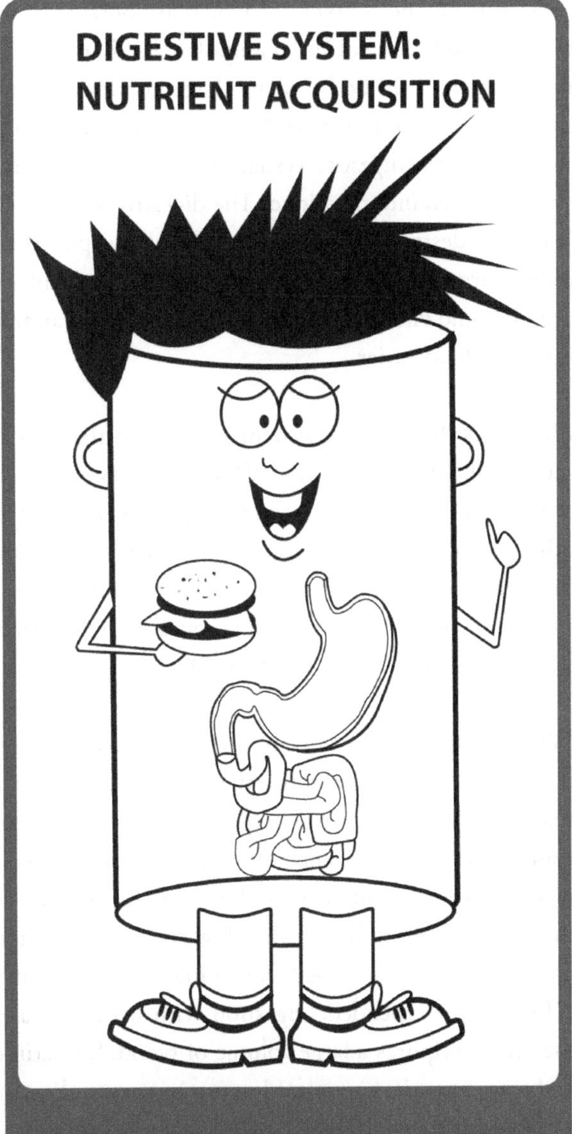

Figure 4.9 - Food is broken down into simple nutrient molecules before being taken into the blood.

The digestive system: The last outside barrier of the body is the digestive system (Fig. 4.9).

Like the lungs, the inside of the digestive system is an environment which is outside of the internal water environment of the body. One can view the inside of the tube which forms the digestive system as a controlled environment, the contents of which are outside the body. However, by surrounding this environment, the body is able to control and manipulate its character. The digestive system is a site of major molecular transfer into and out of the body. We put large quantities of molecules and ions of all types into our mouths. Some of these are large molecules with elaborate structures, and some are small simple molecules like water. The first function of the digestive system is to convert as many of these molecules as possible into a form which can be transferred into the blood. The digestive system attempts to take all of the structure out of big molecules. It tries to destroy the structural information put in the molecule by the plant or animal that made it and to reduce large complex molecules into simple molecules (simple fats, sugars and amino acids). In addition, some minerals and vitamins are converted into forms, and/or are put into environments, which make their transfer into the blood easier.

How does the digestive system accomplish this process of breaking up large, complex molecular structures into small, simple molecules? When large, complex molecules are put into the digestive tube, circular muscles around the tube move the food through this tubular system. As the complex molecules move, the body "squirts" other molecules (protein digestive enzymes) into the tube which break up the large molecules. All of the time the circular muscles around the tube are mixing the food with the new molecules squirted in and moving everything down the tube. In this way as many molecules as possible are broken up and converted into forms which are easy to transfer into the blood.

What is left is dumped out of the body. Molecules are transferred into the body through the large system of blood vessels that closely intertwine around the tube at the level of the small intestine. This system of blood vessels coalesce into a single large blood vessel called the hepatic portal vein which goes directly to the liver. For some molecules, this transfer is passive. These molecules move from a high concentration on the inside of the tube to a low concentration in the blood. Other molecules are so important to the body that it actively moves these molecules "up hill," from a low concentration in the tube to a higher concentration in the blood. When molecules are being moved from a low concentration to a high concentration, high energy molecules such as ATP must be expended. Digestion, with all of its phases from breaking up molecules to transferring them into the blood, requires a large volume of blood. Sometimes the body provides the necessary blood immediately, and sometimes it does not. However, since not digesting your food right away won't kill you, digestion is not a high priority in an immediate sense to the brain.

The digestive system also protects us from foreign invaders such as bacteria. Like the lungs and the skin, the inside of the digestive system is exposed to foreign pathogens. The lining of the digestive system also has a high population of immune cells, which serve as the first line of protection against invaders that try to enter our body through the digestive tract. These cells are called mucosal macrophages. Mucosal macrophages carry out the same function in the digestive system that macrophages carry out in the skin and lungs.

Adjusting the composition of blood—the liver and kidney: With so many molecules coming into the body and so many functions going on, it is necessary for the body to constantly adjust the molecular composition of the blood. The liver and the kidney carry out this function. The liver is a large, reddish organ that lies in the abdominal cavity just below the diaphragm. Although some of the cells of the liver are involved with digestion, and the liver is the primary site for making and storing sugar for the entire body to use, another major function of the liver is helping to control the quality of the blood (Fig. 4.10).

Figure 4.10 - Most molecules that enter the water of the body will at some point pass through the liver.

Because the blood from the digestive system flows directly to the liver, the liver actually serves as a backup barrier to the digestive system. In the period after eating, as much as 15% of all of the blood in the body is flowing from the small intestine to the liver. This blood represents about 70% of all of the blood going to the liver. This blood from the small intestine combines with blood from the hepatic artery and enters an extensive capillary network known as the sinuses. Like all capillary beds the sinuses allow molecules and water to move in and out of the blood. In essence, this provides the liver first access to everything entering the blood through the digestive system. This allows the liver first access to nutrients such as sugar for storage. This also allows the liver to serve as another barrier for pathogens that may have gotten past the macrophages of the intestine. The liver has its own extensive population of macrophages known a Kupffer cells, which adds another layer of protection against pathogens. The liver often removes toxic molecules, like drugs, which enter the blood through the digestive system. In addition, the liver, with its large array of blood vessels, is constantly removing inappropriate molecules such as damaged proteins from the blood. Some of these molecules are destroyed and some are converted to forms which can be removed from the body by the kidney. The liver also adds molecules such as blood clotting factors to the blood. The liver has a large blood supply, and the liver cells spend their time taking some molecules out of the blood and adding others in. Eventually, almost every molecule that enters the blood must go through this screening process in the liver. Because of this, the liver, far more than most groups of cells, is subjected to damage from toxic molecules that enter the blood of the body. For example, the gasoline that entered the body through the lungs will be picked up by the liver.

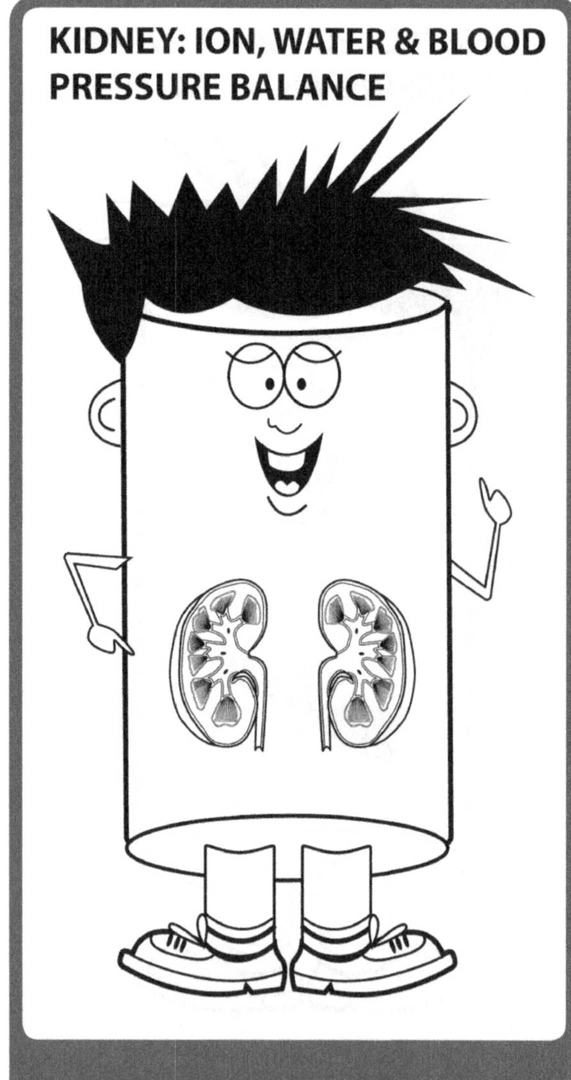

Figure 4.11 - Water, ions, and small molecules are removed from the blood by the kidney and the correct amount of each is returned to circulation.

The other group of cells involved in controlling the quality of the blood is the kidneys (Fig. 4.11). We have two kidneys, which sit in the lower back portion of the abdominal cavity, and are often described as "bean shaped" (because they are!). The function of the kidneys is to control the concentration of a variety of ions and small molecules in the blood. With all of the water and salts entering the body, it is necessary to maintain a balance between the amount of water and the amount of various ions like sodium and hydrogen. The kidney is like a filter. Water, small molecules and ions are removed from the rest of the blood as the blood enters the kidney.

The kidney is constructed of a vast array of microscopic tubes called tubules. The cells of the kidney tubules use a complicated flow system. Energy is used by the kidney to recover some important molecules like sugar. Energy is also used by the kidney to put back into the blood the appropriate amount of the various filtered ions. The water moves back and forth between the tubule and the blood by following the ions and small molecules. The brain controls both how much blood is allowed to go to the kidney and the process by which ions and water are recovered. The brain uses two types of information to control the kidney. The first is the concentration of a variety of ions such as sodium and hydrogen in the blood circulating to the brain. The second is blood pressure. If blood pressure is too low, hormonal signals are sent to the kidney to recover more water. If blood pressure is too high the kidney recovers less water. Because many toxic substances are concentrated in the kidney before they leave the body, the kidney is also often subjected to damage.

Storage tissues—muscle, fat, and bone: By far, the bulk of the cells of the body are part of the muscle, the fat or the bone. These three tissues, along with the liver, store large amounts of particular types of molecules to be used by all of the cells of the body. Within a few hours after eating, digestion is finished and no new molecules are entering the blood from the digestive system. During the period after eating when there were plenty of nutrients entering the body, some groups of cells were storing molecules for their own use and some were storing them for use by the whole body. In contrast to most cells of the body, the brain stores virtually nothing for its own use. If nutrient molecules like sugar are not available to the brain from the blood on a constant basis, then the cells of the brain rapidly begin to die. Because availability of nutrient molecules from outside is variable (because we don't eat all of the time!), the body has a very elaborate system for the storage and release of some molecules. This system involves the liver, the fat, the muscle and the bones. During the "period of plenty" after eating, the liver stores sugar for later use by the entire body. The muscle stores amino acids for the body and the fat tissue stores lipid molecules for the body. The bones are somewhat different than the other three types of storage tissues. The bones serve as a reservoir for the control of calcium. The calcium concentration of the blood is very important because the flow of calcium into cells controls many functions. For example, every time the heart contracts, it is because calcium ions flow into the heart cells. These charged calcium ions (Ca^{++}) enter the cells through the controlled channels in the surface membrane of the heart cells.

MUSCLE: MOVEMENT & AMINO ACID STORAGE

Figure 4.12 - Almost half of the protein in the body is in the musculature.

Depending on the individual, close to one-half of the entire body is muscle (Fig. 4.12). The obvious function of muscle is to do mechanical work (both movement and posture, which allows us to hold our bodies up against the force of gravity). The muscle (in connection with the bones) provides the brain with the capacity to do mechanical work. Muscles do mechanical work by shortening. Most of the mass of a muscle are proteins that use energy from ATP to shorten. The nervous system tells a muscle to shorten and it does so. The shortening of all of these muscles produces one very important byproduct, heat.

The protein mass of the musculature produces most of the heat that the blood moves around the body. Cooler blood flows through the working muscle delivering nutrients and oxygen. At the same time, because the working muscle is warmer than the blood, heat flows downhill from the water of the muscle into the blood.

Figure 4.13 - Fat tissue contains little water, so it is lightweight and a poor heat conductor.

This both prevents the muscle from over-heating and provides heat to maintain the warmth of tissues such as the brain which generate little heat.

In addition to movement and the production of heat, muscle has another very important function in the body. Muscle is a major source of stored potential energy in the body. The carbon part of the amino acids that make up that large protein mass of the muscle is available on demand for making sugar in the liver and kidney. The brain can cause a chemical message to be sent to the muscle, which tells the muscle to break down more protein than it makes. Because the muscle is breaking more protein down than it is making, the muscle has an excess of free amino acids. These amino acids are released into the blood. The blood carries them to the liver and kidney. The same signal that directed the muscle to make amino acids available also directed the liver and kidney to be ready to use these amino acids to make sugar. The liver removes the nitrogen part of the amino acid and uses the carbon part of the molecule to make sugar. The nitrogen part of the molecule is made into urea, which is filtered in the kidney for removal from the body in the urine. In the kidney, the nitrogen removed from amino acids is directly filtered out. The sugar from both tissues is put back into the blood for use by the body. If a person had a sumptuous meal and filled the liver to capacity with stored sugar, then simply sat down and did nothing, all sugar stored in the liver as glycogen would be gone within 12 to 18 hours. However, as sugar stored in the liver in the form of glycogen begins to become depleted, the liver and kidney begin to make sugar mainly from amino acids. If you do not eat for 18–24 hours, all of the sugar found in the blood would be sugar made from the amino acids taken from the proteins of the musculature or from a small lipid molecule from fat.

Fat is the other large source of stored potential energy in the body (Fig. 4-13).

Body fat functions both as a source of stored energy and as insulation. The heat transfer through fat is much lower than the heat transfer through water. Most cells have far higher water contents than fat cells. By distributing fat underneath the skin, the body insulates itself from the cold. If the hot blood from the inside of the body is allowed to move outside of the fat to the surface of the body, then heat is lost. However, if you keep the blood on the inside of the fat, then heat is conserved. Apart from this insulation function, fat also serves as a storehouse for molecules which can be used as fuel. Like muscle, the energy stored as fat is also available on demand. A major user of the energy stored in the fat is muscle. As long as enough oxygen is available, then muscle can function primarily on potential energy produced by condensing the energy in the bonds of lipid molecules. Increasingly, we are beginning to learn that fat (adipose tissue) is also an important source of chemical messages (hormones). These chemical messages are important in regulating appetite, insulin sensitivity and the immune system. Dysregulation of the fat derived signals has been associated with obesity, diabetes and atherosclerosis.

At this point it is appropriate to consider the example of "sugar" diabetes (that is, diabetes mellitus). Diabetes mellitus is a problem of nutrient storage. There are two basic types of diabetes mellitus. Some diabetics are persons who, because of damage to certain cells in their pancreas, do not make enough insulin. Insulin is the chemical message that directs cells to store sugar and amino acids when they are in excess in the blood following eating. This type of diabetes mellitus is referred to as type 1 diabetes (which used to be called "juvenile diabetes" because the disease usually develops during childhood). These people must take insulin after they eat. The other type involves people who at least initially do have sufficient insulin but for some reason their cells, particularly the muscle and fat, are not sensitive to the insulin message. These people are referred to as having type 2 diabetes (previously called "adult-onset diabetes" because it most commonly occurs in mid- to late-adulthood).

When a diabetic eats a sumptuous meal, unlike the normal person, he or she cannot store nutrients. The most important stored commodity is sugar. When the muscle and fat tissues cannot take in sugar, either because of the lack of insulin or the inability to respond to insulin, then the liver is continually bathed in sugar. Because the liver can take in sugar in the absence of the insulin response, the liver takes in large amounts of sugar. Because the synthesis of glycogen from sugar in the liver is coupled to an insulin response, this sugar is used by the liver to synthesize lipid molecules. With time, uncontrolled diabetes can result in the accumulation of large amounts of fat in the liver. This also means that at times the liver can run out of stored sugar (glycogen) for the body's use and the muscle runs out of stored sugar for its own use. All of the body, including the large percentage of the body mass represented by the musculature, must now get its energy from fat.

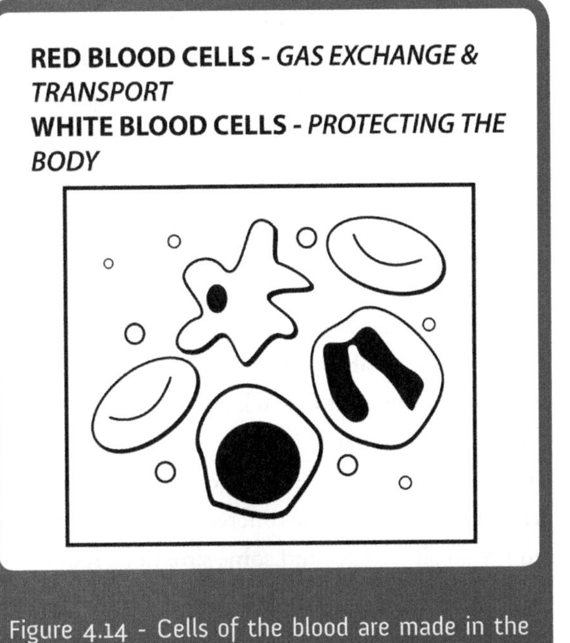

Figure 4.14 - Cells of the blood are made in the bone marrow and are controlled by chemical messages.

Using fat for energy is a rather expensive process in terms of oxygen. Quite often what happens is that the intermediate molecules (the molecules that are produced from the fat on the way to making ATP) build up in the liver. As the intermediate molecules build up they leave the liver and enter the blood. These intermediate molecules are called ketone bodies and can be used by most cells including the brain for energy.

Some cells live in the blood: Besides all of these relatively organized groups of cells there are other important groups of cells that exist in the body. These cells live within blood (Fig. 4.14). There are two basic populations of cells in the blood. One population is red blood cells. These cells serve the functions of carrying oxygen from the lungs to the rest of the body and altering carbon dioxide so it can be transported to the lungs for removal from the body. These are cells that are "born" (formed) from stem cells in the bone marrow. Mature red blood cells do not have a nucleus and therefore lack genetic material (DNA). Because of this they are limited to the capacities, the set of proteins, they were given at "birth." The other population of cells in the blood is white blood cells which are also derived from stem cells in the bone marrow. White blood cells are cells of the immune system. Several different types of immune cells work together to protect the body from infection. Unlike red blood cells, the cells of the immune system have the ability to release chemical messages which coordinate the actions of the various types of immune cells. One message called histamine opens up the capillary barriers of all groups of cells. This allows immune cells to leave the blood and enter the unique environments around all of the groups of cells that do not live in the blood. Because immune cells and red blood cells exist in the blood, their priority is not expressed by blood flow. Their priority is expressed directly by chemical messages, hormones.

The priority of the species: The last groups of cells which we will discuss are those associated with reproduction. These groups of cells are also unique when it comes to priority. Their priority is expressed by blood flow. However, the priority of reproduction is a priority of the species and not one of the individual (Fig. 4.15). If any species of animal does not reproduce, it ceases to exist. Humans which have imbedded in the primitive part of their brain the compulsion to reproduce have combined the process of trading genetic material with the process of making new individuals. We will discuss the processes involved in human reproduction in detail in a later chapter.

Change is stress: Using the term "priorities" implies that there are circumstances in which decisions must be made at the level of the body concerning the distribution of the blood. There are many such circumstances. All of these circumstances are related to the degree of real or anticipated change in the body. Change requires the body to adjust. Change is stress. Any change that the body is required to make is a stress. The body is almost always under some degree of stress, and the hormones that control the stress responses of the body are always present in the blood. The difference between conditions of stress is simply a matter of degree. Some situations require small changes and some situations require very large changes. If you stand up from a sitting position, the body must respond to this change. Your body must adjust its flow of blood so that gravity doesn't pull all of the blood down around your ankles. These adjustments require energy. If you go out on a cold day, your body must respond to the change. The body must adjust blood flow to conserve heat. All of these adjustments require energy. The response to stress involves blood flow and requires energy. In addition to responding to

Figure 4.15 - Reproduction: "Mother Nature" makes us want to participate!

change, the body can also respond in anticipation of change. The brain can initiate a stress response to prepare for a situation. For example, going back to our riddle of the tiger in the jungle: while you are hoping that the tiger doesn't see you, your body is making a massive stress response. This stress response involves blood flow and energy. Hormones and nerve signals speed up your heart. Hormones and nerve signals open the airways in your lungs in order to provide increased oxygen for the increased blood flow. Hormones and nerve signals open the blood vessels to your brain. Hormones and nerve signals close blood vessels to your digestive system (because digesting your food is a slow process and isn't a high priority at this time). Hormones and nerve signals increase the blood flow to your muscles in anticipation of running. Hormones cause the liver to dump sugar, and at the same time begin to replace that sugar with amino acids from the muscle. Hormones mobilize fat because that energy might be needed. Internally, the brain even focuses its attention on the danger. The body is ready to act. However, the health problem of modern society is that we have "killed most of our tigers." We do not follow such preparations for action with action, and usually the body is left with poor digestion and the problem of putting everything back into storage. In addition, many drugs like nicotine and caffeine "work" by causing a stress response. Not many people exercise while smoking and drinking a cup of coffee. Stress responses without action are the cause of many of the diseases prevalent in our society today.

chapter five

a question of pressure

Blood pressure is a common word: Everyone has heard the term "blood pressure." Blood pressure is one of the most common measurements taken on the body. Maintaining the "right" blood pressure is essential for health, and in fact for survival. A very prevalent medical problem today, both in the US and worldwide, is "high" blood pressure. Drugs are usually used to treat this problem. In addition, a side effect of many drugs is an alteration in blood pressure. What exactly is blood pressure and why do we measure it? What do these measurements mean? Why do so many drugs influence blood pressure? How does the body use blood pressure to express its priorities?

Figure 5.1 - The cardiovascular system includes the heart and blood vessels.

The cardiovascular system: The common water environment of the body, the blood, resides in the cardiovascular system. The cardiovascular system includes the blood itself, the heart (a muscular pump which pushes blood around the body), and the blood vessels which are the passages through which blood flows. The purpose of the cardiovascular system is to distribute the blood with the resources that it contains to the tissues, as well as to carry the

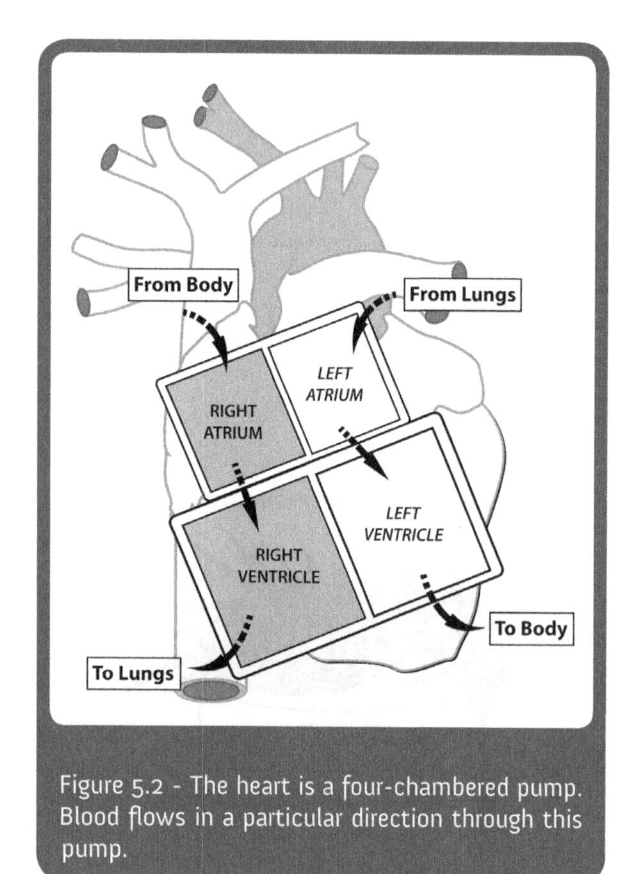

Figure 5.2 - The heart is a four-chambered pump. Blood flows in a particular direction through this pump.

waste products of life to the lungs and the kidney for removal from the body (Fig. 5.1).

At the center of the cardiovascular system is the heart (Fig. 5.2). The heart is a four-chambered pump. At the top of the heart are two smaller chambers called right and left atria. The right atrium receives blood from the body while the left atrium receives blood coming from the lungs. At the bottom of the heart are two larger muscular chambers called the right and left ventricles. The right ventricle pumps blood to the lungs to remove CO_2 and pick up O_2, while the left ventricle pumps oxygenated blood to the body. If you follow the flow of blood around the body, it leaves the heart from the left ventricle, travels through the body, and returns to the heart through the right atrium. From the right atrium it passes to the right ventricle, and then on to the lungs.

In the lungs it gets rid of carbon dioxide and picks up oxygen. It then leaves the lungs and travels back to the heart by way of the left atrium, then to the left ventricle and back out to the body again (Fig. 5.3). Blood leaving the right ventricle going to the lungs has just returned from the body, so this blood is low in oxygen and high in carbon dioxide. Blood leaving the left ventricle going to the body has just returned from the lungs, so this blood is high in oxygen and low in carbon dioxide.

Connected to the ventricles are arteries that carry blood away from the heart. Arteries are channels surrounded by circular muscles (a type of muscle called smooth muscle). Although the brain does not direct the heart to contract, it can regulate how often and how strong the heart contracts. The brain does not cause the heart to contract, it just regulates how it contracts. In addition, by controlling the muscles around the arteries, the brain controls the relative openness of every artery in the body, thus controlling where the blood goes.

The main artery that leaves the left ventricle branches into smaller and smaller arteries that eventually connect with very small blood channels called capillaries (Fig. 5.4). Capillaries are very small blood channels whose walls are only a single cell thick. Capillaries are so small that the cells of the blood must almost pass through them in single file. Exchange between the blood and the water around tissue cells takes place in the capillaries. In capillary beds, molecules, ions and water leave the blood and enter the cells. In addition, molecules, ions and water also enter the blood from the cells of the body through the capillaries. Many of the molecules that enter the blood are waste products of life processes, such as carbon dioxide. The capillaries then connect to very

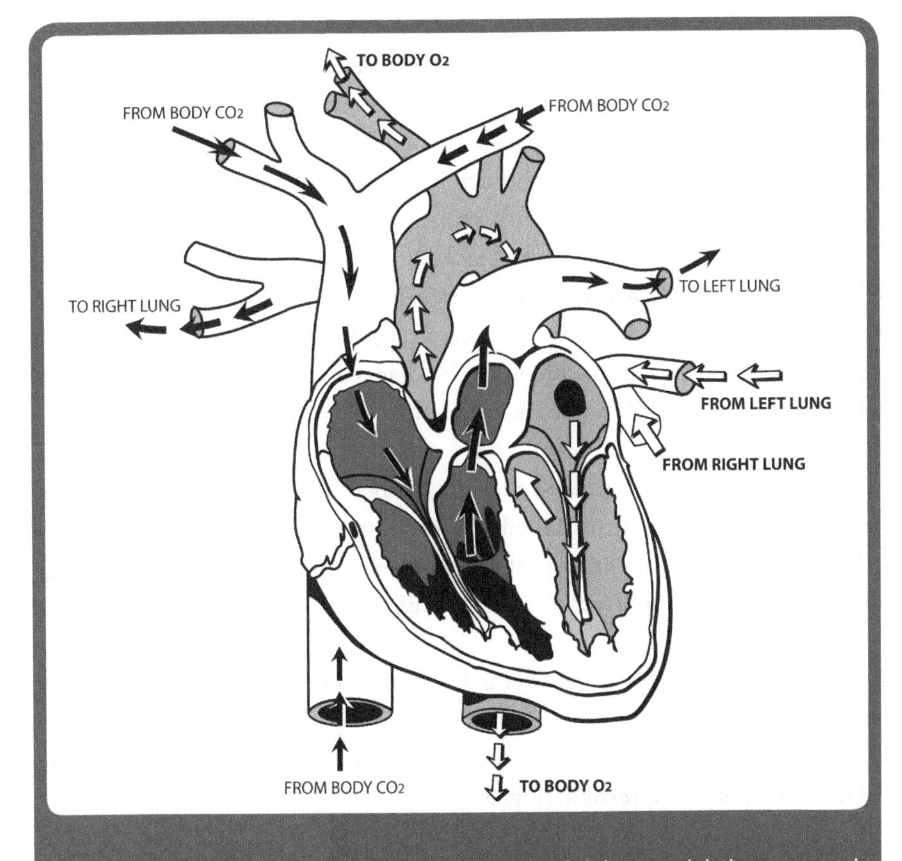

Figure 5.3 - Blood present in the right atrium and right ventricle has returned from the body and has a high concentration of carbon dioxide (black arrows). Blood in the left atrium and left ventricle has returned from the lungs and has a high concentration of oxygen (white arrows).

small channels that begin to join together to make larger and larger channels for the return of blood to the heart. These channels are called veins. Unlike arteries which are surrounded by strong muscles, veins are surrounded by very weak muscles.

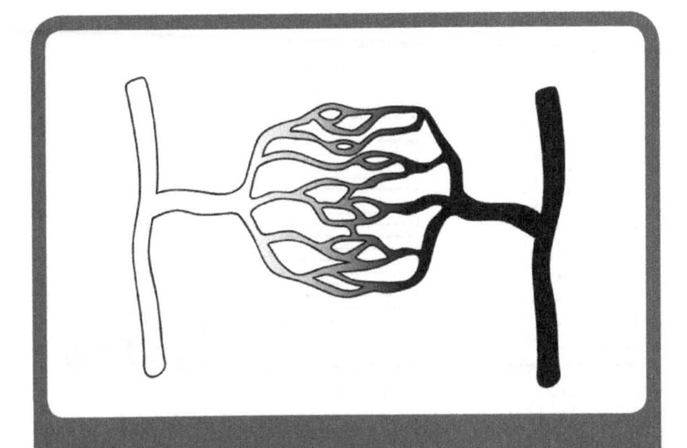

Figure 5.4 - Arteries branch into very small vessels called capillaries, which are only one cell thick. These capillaries branch into larger vessels called veins.

As discussed in the previous chapter, the capillaries, which allow exchange with each unique cell type, have different characteristics. All capillaries are not the same. The differences in capillaries pertain to how exclusive they are with respect to which molecules they allow to leave the blood and enter the environment of that particular type of cell. Although the selectivity or permeability of capillary beds is rather unique to the particular tissue, this selectivity is not a constant. Chemical signals, released primarily by the immune system, can change the permeability of a capillary bed. Such changes in permeability usually occur in localized areas. The sum of the "collection of channels" (arteries) represents potential paths through which some volume of blood can travel. How much of the volume of blood travels through any potential path at any given time depends on the priorities of the brain at that time. The brain can direct arteries in some places in the body to "open up" so that more blood flows to these areas, while at the same time narrowing the arteries that feed other parts of the body so that less blood flows to these areas. The priorities of the body are expressed by the relative distribution of blood to the various paths available to it. The distribution of the blood can be changed quite rapidly only because the blood moves rapidly from the heart around the body and back to the heart.

Blood pressure is the force which moves blood through the body. Blood pressure reflects the capacity of the heart to accomplish the work of moving the blood. The capacity to do this work is the difference between the pressure in the blood vessel when the heart is at rest (between contractions) and the pressure of the heart when it is pushing blood (during a contraction). It is the differences in these two pressures that move blood.

Blood pressure is a force: Pressure is a force. On earth, a dominant pressure is caused by gravity. The effect of this pressure on the gases of the atmosphere is to make the air close to sea level denser than the air at the top of a mountain. There is more air sitting on top of the air at sea level than there is sitting on top of the air high on a mountain. Gases are quite compressible (Fig. 5.5).

The result of air sitting on air sitting on air is to compress the air at the bottom. The air closest to the center of the earth is denser because it has more air sitting on it. The farther you move away from the center of the earth the less dense the air becomes. In an airplane, the air needs to be artificially compressed because the air at high altitudes is not sufficiently dense to sustain life. This is because there simply is not enough of it.

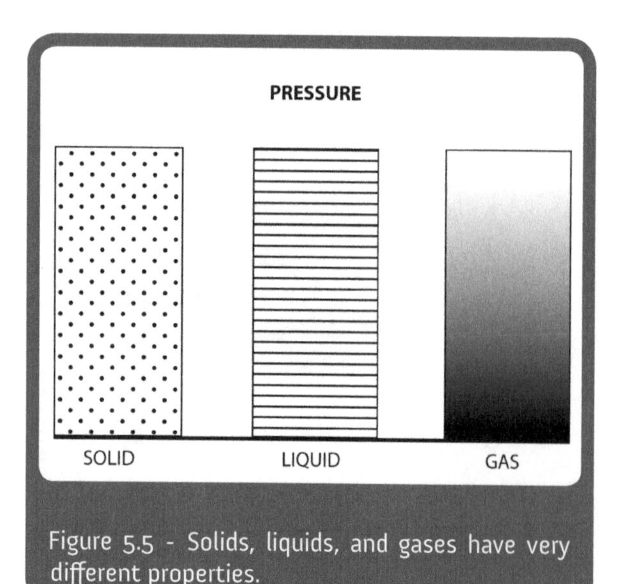

Figure 5.5 - Solids, liquids, and gases have very different properties.

In contrast to gases, solids like cement do not change their density or shape in response to the weight of air or anything else sitting on it. All of the pressure of the weight is distributed at the surface of a solid object. If you hit your head on a bag of air, the force of your head hitting the bag is distributed by compressing and moving the air. In contrast, if you hit your head on a rock, all of the force of your hard head hitting the hard rock is distributed at the surface between your head and the rock. The result is a fractured skull.

Liquids like the water of blood are different than a solid in that they are deformable. That is, liquids can change their shape like a gas. However, liquids are different than gases in that they are not compressible. It is these qualities of a liquid that make them ideal as a vehicle for moving molecules and energy around the body against the force of gravity. The pressure at the surface of the ocean is one atmosphere. This is equivalent to the pressure of 760 mm of mercury, or 14.7 lbs. per sq. inch. All of the air sitting on the surface of the water exerts that much pressure. In comparison, the pressure 32 feet under water is two atmospheres. Just 32 feet of water exerts the same pressure as the entire atmosphere of the earth.

Figure 5.6 - As a bubble of compressible gas rises in a six-foot cylinder of water, it expands as the pressure of the water above it decreases.

If we consider a six-foot tall human with a heart at the level of four feet when he or she is standing up, much of the work of the cardiovascular system involves moving the blood (a liquid) against the force of gravity. Gravity exerts a force caused by the pressure of the liquid at the top of the body sitting on the liquid at the bottom of the body (Fig. 5.6). If you remove that person from the earth and place him or her in the microgravity of space, then the distribution of the liquid in his body changes. It changes because the force of gravity is no longer pulling the liquid of the body towards the earth. Blood pressure is different in space because of the greatly reduced gravity in space. Similarly, if you take a person's blood pressure on earth by putting the cuff around the arm, it is different from the blood pressure measured at the level of the leg. Much of the difference is simply caused by differences in the force of gravity. All of the priorities for distributing blood around the body must be expressed within the context of the force of gravity.

The heart generates pressure: The cardiovascular system of the body is a closed system of channels connected to a pump. The heart is the pump. As we discussed above, this is a four-chambered muscular bag. The top two chambers (atria) are designed to receive blood. The right atrial chamber receives blood from the body. The left atrial chamber receives blood from the lungs. The other two chambers (ventricles) are strong muscular pumps. The right ventricle chamber pumps blood to the lungs. The left ventricle chamber pumps blood to the entire body. When the heart relaxes, blood from the body enters the right atrial chamber and blood from the lungs enters the left atrial chamber. Blood enters the heart when the heart is at rest. The blood flows through both atrial chambers into both ventricular chambers. The blood from the right atrial chamber flows to the right ventricle and the blood from the left atrial chamber flows to the left ventricle. Blood flows in this direction around the body because of one-way valves that are present between each chamber of the heart. These valves open as blood flows in this direction and then close, preventing a "backflow" of blood.

The heart contracts as a wave beginning at the atria spreading down to the ventricles. We said that the brain does not cause the heart to contract. So why does the heart contract? The signal for the contraction actually comes from "pacemaker cells" in the left atria. These cells, in fact all heart cells, are rather unusual in that they have an intrinsic beat. For example, if one were to remove the heart of a chick embryo and separate these cells in a dish containing salt and sugar water, each individual living cell could be seen to be rhythmically contracting. Some cells would be contracting at about 70 times a minute while others would be contracting at about 30 times a minute. The ones with 70 beats a minute would be from the atria while the ones with 30 beats a minute would be from the ventricles. If you pushed two cells with 30 beats a minute together so that they touched, they would beat together in synchrony. If you shoved a third cell into contact with the first two it also would join the synchronous contraction. This could go on until you had a string of cells all in contact, and all beating together at a rate of 30 beats per minute. Now if you were to push over one of the atrial cells that beats at 70 beats a minute, as soon as it established contact, the entire string of cells would begin to beat 70 times a minute. This essentially is what happens in an intact heart: the fast atrial cells control the rhythm of the heart. If one cut every nerve connection from the brain to the heart, the heart continues to beat at its one intrinsic rhythm. All the brain can do is speed up or slow down the heart.

When the heart begins a contraction, blood in the atria is pushed into the ventricles. As the wave of contraction spreads from the atria to the ventricles, they begin to contract. Because there are valves between the atria and the ventricles that do not allow blood to flow backwards, the pressure generated by the contraction pushes the blood out of the ventricles into arteries. The blood in the right ventricle is pushed through the pulmonary artery to the lungs for exchange of gases. The blood in the left ventricle is pushed through the aorta to the body. Blood moves out of the heart only if the pressure generated by the heart's contraction is greater than the "back pressure" of the blood in the arteries. When the heart again relaxes, then blood that was pushed to the lungs by the right ventricle returns from the lungs carrying oxygen and enters the left atrial chamber. At the same time, new blood from the body enters the right atrial chamber. The process begins again.

How do you measure blood pressure? When you put a cuff around a person's left arm and pump it up, as is done when blood pressure is being measured, you block the flow of blood in the brachial artery carrying blood to the arm. When the pressure of the cuff is greater than the pressure generated by the heart to push the blood through the brachial artery, blood flow in this artery stops. As you slowly release the air in the cuff, at some pressure exerted by the cuff around the arm, blood will begin to flow into the brachial artery when the heart contracts. The pressure produced by the contraction of the heart is now just enough to overcome the pressure exerted by the cuff around the arm. When the person taking your blood pressure hears the turbulence of the blood pushing past the cuff, they read the gauge and obtain a number. This number, for example 120, is the pressure in millimeters of mercury exerted by the contraction of the heart together with the resting pressure of the blood in the artery. As you continue to release the pressure of the cuff, more and more blood flows through that artery. The pressure of the cuff is still an impediment, so the flow of blood is turbulent and makes a sound. How much blood flows is a function of the balance of the pressure generated by the heart against the pressure still in the cuff. However, at some pressure the cuff no longer interferes with the flow of blood to the brachial artery. The sound of the blood pushing past the cuff disappears. The pressure of the cuff is at or below the pressure in the artery when the heart is not pumping. This is the resting pressure in the artery, for example 60 mm of mercury. The person's blood pressure in this example is expressed as 120/60 (Fig. 5.7). The top number of a blood pressure (in this case 120) is referred to as the "systolic pressure": it is the pressure generated by the heart contracting. The bottom number (in this case 60) is referred to as the "diastolic pressure": It is the back pressure in the arteries which impedes blood flow.

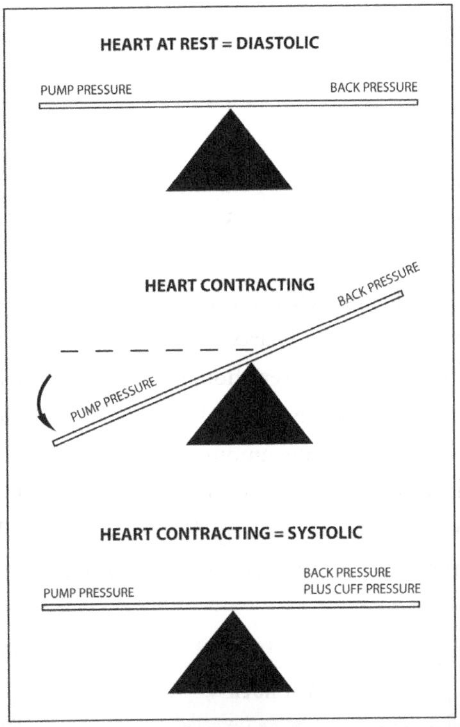

Figure 5.7 - My blood pressure is 120/60: what does this mean? When the heart is at rest, the pressure in the artery of the left arm is 60. When the heart contracts, the pressure is 120. The ratio is the volume of blood which is moving.

The difference between the two numbers is the pressure available to do the work of moving the blood. The resting pressure reflects the resistance to flow in that artery carrying blood to the arm. When the cuff was in place and pumped up above 60 mm of mercury, the volume of blood flow through that artery in the arm was reduced. When the cuff exerted a pressure above 120, the blood flow stopped. What happened to the blood volume that did not go to the arm? It simply went down other arterial paths with less resistance to flow. There is, in fact, a predictable relationship between the factors controlling the flow of blood in any artery:

$$P = V \times R$$

where:

P = pressure

V = volume of blood flow

R = resistance to flow

What the blood pressure really indicates is the volume of blood flowing in the artery since the relationship can be rearranged as follows:

$$V = P/R$$

This equation is really exactly the same as the equation you probably learned in high school physics when you studied electricity (Fig. 5.8)!

This ratio also reflects how much effort the heart must expend to accomplish the work of moving the blood. For example, the hearts of two people with blood pressures of 120/60 and 160/80 are moving the same volume of blood, even though the person with 160/80 has somewhat "high blood pressure." Because the back pressure is increased, the force of contraction (reflected by the top number) also increases to compensate and move the same amount of blood. High resistance in arterial paths (such as occurs when athrosclerotic placques form in arteries narrowing them) leads to an enlarged heart and pathologies associated with high blood pressure. An enlarged heart can eventually lead to congestive heart failure, a situation in which the heart cannot generate enough force on contraction to empty.

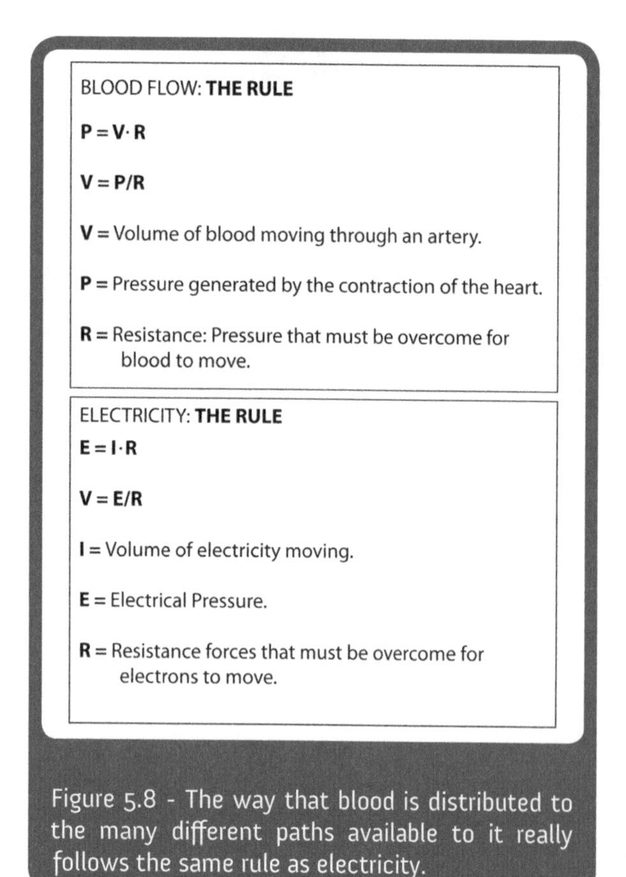

Figure 5.8 - The way that blood is distributed to the many different paths available to it really follows the same rule as electricity.

Blood distribution is dependent upon relative resistances: As we mentioned, arteries are vessels with walls made of a strong circular muscle. When the muscle contracts the diameter of the vessel becomes smaller and resistance to flow increases. Conversely, when the muscles of the wall relax, the diameter of the vessel becomes larger and the resistance to flow decreases. The volume of the blood is distributed among the arterial paths based on relative resistances (openness).

The above equation simply says that if you increase resistance to flow, you must reduce the volume to keep pressure constant. Similarly, if you increase resistance to flow and keep volume constant, pressure goes up. This equation describes the flow of any liquid through a tube. For example, if you turn on a garden hose, the water flows out of the hose. The movement of the water out of the hose releases the pressure behind the flow. How much water flows out is the volume, and the diameter of the hose is the resistance to flow. If you put your finger over the hose, partially blocking the end of the hose, then you are increasing resistance because you are reducing the size of the opening. The same amount of water still flows out of the hose, but it squirts farther because the pressure is increased due to the increased resistance. If you put a small hole in the side of the hose, some of the water now flows out of the end of the hose and some water flows out of the hole on the side. How much of the water flows out of the end of the hose and how much of the water flows out of the hole in the side of the hose will be a function of the relative size of the two holes. If you now put your finger over the end of the hose again and reduce the size of the opening, more water will come out of the hole in the side of the hose because you have increased the resistance to flow out the end whereas the resistance to flow through the hole in the side is unchanged. In fact, if you had many holes all of different sizes through which the water could flow, how much of the available volume that would flow through each would be a function of the size of each hole relative to the size of all others. More of the water would flow through the larger holes with less resistance and less water would flow through the smaller holes with more resistance.

This is the situation in the cardiovascular system of the body. The heart pumps thereby creating a pressure to push the blood volume through the many arterial paths available to it. How much of the volume goes down each arterial path is a function of the resistance of each path relative to all others. The resistance of each artery can be changed by changing its diameter due to contraction of its surrounding muscle layer. In the case of the heart, all of the blood going to the body leaves through a common arterial path (the aorta). The common path begins to branch. The first branch is a very important path known as the coronary artery. This branch and all of the smaller vessels that branch off of it deliver blood to the heart muscle itself. If one of the branches of this artery gets "clogged up" with fat, blood flow to part of the heart muscle is disrupted. A type of heart attack known as a "coronary" occurs. If blood flow through the coronary artery is reduced due to narrowing, a type of pain called angina is felt in the left shoulder.

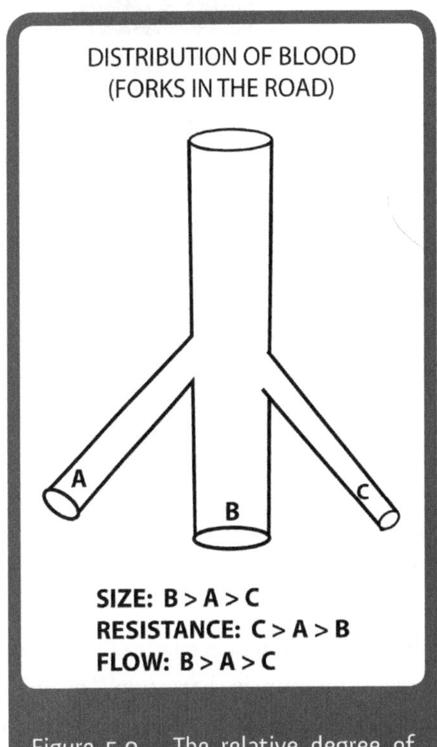

Figure 5.9 - The relative degree of opening of arteries determines where blood is distributed.

Past the coronary artery many branches leave the major path leaving the heart. At each branch point in the path, some of the volume goes one way and some of the volume goes the other way. How much of the volume goes each way will depend on the resistance to flow of each path relative to the other (Fig. 5.9). Branching paths split and the distribution of the volume of the flow is determined in the same way.

Gravity is important to blood flow: The heart is not the only generator of pressure in the system. Gravity generates much of the pressure in the system. When a path reaches a branch where the volume can go down towards the earth instead of up, then the

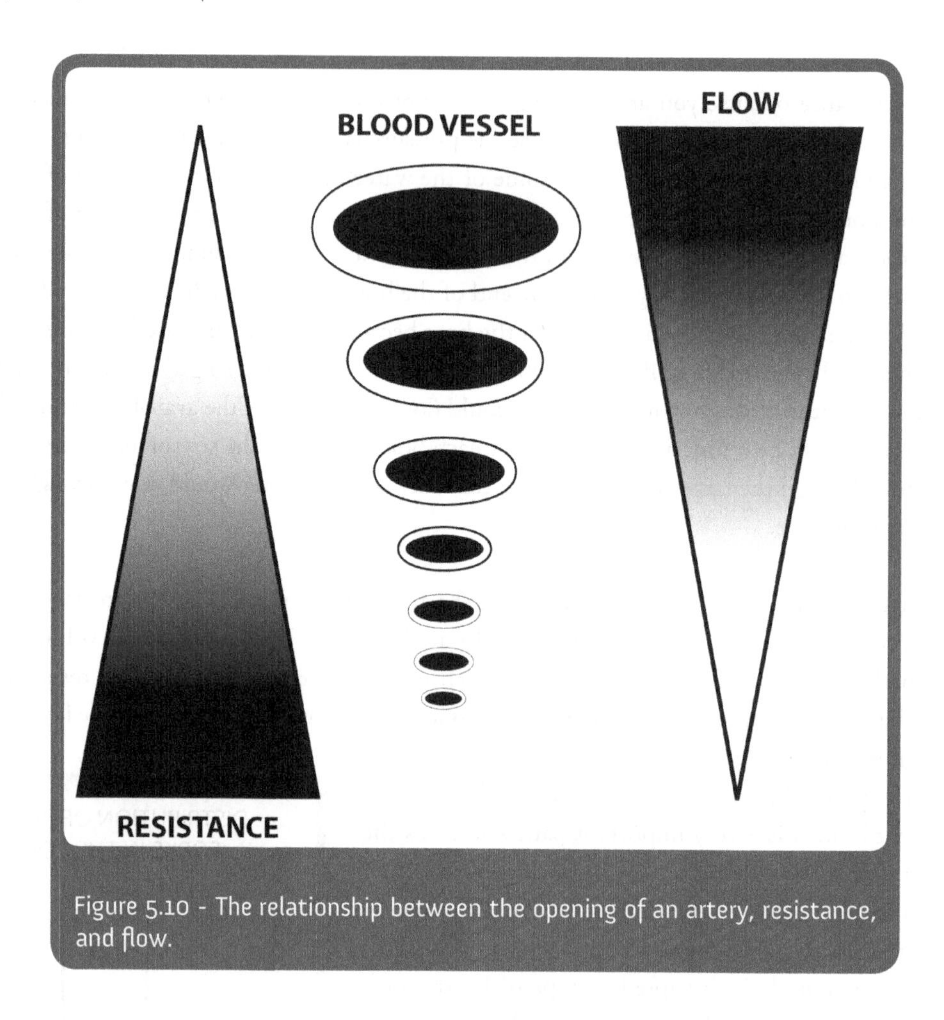

Figure 5.10 - The relationship between the opening of an artery, resistance, and flow.

force of gravity is added to the pressure for the volume to go down and subtracted from the pressure to go up. The only way that the body can prevent all of the blood volume from going down is to increase the resistance in the paths going down towards the earth and to decrease the resistance in the paths going up away from the earth. The mechanism by which resistance to flow is increased or decreased is the same as with the garden hose. When you decrease the diameter of the arterial path, you increase resistance. When you increase the diameter of the arterial path, you decrease resistance (Fig. 5.10). In this way the priorities of the brain for the distribution of the volume of the blood are expressed as the relative opening and closing of the many possible arterial paths that the blood can take at the many branch points available.

All of these changes in relative resistances are taking place within the context of changes in the two factors that generate pressure in the system. The pressure generated by the heart changes in both how often and how much pressure is generated. "How often" is controlled by heart rate, and "how much" is controlled by the strength of the heart's contraction. The pressure generated by gravity changes with the orientation of the body to the earth. If you lie down, then the direction and the magnitude of the pressure of gravity are different than when you are standing up.

Volume can also change: Just as pressure and resistance undergo controlled changes, so does the third factor in the equation: blood volume. There are two ways by which volume changes. One is short-term and rapid, and the other is longer-term and slower. The short-term, rapid mechanism is expressed in local areas of the body. For example, during an acute stress response the body can raise the pressure in the arterial paths going to the brain, heart and muscles by greatly reducing the volume of blood distributed to other areas such as the digestive system. Pressure increases in the arterial paths to the brain, heart and muscles (even though resistance in them is decreasing) because more volume is being directed into them. This is accomplished by greatly increasing the resistance in the arterial paths going to other areas such as the digestive system. It is, in fact, this mechanism along with the mechanism of temperature control that causes people to sometimes have heart attacks while shoveling snow. When a person exercises, the heart, the muscles, and the brain need more blood. Blood is mobilized into the arterial paths leading to these tissues. When the body wants to conserve heat on a cold day the body allows less blood to flow to the skin. This brings even more blood to the interior arterial paths of the body. When you exercise on a cold day this blood flows into the arterial paths of the heart, muscle and brain. The combined effect of all of this increased volume in these paths is a large increase in the pressure in these arteries. If a person is susceptible to a heart attack, this increase in pressure often causes it to happen.

The second mechanism of volume control is associated with the kidney. However, volume control by the kidney is also associated with the control of the concentration of ions in the blood. The concentration of anything in the body, including ions, is a function of the number of molecules and the amount of water. If the body needs to raise the concentration of an ion, it can decrease the amount of water or increase the number of ions. Conversely, if the body needs to decrease the concentration of an ion then it can increase the amount of water, or it can decrease the number of ions. The kidney is a filtration system for the blood. More than a quart of fluid is filtered through the kidneys every minute. In the filtration process, small molecules such as sugar and ions, along with water, leave the blood and enter a separate series of tubes called kidney tubules. As soon as this water is filtered out of the blood, the cells of the wall of the tubule begin to very selectively return some molecules and ions back to the blood. Sugar is returned, some ions are returned, and water is returned. By returning more or less water to the blood, the kidney can have a large effect on the fluid volume in circulation.

Most of the control of the kidney is based on a balance of two factors. The first is the sodium ion concentration in the water of the body. Sodium ions are so important to the functioning of the body that molecular energy in the form of ATP is used to actively transport sodium out of the kidney tubule and back into the blood. If the sodium ion concentration is too high, then fewer sodium ions will be transported. If it is too low, then more sodium ions will be recovered. The second factor is blood pressure. If blood pressure is too high, then less water will be recovered; if it is too low then more water will be recovered. The balance between these two processes derives from the fact that the system cannot substantially recover sodium ions without recovering water (Fig. 5.11).

Generally, these two factors are not in opposition. However, they are not always compatible with the optimum regulation of the water volume of the body. This is because recovering water provides a limit for recovering

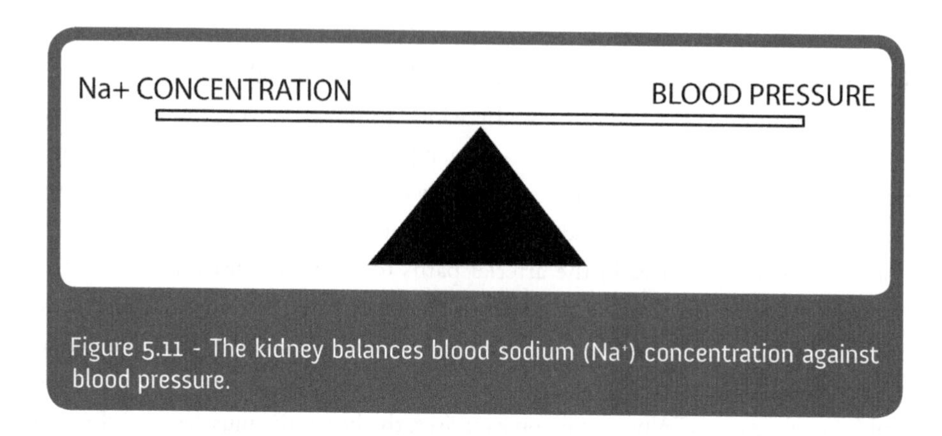

Figure 5.11 - The kidney balances blood sodium (Na^+) concentration against blood pressure.

sodium ions. If the kidney tries to recover too many sodium ions, then it increases water volume to the point that the increased pressure caused by the increased volume of water begins to inhibit the recovery of water, which in turn inhibits the further recovery of sodium ions. This happens because water recovery is highly promoted by a hormone from the pituitary gland which responds to signals concerning both sodium concentration and blood pressure. Fundamentally, the way in which this hormone works is that it controls the number of channels for water to flow through between the kidney tubule and the blood. The more hormone molecules that are present the more channels available for water. The water is simply trying to follow the sodium ions. As the blood pressure goes up, the amount of this hormone, which is known as the antidiuretic hormone (ADH), decreases. As the amount of the hormone decreases, fewer channels are available for the water to follow the sodium ions back into the blood. This hormone attempts to prevent the loss of water from the body only up to the point where blood pressure is too high. The converse situation is also part of the balance: if the kidney tries to recover too much water because blood pressure is too low, then the increased water dilutes the concentration of sodium ions in the body. When the concentration of sodium ions begins to drop, it causes the release of another hormone from the adrenal cortex. This hormone, mineralocorticoids (aldosterone), is a type of steroid hormone. Mineralocorticoids promote the recovery of sodium ions by the kidney. Promoting the recovery of sodium ions promotes the recovery of more water by the kidney.

This cycle of water recovery promoting the recovery of sodium, and sodium recovery promoting the recovery of water, is fine as long as the system properly defines when blood pressure is "too high" or "too low." However, human bodies for one reason or another often have difficulty properly defining "too high" and "too low." A medical solution is the use of a class of drugs known as diuretics. Sometimes non-medical people refer to these drugs as "water pills." Diuretics inhibit the recovery of sodium ions, and therefore increase water loss from the body by increased urination. Diuretics, in combination with a low sodium diet, are often an approach used to reduce blood pressure. This combination provides fewer sodium ions to the body and inhibits the recovery of sodium ions filtered in the kidney. The result is the recovery of less water volume in the kidney, and consequently a lowering of blood pressure.

Problems with blood pressure: Although blood pressure being too high (hypertension) or too low (hypotension) is a somewhat individual thing, much above or much below the normal range can cause medical problems. The upper limit of what is considered normal blood pressure is 120/80. Pre-hypertension is classed up to 140/90 and beyond this is considered as hypertension. As discussed above, the two variables impacting blood pressure besides the strength of contraction of the heart, are volume and resistance. Volume to a great degree is regulated by a combination of kidney function and diet. Excessive salt in the diet forces the brain to balance concentrations with blood pressure. Since the brain never tolerates a blood sodium concentration above normal even for a brief period of time, the balance always allows blood pressure to go up. High salt diets raise your blood pressure. The other factor, resistance, relates to the opening and closing of arteries. A sympathetic response ("stress response"), whether it is caused by exercise or anger will raise your blood pressure above the "normal" limits. If it is caused by exercise then it should return to "normal" within a reasonable time after the person stops exercising. If the sympathetic response is caused by anger or other chronic neurogenic stressors then this can be the cause of the hypertension. The other impact on resistance is flexibility of the arteries. The incidence of hypertension due to decreased flexibility of arteries increases with age. "Hardening of the arteries," atherosclerosis, is a condition in which plaques (deposits) containing cholesterol and other substances build up in arteries and both narrow them. These plaques both increase resistance and reduce the artery's ability to relax and open.

Hypotension, low blood pressure, is more difficult to define in a chronic sense, but can occur quite commonly in an acute sense. Well-trained athletes at rest can have blood pressures such as 100/50 which is well below what would be considered normal. However acute drops in blood pressure, which can be quite life threatening, are not uncommon. When humans began to stand up and walk on their hind legs this created a problem for blood distribution. When a person stands up, major adjustments must occur in vascular resistance, or gravity will pull all of the blood towards his or her feet and away from the head. If these adjustments do not occur then blood supply to the brain is compromised and the person loses consciousness (faints). There are a variety of conditions and drugs such as those used to treat high blood pressure that can interfere with these adjustments and cause the person to pass out when standing up. The condition is known as postural hypotension. Volume problems can also cause hypotension. Although severe dehydration can cause low blood volume, more commonly, loss of blood either internally or externally causes low blood volume and hypotension. Finally, hypotension can be caused by both a loss in vascular tone and volume. Acute

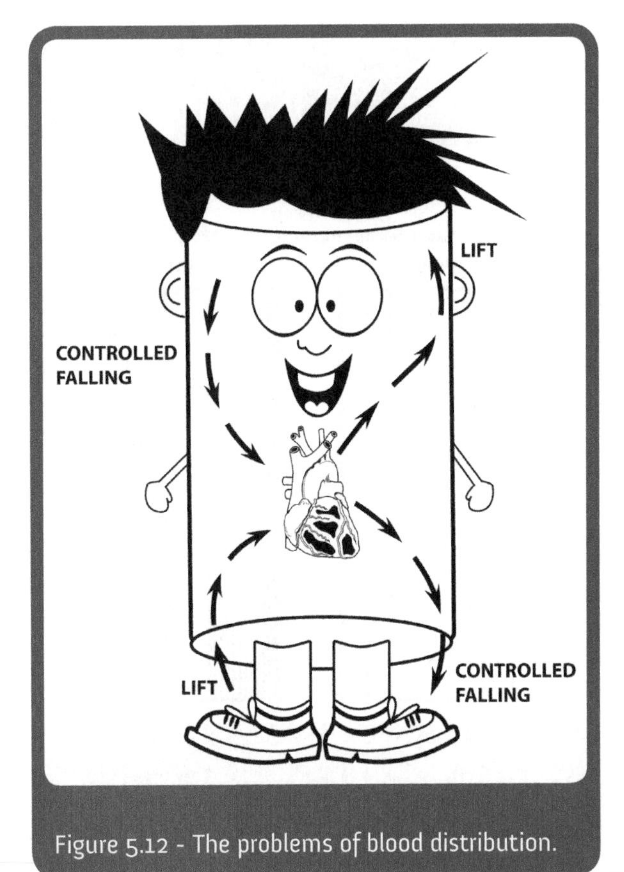

Figure 5.12 - The problems of blood distribution.

Figure 5.13 - Like sinuses, fluid cannot leave the straw unless the space behind it is filled.

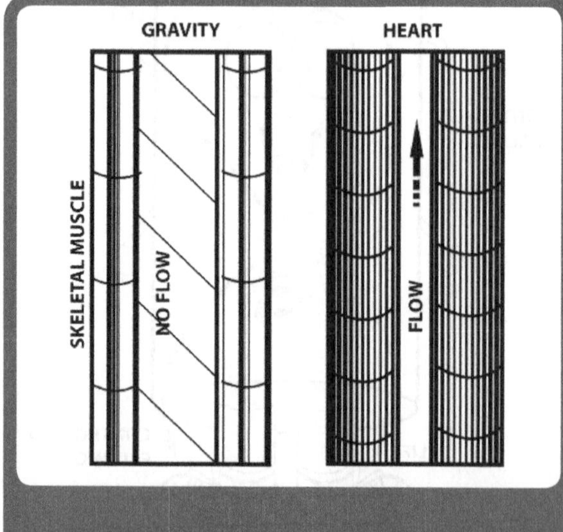

Figure 5.14 - Skeletal muscles contract and squeeze veins, which then pushes blood towards the heart.

allergic reactions can cause life threatening hypotension. This condition is known as anaphalactic shock. In this case a systemic immune response causes both a loss in vascular tone and opens up the capillary beds and allows water to leave circulation and enter the space around cells reducing blood volume. This loss of both vascular tone and reduced blood volume can be quite deadly.

Returning blood to the heart: There are two aspects to the problem of returning blood to the heart (Fig. 5.12).

One aspect deals with the blood that went up to the brain. The second aspect of the problem is the blood that went down toward the earth. Humans, being animals that walk on two legs, must generate enough pressure to return most of the blood of the body to the heart against the force of gravity. Volume is a problem because there is often too much of it. The human body has the problem of lifting much of its blood volume back to the heart against the force of gravity. Returning the blood to the heart is a different process from that of distributing the blood to the tissues of the body. As blood leaves the heart, a significant but relatively small fraction of the blood volume must go to the head, against gravity. This is a special problem which is solved by the judicious control of resistance and pressure. However, most of the blood leaving the heart simply drops from the heart to the rest of the body. Resistance, the relative opening and closing of arterial paths, is used to direct more or less blood to different areas along the way. However, returning the blood to the heart is a difficult fight against gravity.

The venous return system has two problems. First it must prevent the blood that went up to the head from returning to the heart too rapidly. Second, it must lift the blood that went down all the way back up to the heart. The problem of preventing the blood that went up to the head from returning too fast is solved by the sinuses. The sinuses serve like a "brake" on the rate of return of blood from the head. These paths for blood on the way back to the heart from the head slow down the return of blood because they are a series of paths with walls of rigid bone. Gravity can pull on them and they don't collapse and get smaller. Sinus pathways do not respond to the force of gravity. There is no change in the resistance to flow. Gravity pulls on the volume of blood. However, liquids are not compressible nor are they expandable. Therefore, gravity can affect neither the size of the sinus tubes nor the volume of the blood in them. Both remain constant, and as indicated by the equation, pressure then remains constant. Because of this, gravity can pull on the sinuses, but for the blood to flow from the sinuses to the heart, blood must be added from the brain to the sinuses. Blood returns from the sinuses in the head to the heart just about as fast as the blood enters the sinuses from the brain. The value of the sinuses as a break on blood leaving the head can be understood by considering the game children play while fooling with a straw and a soft drink. Everyone has seen a child put a straw into a soft drink, put his or her finger over the end and lift the straw out (Fig. 5.13). The straw will remain full of fluid until the finger is removed. When the finger is removed from the end, the fluid runs out on the table because air is allowed in behind the fluid.

Return of the blood that went down towards the earth is a much greater problem. Blood is pushed and sucked back to the heart by several forces. First, blood is pushed back to the heart by the contraction of skeletal muscles. The contraction of skeletal muscles associated with movement of the body, especially in the legs, squeezes and pushes on the venous paths. Venous paths are more sensitive to the squeezing and pushing of the skeletal muscles than are arterial paths because they are thinner and less rigid than arteries (Fig. 5.14). When the veins are squeezed and pushed on, the blood in them is shoved out of the way. Since the venous paths contain valves that only allow blood to flow one way (towards the heart) the only way the blood can move is toward the heart. Also because of these one-way valves, when the muscle stops pushing on the venous path, the blood does not fall back down with gravity. The second force aiding in the return of the blood to the heart is the negative pressure created by expansion of the chest during breathing. When the chest expands during breathing, it creates a negative pressure that sucks air into the lungs. This negative pressure also causes the large venous paths in the chest to balloon out a little bit. When these veins balloon out, they suck blood up towards the chest. Again, one-way valves prevent the blood that has been lifted up from falling back towards the earth. The last force for returning blood to the heart is created by the heart itself. Blood leaving the heart through the arterial paths must be replaced when the ventricle relaxes. The relaxation of the ventricle creates a negative pressure which sucks blood into the heart. This negative pressure is transferred down the major venous paths and pulls blood towards the heart. Finally, as people often find out as they get older, one can aid the return of blood to the heart. This can be done by simply raising the feet, which reduces the force of gravity on the process. By sitting down and putting your feet up you reduce the amount of work necessary to return blood to the heart.

What is the purpose of blood flow? The movement of this large volume of blood has a purpose. The purpose is to deliver blood to the capillary beds of the body. Exchange between the water environment of the blood and the water environment directly around tissue cells takes place in the capillary beds. There is then an

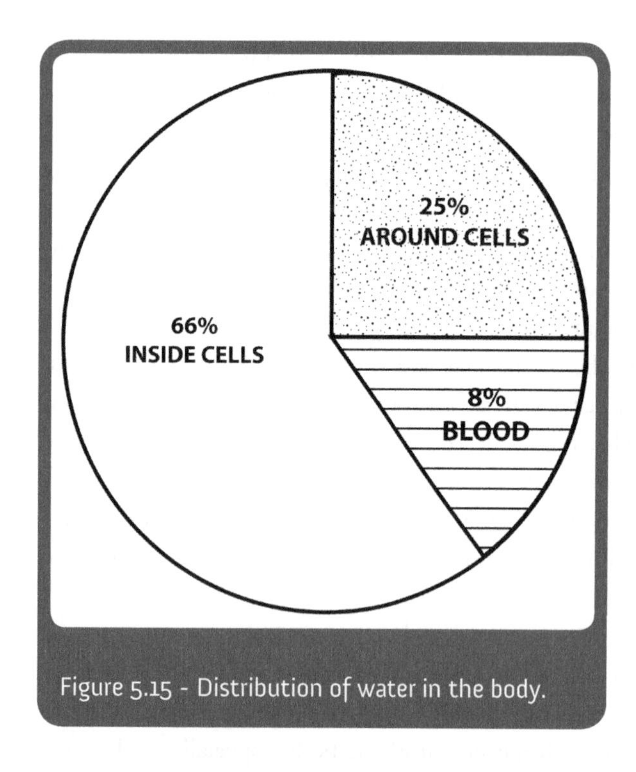

Figure 5.15 - Distribution of water in the body.

exchange between the water environment around the cells and the water environment inside cells. To put these three water environments into perspective, from 60%t to 70% of the entire weight of the body is water. About 66% of this water is inside cells. Twenty-five percent of this water is around cells, and the last eight percent of the water of the body is in the blood (Fig. 5.15). This means that a 150 lb. person has a little less than a gallon of blood in circulation. About every minute, a volume of water equivalent to the entire volume of blood is exchanged in the capillary beds of the body. What this means is that at the same time that the volume of blood is flowing from the heart to capillary beds and back to the heart, there is a large flow between the blood and the water environment around cells. The net flow is out of the blood into the water environment around cells. To compensate for this net outflow from the blood there is a drainage system, known as the lymphatics, which returns the excess water in the extra-cellular water environment (the water around cells) to the venous paths for return to the heart.

Why do so many drugs influence the cardiovascular system? So many drugs influence the movement of blood in the body because directing the distribution of the blood in the body is very important to the body. Blood is important because it is the mechanism for dividing up the limited supply of nutrients. Blood is necessary for controlling body temperature. Blood is necessary to the control of the ion concentration of the body fluids. Blood is necessary to control blood pressure. Everything related to blood is necessary. The distribution of blood is a constant "emergency" in the body, and the body always maintains priorities concerning the distribution of blood. Because of this, drugs that influence one function or another have either a direct or indirect effect on the distribution of the blood.

chapter six

barriers of the body

What do outside barriers do? The lungs, the skin, and the digestive system form the outside barriers of the body. It is these structures that separate an individual human from the world. Humans are always exchanging materials with their environment. They must take in food and water from the environment to both replace water loss and to provide substrates that can be broken down by biochemical processes to extract cellular energy. They must take in oxygen, which is a necessary component for extracting energy by the cell. In turn, they must remove carbon dioxide from their bodies, which is a toxic waste product of the energy generating pathways. In addition, the body must be maintained within a certain temperature range in order to function properly. Most of the molecules and heat necessary for maintaining the quality of the internal water environment of the body flow in and out through the cells of the lungs, the skin and the digestive system In order to maintain the stability of the molecular composition of the internal water environment, there must be a constant flow of molecules into and out of the body. Similarly, in order to maintain the body at the appropriate temperature, there must be constant control of heat loss from the body.

The quality of being warm-blooded (of maintaining a constant internal temperature relative to the changing temperature of the outside world) allows an animal a very high level of year-round activity. However, it is also very expensive with respect to molecular flow, and very difficult with respect to controlling heat flow. The basic problem is one of maintaining a high enough molecular flow to support the heat production which is necessary for temperature regulation. Therefore, a human must eat a lot of food and breathe a lot of air simply to maintain his or her body temperature. The breakdown of food molecules, which requires oxygen, produces condensed high-energy molecules like ATP. A byproduct of the use of this condensed energy, especially by the muscles of the body, is heat. The use of metabolic energy is not 100% efficient. In fact, more than one-half (62%) of the energy contained in the food molecules is lost to heat. Land animals that don't maintain a relatively high constant temperature (cold-blooded animals) must settle for periods of lethargy and unconsciousness. The quality of warm-bloodedness allows for a constant awareness of the world and a high mobility in the world.

In general, humans are best suited to exist in a world that is somewhat cooler than their body temperature. Humans bring large amounts of food and large amounts of oxygen into their bodies. Much of these resources are used to generate high-energy molecules by the various types of muscles to generate large amounts of heat. The temperature of the body is maintained by controlling the loss of this heat from the body (Fig. 6.1). The warmer it is outside of the body, the faster we lose heat. The colder it is outside, the slower we lose heat. Although some of the heat is lost through the lungs, the digestive system, and the kidney, heat control primarily involves the skin.

The skin: The surface of the body is covered by a highly dynamic group of cells which constitutes both the immediate surface of the body and the cells which will replace that surface. The skin provides a protective barrier for the body. It prevents water loss. It also prevents the entry of potentially infecting micro-organisms (this will be discussed in more detail in a later chapter). The surface of the body is in a constant process of renewal. There is a very active stem cell population in the skin that is constantly proliferating (reproducing new cells) and differentiating (becoming new skin cells). This process of renewal occurs at a rate sufficient to keep the body covered. Furthermore, there can be adjustments in the rate of replacement of skin cells to compensate

Figure 6.1 - The degree of heat lost through the surface of the body helps maintain body temperature.

for damage. This of course does not mean that damage cannot exceed the capacity of the skin to renew itself. It simply means that damage accelerates the renewal process. Among the skin cells live a relatively few hairs and a few channels which deliver water from the sweat glands to be poured over the surface of the skin. Also among the skin cells is a vast network of blood vessels which carry nutrients and waste to and from the capillary beds of the skin. This network of blood vessels also carries heat generated by the muscles to the surface of the body. In addition, in this community of cells at the surface are nerves which carry information concerning touch, pain, and temperature to the spinal cord and ultimately to the brain. Like everything else in the body, the brain keeps track of body temperature. It monitors the temperature of the blood going through the brain, and it monitors the temperature of the surface of the body. The major purpose of all of this monitoring is to control the temperature of the environment of the brain.

Water, the environment of the body, has two special characteristics with respect to heat. The first is that it has a high heat capacity, and the second is that it has a high heat transfer. The consequence of the high capacity is that it can carry a relatively large amount of heat without changing form. The consequence of the high heat transfer is that heat moves rapidly from one place to another in water. Therefore, the blood flowing around the body carries a large amount of heat. If the water environment in a particular area of the body is warmer than the blood, then heat is transferred from that area to the blood. Conversely, if the water in a particular area of the body is cooler than the blood, the heat is transferred out of the blood to that area.

There is a group of cells in a part of the brain called the hypothalamus which both monitors the temperature of its own water environment and maintains the temperature set point of the body (Fig. 6.2). This group of

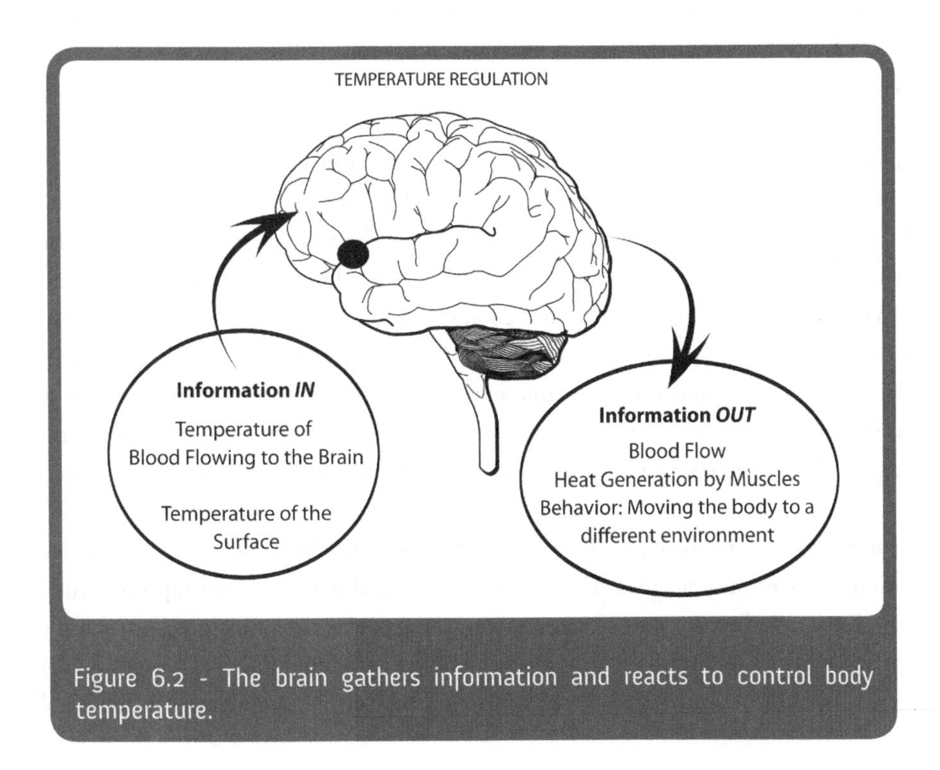

Figure 6.2 - The brain gathers information and reacts to control body temperature.

cells acts to maintain the body temperature around 99° F. Body temperature can actually normally fluctuate about a degree during the day. However, body temperature can also increase several degrees associated with disease. In this case, chemicals released by infecting micro-organisms or chemical messages released by cells of the immune system can act on the temperature regulating cells of the brain and increase the temperature set point of the body. In this condition, called a fever, the temperature regulating cells of the brain cause nerve signals to be sent out to the body reducing heat loss and generating more heat. The body temperature elevates to the new artificially high set point.

In addition to monitoring the temperature of the blood circulating to the brain, the brain receives information about the temperature of the surface of the body. The brain uses the information about the surface temperature for a very obvious but important aspect of temperature regulation: the brain can use skeletal muscles to move the body to a warmer or cooler environment. However, the information about blood temperature gathered by the hypothalamus is used to help adjust blood flow to the surface of the body. The temperature-sensitive cells of the hypothalamus have the capacity to set in motion signals through the autonomic nervous system which change the flow of blood in the body and thus change the flow of heat. By controlling the flow of heat in the body, the brain attempts to control the temperature of its own water environment. If the temperature of the water environment of the brain increases or decreases by more than five or six degrees, then the situation is life threatening. For example, a temperature of 106° Fahrenheit, if maintained for very long, will cause permanent brain damage and eventually death. Conversely, if the temperature of the brain is decreased by about15 degrees, then the brain has dysfunctioned sufficiently such that the brain can no longer warm the body up. However, in such hypothermic situations, if an outside source of heat is used to warm the body, even from much lower temperatures, the result is usually survival without substantial damage. In essence, the brain can survive but not function if too little heat is available but it can neither survive nor function if too much heat is present.

Air (the gas mix around the body) has both a lower heat capacity and a lower heat transfer than water. As the air around the body gets closer to the temperature of the body it becomes more and more difficult to remove the heat which is produced by the muscles of the body. When the temperature of the air is higher than the temperature of the body, heat removal quickly becomes an acute life-threatening problem for that person. Conversely, as the temperature of the air becomes cooler and cooler relative to the temperature of the body, the problem becomes one of preserving the heat generated by the muscles in order to maintain the core body temperature. The single most important principle governing the movement of heat is that it always flows "downhill": that is, heat always moves from the warmer place to the cooler place. When the temperatures of two places are equal, there is no potential for movement and no net flow of heat in either direction.

Why is temperature important to the body? Heat is important to the water environment because it represents the random movement of molecules. The myriad of chemical reactions that represent life occur because molecules bump into each other. As water becomes warmer and warmer, collisions between molecules occur both more often and with more force. For example, sometimes individual oxygen molecules in the population of oxygen molecules bump into individual hemoglobin molecules in the population of hemoglobin molecules

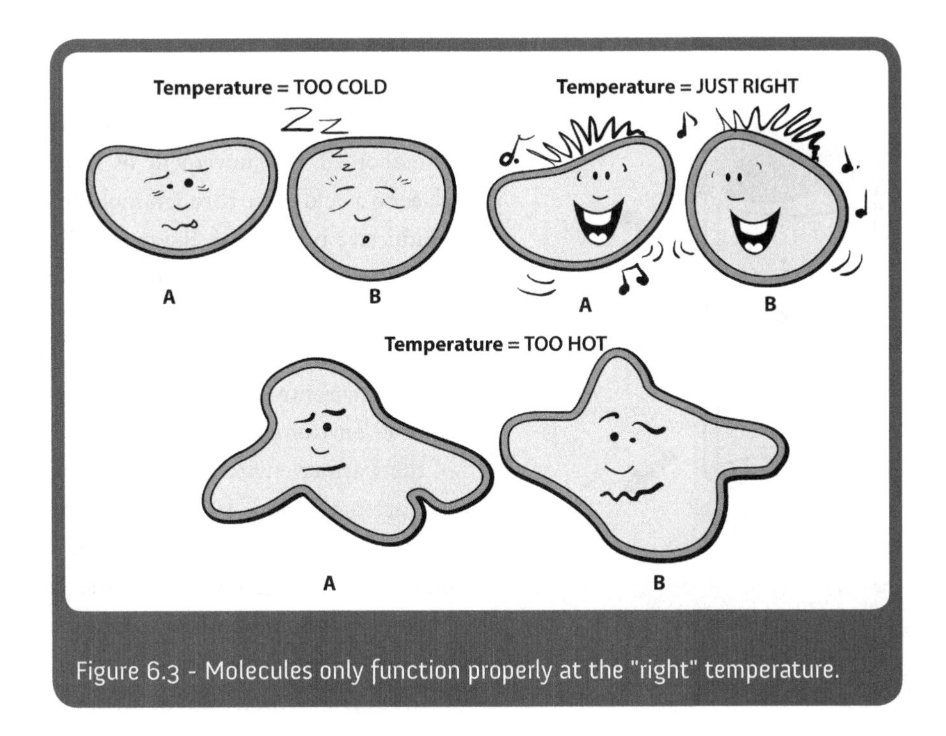

Figure 6.3 - Molecules only function properly at the "right" temperature.

and the two stick together in the special way necessary to deliver oxygen to the tissues of the body. Sometimes a hormone molecule in the population of hormone molecules bumps into a receptor molecule in the population of receptor molecules and they stick together in the special way necessary for one group of cells to tell another group of cells what to do. The same could be said for virtually every reaction in the body. How often collisions between molecules occur and how often the collision between molecules results in the special molecular reactions that run the body depends on temperature. The warmer the water the more frequently collisions occur. The stronger the force of the collision the more likely the two will interact with each other in the special way necessary for molecular reactions to occur. Of course, the role of temperature has an upper limit of value. This upper limit of value, which is a few degrees above normal body temperature, is the point where the heat begins to destroy molecules (Fig. 6.3). The temperature of the brain is controlled because too much heat is lethal, and too little heat does not allow the brain to carry out its reactions with sufficient frequency to function.

There are three processes by which heat enters and leaves the body through the skin (Fig. 6.4). The first process is radiation. Radiation is the phenomenon of the movement of heat from a warm object to a cooler object without molecular contact. Radiation is the flow of heat in the form of energy waves from one place to another. The characteristics of the movement of energy in this form are the same as those of visible light. It can be blocked by putting an object in between. Also, dark dull surfaces absorb the radiation energy, and light shiny surfaces reflect the radiant energy. Radiation is the mechanism by which energy is transferred from the sun to the earth. Dark colored, dull surface clothes are warmer on a sunny day because they absorb radiant energy. Light colored, shiny clothes are cooler because they reflect the radiant energy.

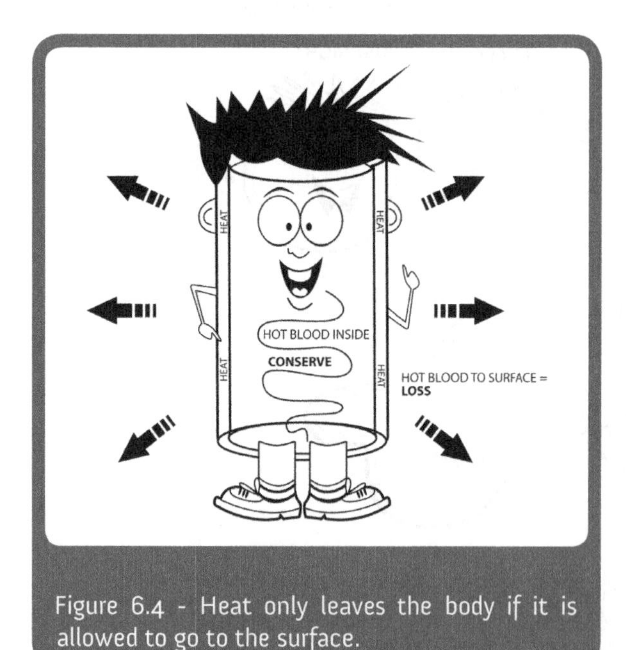

Figure 6.4 - Heat only leaves the body if it is allowed to go to the surface.

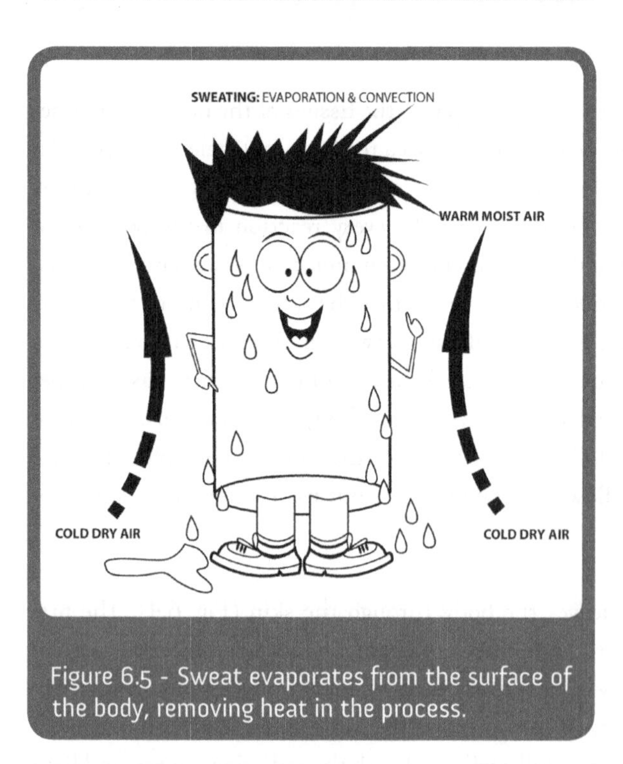

Figure 6.5 - Sweat evaporates from the surface of the body, removing heat in the process.

The second process by which heat enters and leaves the body is by conduction. Conduction is the phenomenon of the movement of heat from a warm place to a cold place through molecular contact. The conductive transfer of energy involves the transfer of energy by molecular collision. Molecules in the warmer place are moving more rapidly in a random fashion than molecules in a cooler place. When a cooler environment is brought into contact with a warmer environment, the rapidly moving molecules of the warmer environment collide with the slower moving molecules of the cooler environment. The result of the collision is that the faster moving molecules move slower, and the slower moving molecules move faster. Some energy has been transferred from molecules in the warmer higher energy environment to molecules in the cooler lower energy environment. In the body, conduction transfer occurs between different water environments and to some extent between the surface of the skin and the air immediately next to the surface.

The third process by which heat is lost from the body is through evaporation (Fig. 6.5). This last mechanism of heat loss from the body involves the movement of air around the body. The air layer next to the body changes very rapidly. Air, warmed by the body, constantly rises up and away from the body; this movement upwards of warm air constantly pulls cool air in next to the body. Because of this there is a constant convection current of air flowing up, along the surface of the body. This convection current allows for evaporation. In the sweat glands, the heat of the blood is transferred by conduction to the water in the sweat. When the sweat is poured over the surface of the body, it evaporates and is carried away by the convection current of air. The result is that the heat energy in the water of the sweat is carried away from the body. Loss of energy through the evaporation of sweat is greatly affected by the humidity (water content) of the air. Humidity, like energy, flows downhill. Water vapor in the air flows from environments of high concentration to environments of low

Figure 6.6 - On a hot day you feel warm because hot blood is sent to the surface of the body so that heat can be lost. On a cold day you feel cold because the hot blood is maintained away from the surface of the body.

concentration. If the humidity is low in the air, the drive for evaporation is high. If humidity is high, the drive for evaporation is low.

Given these three processes of transferring heat at the surface of the body, the problem of temperature regulation really becomes one of controlling the flow of heat to and from the surface of the body. This is accomplished by simply controlling the flow of blood to the surface of the body. When the body needs to lose heat, the resistance to blood flow to the surface of the body is reduced by opening wider the arterial paths to the surface.

More blood carrying heat flows to the surface and heat is transferred by the above three processes from the body to the environment. In contrast, if it is cold outside and the body needs to conserve the heat produced by the muscles, the resistance to blood flow to the surface is increased by closing down the arterial paths to the surface. The result is that less blood carrying heat flows to the surface, and therefore less heat is transferred to the environment from the body (Fig. 6.6). Because our skin contains special cells that sense heat, we feel warm when more blood flows to the surface and we feel cold when less blood flows to our body surface.

Another way to conserve heat is to cover the surface with a relatively undisturbed air layer. Most mammals accomplish this with fur, and birds accomplish this with feathers. In both cases, the fur and the feathers can be puffed up to increase the size of the unstirred air layer or flattened down to decrease the insulating air layer. Naked humans accomplish this with clothes. Air-trapping clothes help maintain body heat, whereas clothes which are porous facilitate the loss of heat from the body. Because of the acute importance of body temperature, heat flow is a major priority of circulation and any condition which disturbs the processes by

Figure 6.7 - Alcohol interferes with temperature regulation.

which the body controls circulation impacts temperature regulation. For example, alcohol interferes with the control of blood flow to the surface of the body. After a few drinks, an individual's body begins to lose the ability to restrict blood flow to the surface of the body. When this occurs, the body is losing the ability to conserve heat. A few drinks on a cold day makes one feel warm. The feeling of warmth is caused by hot blood from the interior of the body warming up the temperature-sensing nerves at the surface. Truly, "cold hands mean a warm heart." However, in reality, the body is actually becoming cooler because that heat sent to the surface is lost from the body. The combination of alcohol and cold weather can, and often has, lead to death due to hypothermia (Fig. 6.7). Conversely, the drug cocaine has the opposite effect. Cocaine, by inducing a strong acute stress response, restricts blood flow to the surface of the body. The consequence is that the body temperature can rise above normal.

Many problems for which skin is treated involve the immune system. Living within the skin are cells of the immune system known as macrophages that were born in the bone marrow and migrated through circulation to the skin. These cells constitute the first line of defense against microbe invasion. These cells have the ability to recognize specific structures associated with bacteria and viruses. When they encounter such structures they both engulf (eat) the microbe containing the structures and they send out chemical messages that attract other immune cells that will continue the attack in a very specific way. Another way that macrophages can be activated is by simple damage to the skin. This is because, by breaking the outside barrier of the body, damage invites invasion. The skin is exposed to many foreign structures. When these foreign structures take the form of infectious microbes or viruses, the immune system response is generally constructive because it is protective. However, often such immune responses occur to foreign structures that are rather benign. The responses of the immune system to these non-infecting foreign structures are referred to as allergies. In the case of allergies the damage is caused by the immune response.

Figure 6.8 - The inside of the lung ends in many small closed sacs called alveoli.

Immune responses of the skin which are due to infection sometimes require drug intervention. Although systemic antibiotics (antibiotics which are taken internally) can be used for such purposes, the most commonly used approach is to apply the antibiotic to the affected area of the skin. For example, it is not uncommon to apply an antibiotic ointment, such as Neosporin, to a cut to prevent bacterial infection. Likewise, creams or powders containing antifungal agents can be applied to areas of the skin infected by a fungus such as occurs with athlete's foot. In contrast, immune responses of the skin which are due to allergy can be treated by suppressing the immune system. Creams containing corticosteroids (such as hydrocortisone cream) are often used for this purpose. Such creams are currently available without prescription. The major problem is that to the untrained person it is often difficult to distinguish between infection and allergy.

The lungs: The lungs are the second of the three outside barriers of the body. Although heat and water move into and out of the body through the lungs, the major transfer being carried out in the lungs is that of gases. The air we live in is composed of several different gases. Depending on the quality of the air, it is about 78% nitrogen, 21% oxygen, about four-hundredths of one-percent carbon dioxide, and the remainder is other gases like helium and argon. The problem of the body is to bring in enough oxygen to be used to generate enough energy to keep all of the myriad reactions necessary for life going. As mentioned above, warm-blooded animals like humans are very high energy users, so they require a constant supply of oxygen to generate this energy from food they take in. The second aspect of the problem is to remove carbon dioxide, which is a waste product of the process of generating energy with oxygen.

Figure 6.9 - Air enters and leaves the lungs because the chest cavity expands and contracts.

The lungs are passive dead-end sacs that connect to the mouth and nose through airways called the bronchial tubes. These airways are sometimes called the bronchial tree because it consists of a major trunk path which splits into smaller branches that in turn split into even smaller branches. At the level of the smallest branches are little sacs, called alveolar sacs. These little sacs are surrounded by vast capillary beds. The gas exchange of the body takes place between the air in the sacs and the blood in the surrounding capillaries (Fig. 6.8). To help this exchange, the walls of the sacs are coated with a substance very much like soap or the phospholipids which form the membranes separating the water compartments of the body. This soapy substance (called a surfactant) reduces surface tension and brings the humidity of the air closer to the walls of the sacs.

The gas exchange is facilitated when all of the components of the system are brought closest together at the place of interface.

Air enters the lungs passively when the chest cavity expands and leaves when the chest cavity gets smaller (Fig. 6.9). In normal breathing, we inhale because the brain, by way of motor nerves, tells the diaphragm muscle (a muscle "sheet" that forms the bottom of the chest cavity) and muscles between the ribs to contract, expanding the chest cavity. For deeper breathing, muscles associated with the ribs contract and pull up expanding the chest even more. Normally we exhale because the diaphragm relaxes, reducing the size of the chest cavity. For heavy deep breathing the brain tells other muscles to contract and pull the ribs up. The process of breathing does not occur unless there is specific direction from the brain. For this reason people with broken necks above the point where the nerves that innervate these muscles leave the spinal cord cannot breathe by themselves. This is also why a small hole in the chest (a pneumothorax) can prevent a person from breathing. In the case of a pneumothorax, when the chest cavity expands, air is sucked into the chest cavity around the lungs through the hole and not into the lungs through the airways.

The most important principle necessary for understanding how gas exchange occurs is that gases can dissolve in water. What regulates the movement of a gas into or out of a water environment is how much of that gas is in the air next to the water and how much of that gas is already dissolved in the water. Like energy, gases move downhill. That is, a particular gas like oxygen or carbon dioxide will move from a high concentration to a low concentration until a balance is reached.

The initial exchange of oxygen and carbon dioxide in the lungs occurs because the air entering the lungs contains a high concentration of oxygen and a very low concentration of carbon dioxide. Similarly, the blood entering the lungs from the heart contains a high concentration of carbon dioxide and a low concentration of oxygen. When the two meet in the alveolar sacs, then both gases flow downhill in opposite directions.

Carbon dioxide flows from the high concentration in the blood to the low concentration in the air of the sac. At the same time, oxygen flows from the high concentration in the air of the sac to the low concentration in the blood. The gases in the air and the gases in the water of the blood move in and out depending on their relative concentrations.

However, only a relatively small fraction of the carbon dioxide and oxygen in the blood exists as the free gas dissolved in the water of the blood. The body needs far more oxygen and produces far more carbon dioxide than can be carried simply dissolved in the water of the blood. As the oxygen flows into the water of the blood along its downhill concentration gradient, it continues to flow with this gradient into the inside of the red blood cell. Inside the red blood cell there is a protein molecule called hemoglobin (Hb) which binds oxygen and essentially takes it out of the "concentration game" (Fig. 6.10).

$$Hb + O_2 \leftrightarrow HbO_2$$

Every oxygen molecule bound to hemoglobin is not viewed as free oxygen in this concentration game. Every oxygen molecule that enters this reservoir of bound oxygen reduces the free concentration of oxygen in the water of the red blood cell. This allows more oxygen to leave the water of the blood, which in turn reduces the oxygen concentration in the water of the blood. This allows more oxygen to leave the air of the alveolar sac and enter the water of the blood. The hemoglobin protein in the red blood cell serves a reservoir, or sink, for carrying oxygen. A very important characteristic of this reaction is that it is very sensitive to the hydrogen ion concentration $[H^+]$ of the water around it. Within narrow limits, increasing the hydrogen ion concentration of the blood reduces the ability of hemoglobin to bind oxygen, and decreasing the hydrogen ion concentration

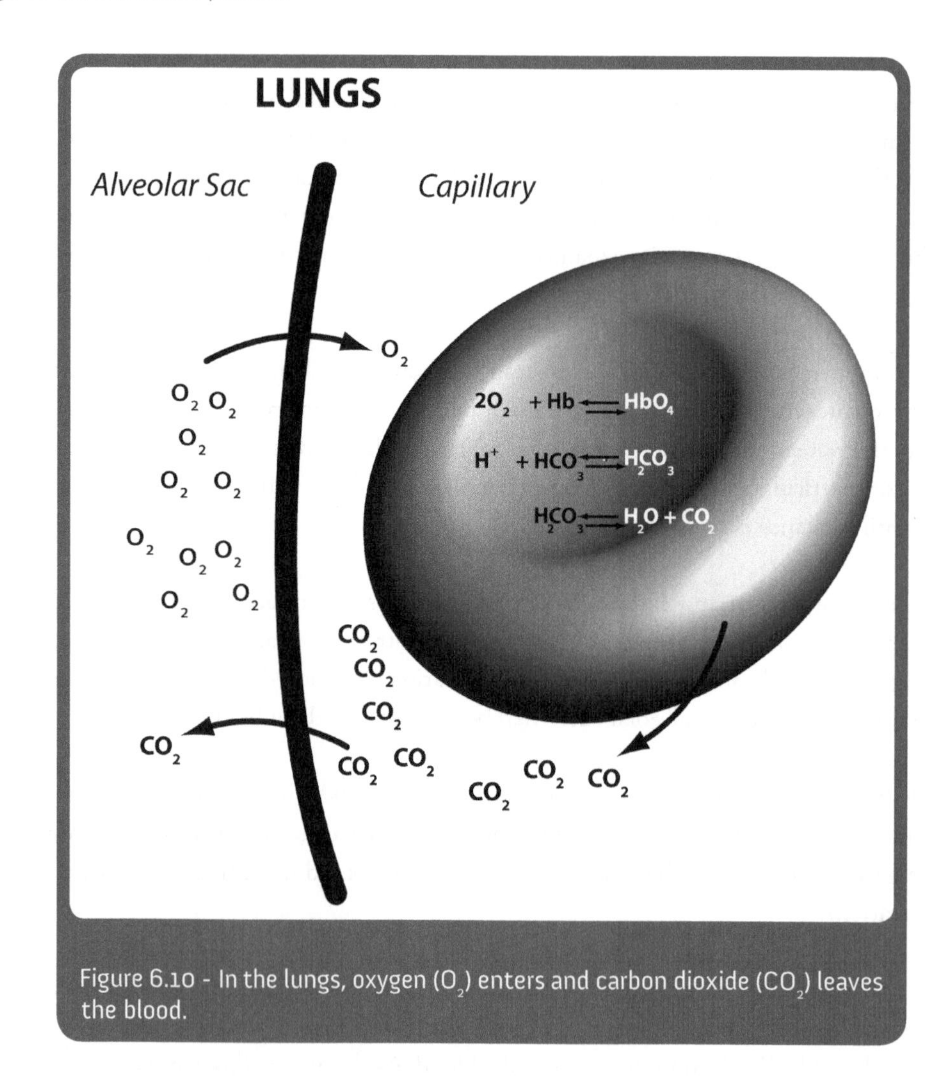

Figure 6.10 - In the lungs, oxygen (O_2) enters and carbon dioxide (CO_2) leaves the blood.

increases the ability of hemoglobin to bind oxygen. The importance of the hydrogen ion sensitivity of this binding reaction relates to the mechanism by which carbon dioxide is carried.

Just as there is a reservoir to carry oxygen in the blood, there is also a reservoir to carry carbon dioxide in the blood. This reservoir is created by another protein in the red blood cell. This protein, carbonic anhydrase, is a special type of protein called an enzyme. Enzymes cause reactions between molecules to occur.

The enzyme carbonic anhydrase makes carbonic acid by combining carbon dioxide and water. This reaction is shown below.

carbonic anhydrase:

$$CO_2 + H_2O \leftrightarrow H_2CO_3$$

The arrows, which point in both directions, indicate that the reaction can flow in either direction depending on the relative concentrations of carbon dioxide and carbonic acid. High carbon dioxide relative to carbonic acid drives the reaction towards forming carbonic acid. Conversely, high carbonic acid relative to carbon dioxide drives the reaction towards forming carbon dioxide and water. However, there is another step in the reaction sequence. That step involves the behavior of acids in water (Chapter 2). Like all acids, carbonic acid has a tendency to lose hydrogen ions. Therefore, as soon as the carbonic acid is made in the red blood cell, it participates in a second balanced reaction in the water of the red blood cell. This reaction is shown below.

$$H_2CO_3 \leftrightarrow H^+ + HCO^{3-}$$

Again the arrows are drawn in both directions indicating that the reaction can flow spontaneously in either direction, depending on relative concentrations of the molecular forms on either side. A high relative concentration of H^+ or HCO^{3-} drives the reaction towards H_2CO_3. A high relative concentration of H_2CO_3 drives the reaction towards H^+ and HCO^{3-}. At the lungs, the carbon dioxide which is dissolved in the water of the blood flows down its concentration gradient into the alveolar sacs. This reduces the carbon dioxide concentration of the water of the blood, causing the carbon dioxide in the water inside of the red blood cell to flow out of the cell down its concentration gradient.

This reduces the concentration of carbon dioxide in the water inside the red blood cell which in turn pulls the flow of the carbonic anhydrase reaction towards forming more carbon dioxide. This reduction in the concentration of carbonic acid molecules with bound hydrogen pulls the flow of the second reaction to the left, which therefore reduces the concentration of both H^+ and HCO^{3-}. Thus, the hydrogen ion concentration is reduced at the same time that the hydrogen ion sensitive binding of oxygen to hemoglobin is taking place.

When the blood leaves the lungs and returns to the heart, it has a lower hydrogen ion concentration than it had when it entered: it is carrying far more oxygen and far less carbon dioxide. The heart then pumps this blood to the body. This blood flows to a capillary bed in some tissue. This tissue has been generating high energy molecules and producing carbon dioxide. Oxygen concentration is low in the water environment around the cells and carbon dioxide concentration is high. Carbon dioxide flows into the blood, driving its two reactions to the right forming carbonic acid and hydrogen ions. Oxygen flows out of the blood pulling its reaction to the left, yielding more free oxygen to diffuse into cells. In addition, the release of oxygen from hemoglobin is facilitated by the reduced ability of hemoglobin to bind oxygen because of the higher hydrogen ion environment created by the carbonic anhydrase reaction. Blood enters the venous paths for return to the heart with a higher hydrogen ion concentration, a much higher carbon dioxide concentration and a much lower oxygen concentration. At the lungs, the entire sequence begins again.

Controlling respiration: Because of the acute importance of gas exchange to the body, the various aspects of the process are highly regulated (Fig. 6.11). The primary signal being monitored by the brain is the hydrogen ion concentration of the blood NOT the oxygen concentration. In the carotid arteries, which are the major arterial paths to the brain, there are sensors which monitor the hydrogen ion concentration of the blood.

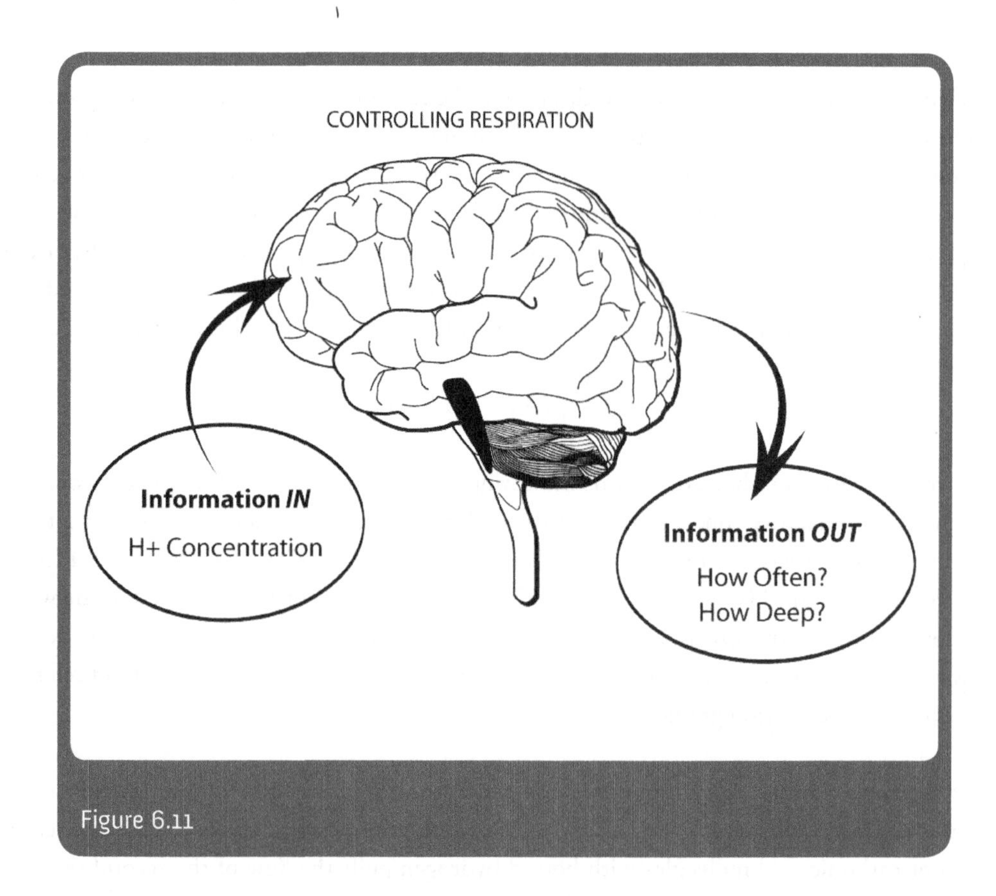

Figure 6.11

The information concerning the hydrogen ion concentration of this blood on its way to the brain is sent directly by nerves to the brain. If the hydrogen ion concentration is too high, indicating too low of an ability of hemoglobin to bind oxygen, the brain sends signals by way of nerves to the muscles involved in breathing. These signals increase the rate and depth of breathing. The result is the blowing off of carbon dioxide which reduces blood hydrogen ion concentration. Conversely, if the hydrogen ion concentration is too low then the individual does not "feel" like taking a breath. The result is the conservation of carbon dioxide and hydrogen ions.

In addition, because energy is the key to dealing with stress, and oxygen is the key to generating large amounts of high energy molecules, there is a stress override of this system. Faced with a "dangerous situation" and the real or potential need for more energy, the brain can accelerate the gas exchange process. Using a combination of nerve and hormonal signals, rate and depth of breathing is increased, the airways are opened wider and more blood is pumped from the heart to the lungs. The entire process of gas exchange is very sensitive to a large variety of drugs. The influences range from affecting the hydrogen ion concentration of the blood to turning up or turning down the stress responses of the system.

As with the skin, the most common problems for which the lungs are treated involve the immune system. Like the skin, the lungs are exposed to foreign structures some of which cause infection and some of which cause allergy or just irritation. Both allergy and chronic irritation can cause chronic lung inflammation. Smoking

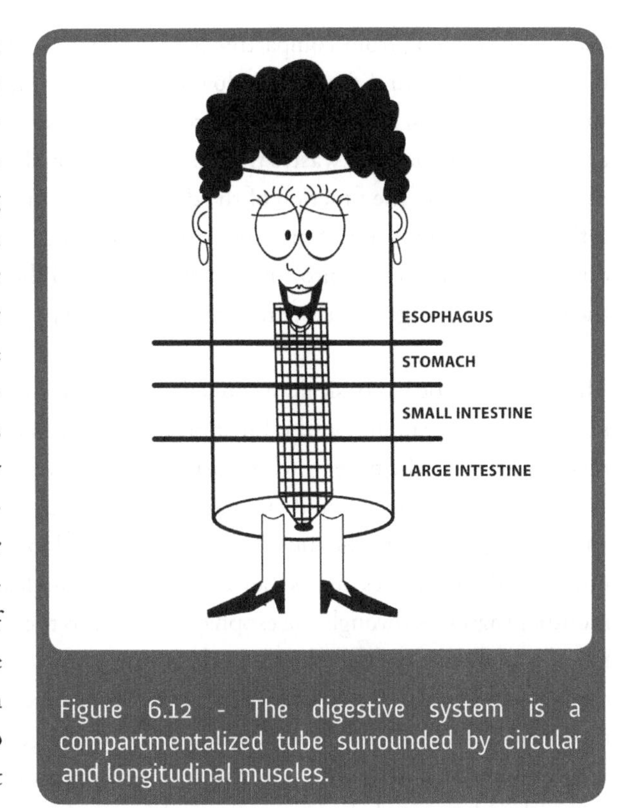

Figure 6.12 - The digestive system is a compartmentalized tube surrounded by circular and longitudinal muscles.

and allergy (hay fever and asthma) are the two most common causes of chronic lung inflammation. Over time such chronic inflammation can lead to permanent damage to the lungs. In response to exposure to foreign structures, cells of the immune system (lung macrophages) release chemical messages. Like in the skin these chemical messages both attract other cells of the immune system and alter the character of the capillary beds. The lungs have very extensive capillary beds associated with the exchange of gasses. One of these chemical messages, histamines, is particularly associated with the change in capillary permeability. Histamine increases capillary permeability, allowing water to leave the blood and enter the space around cells. The histamine release associated with immune responses causes congestion of the lungs. In addition, such congestive responses are often associated with blockage and/or constriction of the airways. However, in some people, due to largely unknown reasons, the airways may constrict without congestion. This is referred to as an "asthma attack." Because gas exchange is so critical to the body, the primary treatment approach is the same as the case of both infection and allergy. That approach is to reduce congestion with antihistamines (drugs which block histamine action) and to open the airways by either facilitating or directly initiating an adrenalin response in the lungs. Adrenalin is a natural chemical message that opens the airways. For example, many medicines that contain antihistamines also contain another class of drugs (xanthines) that blocks the turn-off of natural adrenalin responses. Caffeine has such an effect. However in acute situations, adrenalin can be directly applied to the airway by use of an inhaler. This is a major form of "respiratory therapy" used during acute asthma attacks.

The digestive system: The third barrier between the outside world and the inside water environment of the body is the digestive system. With the exception of gases, the major molecular flow into the body occurs through the digestive system. Sugars, amino acids, fats, vitamins and minerals enter the body by way of the digestive system, Solid wastes that the body cannot process are removed from the other end of the digestive tract In addition, the digestive system is a major vehicle for introducing drugs into the body.

Fundamentally, the inside of the digestive system is a water environment that is controlled by the body but is treated as being outside of the water environment of the body. The digestive system is simply a compartmentalized tube that runs through the body and is open at both ends (Fig. 6.12). The complex molecular structure of carbohydrates, proteins and fats enters the system at the mouth along with vitamins, minerals and water.

As they pass along from compartment to compartment within the various regions of the digestive system, the complex structures are broken down into smaller molecular forms. When they are in the simple forms of sugar, amino acids, fatty acids and glycerol they are absorbed along with the minerals and vitamins into the circulatory system. Solid wastes that can't be used by the body are eliminated at the opposite end of the tube through the anus. The function of the digestive system involves the coordination of three processes. The first is the mechanical process of moving material from one end of the tube to the other. The second is the breaking down of large molecules into smaller molecules, and the last is absorption of the small molecular units from the inside of the digestive system into the circulatory system.

The digestive tube has basically a similar structure from mouth to anus. However, there are regional variations in structure, depending on the function of a particular part of the digestive tube. Generally, there is a fine membrane lining the tube which is surrounded by two layers of muscle. One layer of muscle runs lengthwise, and the other is a segmented circular layer of muscle. These muscles are of a type called "smooth muscle," and are very similar to the muscles that surround blood vessels. Nerves from the autonomic nervous system innervate these muscles just as they innervate the muscles around blood vessels. Food enters the system by the mouth, progresses through the esophagus, and into the stomach. After the food is converted into the "proper form," it is allowed to leave the stomach through a valve and enters the small intestine. As the food progresses through the small intestine its form is further converted and it is absorbed into circulation. The non-digested remainder leaves the small intestine and enters the large intestine. In the large intestine the body recovers the large amount of water used in the digestive process back into circulation. The final undigested or indigestible material with most of the water removed leaves the body through the anus. The mechanism of all of this movement is the rhythmical contraction of the muscles of the tube. These rhythmic contractions mix the food with the material squirted into the tube by the body and push it along. Digestion is the process of breaking large complex molecules into smaller simpler molecules. Different aspects of digestion take place in different compartments of the tube.

In the mouth, saliva is mixed with the food. Saliva is basically water containing protein enzymes that break large carbohydrates into smaller carbohydrates and sugars, as well as some large fats into smaller fats. The process of chewing breaks up the food and mixes it with the water and enzymes. The result is that the food is softened and carbohydrate and fat digestion begins. Swallowing involves the action of skeletal muscles which are controlled by motor nerves. The primary difference between the control of smooth muscles which are innervated by the autonomic nervous system and skeletal muscles which are innervated by motor nerves is one of control. The control of smooth muscle contraction by the autonomic nervous system is usually outside of the conscious control of the individual. In contrast, the motor control of the contraction of skeletal muscles is generally under the conscious control of the individual.

However, like all generalizations, exceptions are abundant. People can learn to control the contraction of smooth muscles through conditioning, and a type of skeletal muscle movement called a reflex is generally out of the conscious control of the individual. When food is pushed to the back of the throat, swallowing is an automatic reflex (Fig. 6.13).

A flap leading to the lungs closes to prevent food from moving into the airways, and food moves into the esophagus, a tube connecting to the stomach. Food moves through the esophagus into the stomach. In the stomach, more water, hydrochloric acid (HCl) and new digestive enzymes are added to the mixture. These new enzymes begin to digest proteins. The HCl serves two purposes. First it destabilizes bonds between the amino acids that form the protein, which helps break the bonds. Second it activates, or "turns on," the enzymes involved in breaking particular bonds between certain amino acids. The food stays in the stomach mixing and digesting for some time. How long the food remains in the stomach will depend on the state of the individual and the composition of the food. Food containing protein and fats stay longer. In addition, stress prolongs the time necessary for digestion (see below). The result of the action of the stomach is to begin protein digestion and to essentially liquefy the food. When the food is in the proper liquefied form, it is allowed to leave the stomach and enter the small intestine.

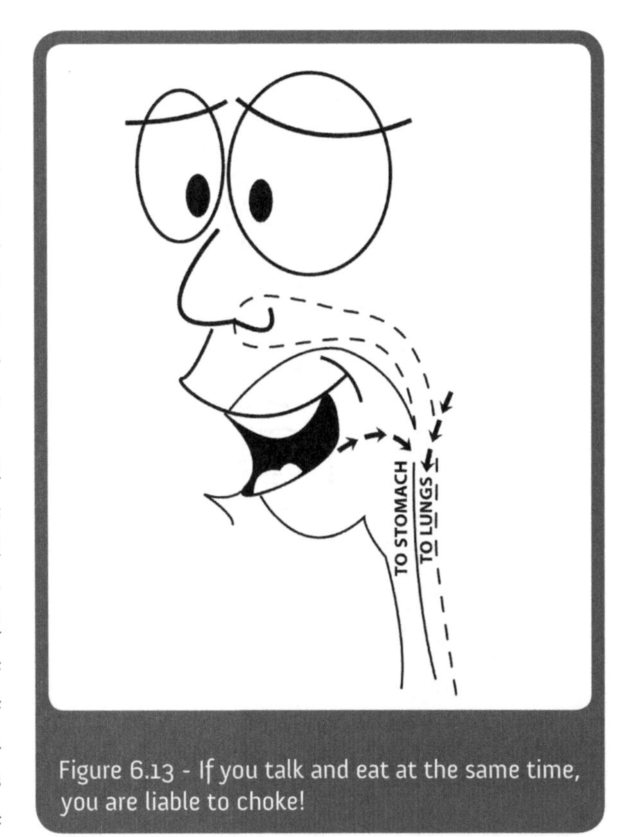

Figure 6.13 - If you talk and eat at the same time, you are liable to choke!

The small intestine is a actually a very long, highly coiled part of the digestive tube, which greatly increases its surface area. In the small intestine final digestion and absorption occur. However, first the acid which was put into the contents in the stomach must be neutralized. Neutralizing the acid involves putting into the contents of the digestive tract a molecule, bicarbonate, which combines hydrogen ions (H^+) with hydroxyl ions (OH^-) to form water (H_2O). The effect of this reduction of hydrogen ion concentration in the liquefied food is to protect the lining of the intestine from the acid, to turn off the protein-digesting enzymes from the stomach, and to activate the digestive enzymes of the small intestine. Like hemoglobin, discussed above, digestive enzyme activity is controlled by the hydrogen ion concentration. In the small intestine, the proteins that can be digested by the body are reduced to amino acids by enzymes made in the pancreas. The pancreas is a relatively small gland which lies under the stomach and connects through ducts to the small intestine. It is important to note that all proteins cannot be digested by the human digestive system. Similarly, the carbohydrates that the body can digest are broken down into simple sugars like glucose by other enzymes made in the pancreas. However, fats are more of a problem. Because of their lack of charge, fats are not soluble in water (Chapter 2). Therefore, before they can be digested they must be brought into contact with the fat-digesting enzymes in the water. This is accomplished by introducing bile salts into the contents of the digestive system. Bile salts are formed in the liver (primarily from cholesterol) and are stored in the gall bladder (a small sac which lies largely under the liver). Bile is moved from the gall bladder to the intestine by another duct which connects

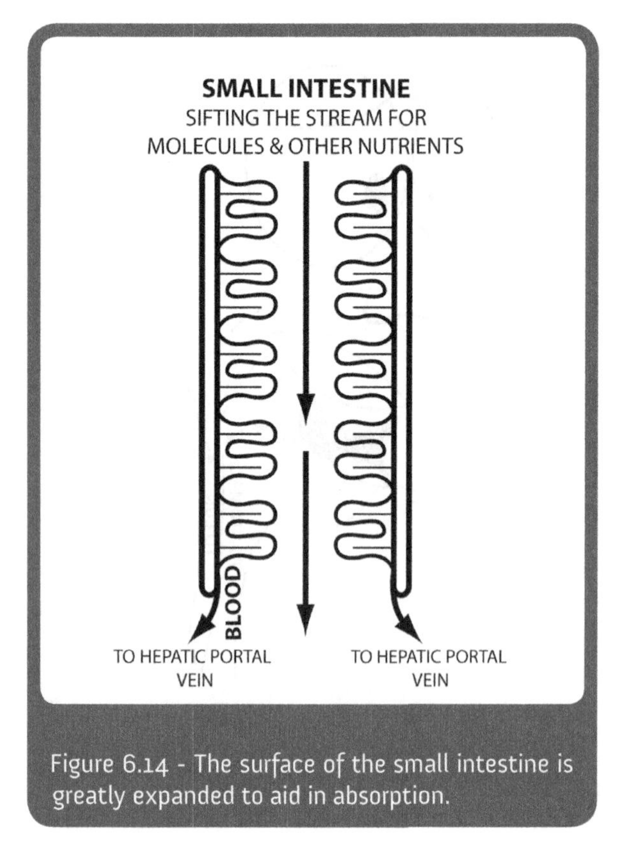

Figure 6.14 - The surface of the small intestine is greatly expanded to aid in absorption.

these structures. In water, bile salts are large ions with a low charge density. Because of their charge, they can interact with the water. However, because of their large size relative to the amount of charge they contain, their effect is to greatly reduce the electrically polar order of the general water structure (Chapter 2).

The effect is to decrease the repulsive relationship between the fats and the now less structurally ordered water environment. This allows greater mixing between the fat and water. As a consequence, the fat digesting enzymes from the pancreas are able to get at the fat and reduce them to fatty acids (Chapter 2) and a three- carbon molecule called glycerol.

The last part of the process is absorption. This is the process by which the small units of the digested carbohydrates, proteins and fats as well as vitamins and minerals are taken into the water of the blood. Structurally, the lining of the small intestine is quite elaborate. The lining of the small intestine is not simply a smooth-sided tube. Rather, it consists of millions of tiny fingers (called "villi and microvilli") which project into the passing fluid (Fig. 6.14). Inside these projections are capillary beds which bring the flow of blood very close to the fluid inside the intestine. The value of the projections is that they greatly expand the surface through which molecules can be absorbed. Amino acids, sugars, minerals, vitamins and some simple fatty acids are taken up by the cells of the lining and then released to the blood.

Depending on the particular substance, absorption is either passively downhill with a concentration gradient, or a combination of passive movement and active transport. Active transport is that process of using metabolic energy in the form of ATP to move molecules. This is generally used to absorb larger quantities of vital nutrients. The bile salts are also recovered back into the blood.

The blood from the digestive system is run through a special circulatory path directly to the liver where it is screened for toxic substances like drugs. This special circulatory path is called the hepatic portal vein. Some of the nutrient molecules are taken up into liver cells, and the rest are released to the lymphatic drainage system and enter venous return to the heart. Most of the bile salts are also recovered back into the liver where they were made. In this way, the liver can be considered as a backup barrier behind the digestive system. However fat takes a different route into general circulation. Large fats are formed into a molecule called triglycerides which consist of a three-carbon glycerol molecule with three fatty acid molecules bound to it (Chapter 2).

Because these large fat molecules are not soluble in water, many of them are brought together and packaged with a thin protein coat. The thin protein coat gives the outside of the structure charge so that it can travel in the water environment of the body. These structures, called chylomicrons, are released into the extra-cellular sewer system known as the lymphatics and travel by way of this path to the venous return system of circulation. With a pass through the heart, they are delivered to other tissues for storage and use (Fig. 6.15). Most molecules which enter the body through the digestive system ultimately, in some form, are destined to leave the body through the kidney.

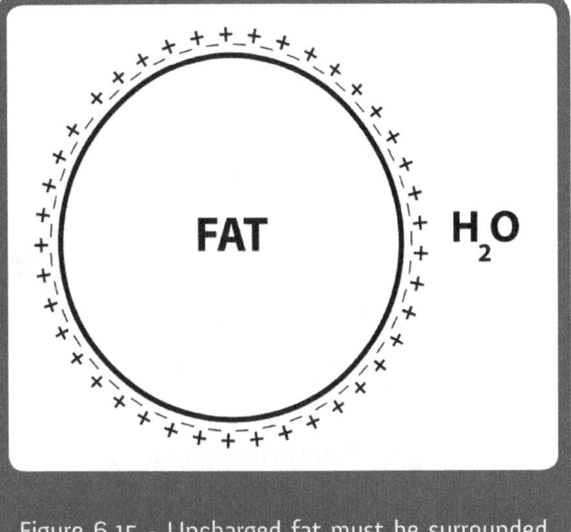

Figure 6.15 - Uncharged fat must be surrounded by charged protein to travel in the water of the blood.

Common problems in the digestive system are usually related to one of two imbalances. One problem is excess acid and the other is constipation. Excess acid commonly causes "heartburn" and less commonly causes ulcers. Heartburn is an irritation of the esophagus caused by the acid containing contents refluxing up (or moving back up) out of the stomach. A more chronic and serious form of this problem is called "acid-reflux disease." Ulcers are weakening or holes in the lining of the digestive tract. Although they can be found in many places, ulcers most often occur at the beginning of the small intestine. If the acid from the stomach is incompletely neutralized then the digestive enzymes from the stomach can stay active and digest the rather delicate lining of the small intestine. Often excess acid secretion is associated with the use of drugs such as caffeine or alcohol. Excess acid secretion is also associated with stress. In either case the solution is to reduce the amount of stomach acid. One method of reducing stomach acid is the use of drugs called antacids. Antacids are compounds that neutralize hydrogen ions. Such compounds act in the same manner as the bicarbonate, which the body secretes into the contents of the small intestine. One side effect of these drugs is associated with their excessive use. Because many of these compounds are subsequently taken into circulation they can influence the pH balance of the blood. A second side effect involves a process called acid rebound. As we shall see later, often if a drug "pushes" some process in one direction the body will respond by "pushing back" in the opposite direction. Acid rebound is such a situation. Antacids reduce stomach acid and the system responds by producing more acid. A second approach to dealing with excess stomach acid is to block the signal directing the cells lining the stomach to produce acid. Histamine is the chemical message sent from nerve cells to special cells in the lining of the stomach directing them to produce and release hydrochloric acid into the contents of the stomach. This is the same chemical used by the immune system to increase capillary permeability. It is also one of the chemicals used in the brain for sending messages between nerve cells. Fortunately for selectivity, the receptors for histamine in the lining of the stomach are somewhat different than those used for signaling in the brain and the immune system.

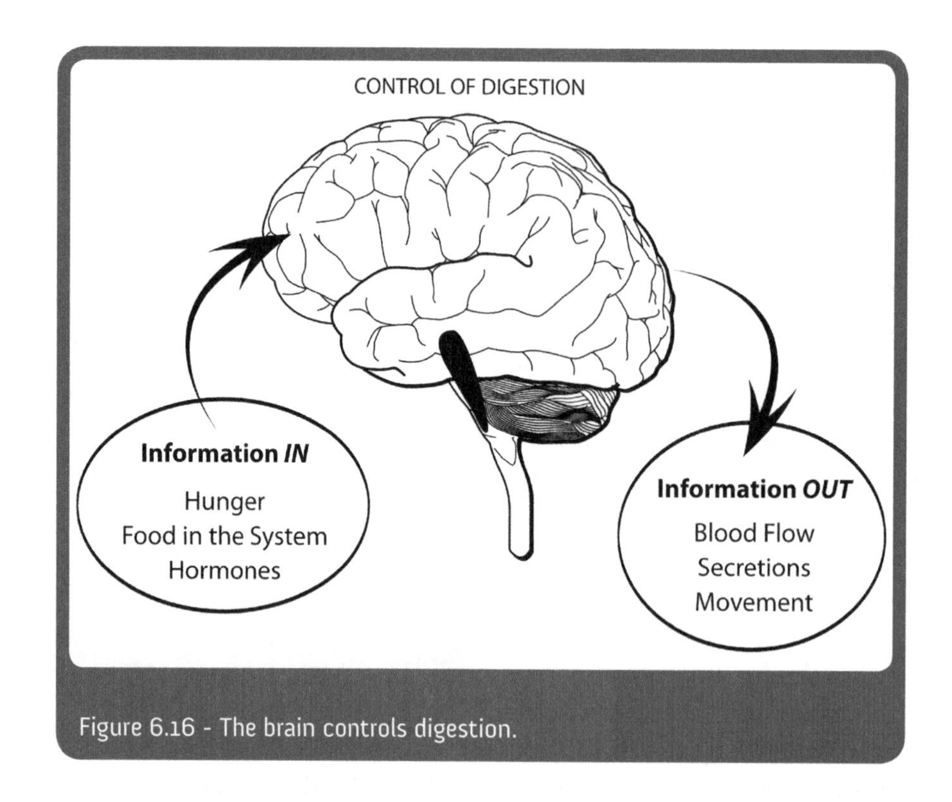

Figure 6.16 - The brain controls digestion.

Antihistamines are available that have reasonable selectivity for the histamine receptors associated with the release of stomach acid.

Constipation is caused by excessive dehydration of the material at the end of the large intestine, the colon. As discussed above, as material transits through the large intestine the large amount of water which was introduced for the purposes of digestion is recovered back into circulation. If too much water is removed, the material that reaches the end is too hard. The most common solution to this problem is dietary. Specifically, one adds "bulk" to the diet. Bulk refers to materials such as bran which are essentially indigestible and absorb water. By analogy, one might consider them like a sponge. They absorb water and thus more water is retained in the feces. In actuality, not withstanding extreme cases, the distinction between the range of normal bowel function frequency and constipation is often lost in social perception and advertising imperatives.

Often both the problem of excessive acid in the stomach and insufficient water in the colon are secondary to some other problem. Quite commonly these other problems are related to stress, drugs, or both. The excess build-up of acid in the stomach as well as the excess removal of water in the large intestine is closely related to "how fast" material moves through the system. How fast such movement occurs is a function of the tone in the smooth muscle that surrounds the digestive system. There are a large number of drugs that have a side effect of reducing the activity of these muscles. Because of the commonality in function between intestinal smooth muscle, vascular smooth muscle and skeletal muscle, and because of the commonalities in function between all muscles and nerves, selectivity is a great problem. For example, anti-arrhythmic drugs for the heart have as a side effect the reduction in muscle tone in the digestive system.

Like the skin and the lungs the stem cell populations of the digestive system are quite actively proliferating, differentiating and replacing the lining of the digestive system. In addition, the digestive system also contains immune cells that help protect our bodies from invading microorganisms. Because the intestines are a very large interface with the outside environment they contain a very large macrophage population. Intestinal macrophages reside under the surface layer of the small and large intestine and function to recognize and destroy pathogens that enter through the digestive system. Because the blood that flows around the small intestine is sent via the hepatic portal system to the liver, the macrophage population of the liver serves as a back-up for the macrophage population of the small intestine.

Like most processes in the body, the functions of the digestive system are controlled either directly or indirectly by the brain (Fig. 6.16). The brain "senses" when we need more food (there are both hunger centers and satiety, or fullness, centers in the brain that control how much we eat). The brain also senses when more water needs to be taken into the body, and "tells" us to drink more if necessary. The brain controls the contraction of smooth muscles around the digestive tract, as well as blood flow to it. It controls the amount of secretions that are added to the digestive tract in various locations. It also controls the amount of water that is taken back up into the body from the digestive tract.

Information flows into the brain, is processed there, and flows back out by way of both direct nerve connections and by the release of chemical messages.

The nature and priority of outside barriers: The basis for the adaptive capacity of the individual is the control of molecular and heat flow between the body and the outside environment. The fundamental mechanisms of the body evolved in a far less stable world than presently exists for most humans in this society. Our systems evolved in a world of feast and famine, within the context of very little control of environmental temperature. To understand the demands on our barriers, it is necessary to consider an individual who was very active in the warmth and plenty of the summer and huddled against the cold during the famine of winter. All of this environmental change occurred within a world in which survival depended on making immediate responses to life-threatening situations.

Gas exchange and temperature control are acute processes which must adapt rapidly to situations of change. Because of this there are rapid adjustments in the functional capacity of the respiratory processes of the lungs and the thermoregulatory processes of the surface of the body. In contrast, digestion is a chronic process. More or less food is available at any one time, and the objective of molecular flow through the digestive system is to make sure enough is available all of the time.

The unpredictable nature of availability has led to a system that attempts to extract for storage purposes the maximal nutritional value of the food. Digestion is not an acute process because stored nutrients are generally available in the body. The consequence of this is that, in general, digestion is a low priority in an immediate sense. In the same stress situations where respiration and thermo-regulation are turned up, digestion is turned down. The chronic process of digestion is simply left to calmer times.

There is one common problem shared by all three barriers. That problem is damage. Because barriers interface with the outside world, they are subjected to toxic and other damaging situations more often than most types of cells. As a compensation for damage, they share the common characteristic of a rapid rate of stem cell proliferation and differentiation. The rapid proliferation and propensity for damage also increases the probability that cancer will develop from the stem cells of the barriers. For this reason, cancers of the skin, lungs, and digestive system are very common.

chapter seven

nutrition and metabolism

Humans have to eat: In general a normal healthy human being can survive for minutes without air, for days without water, and for weeks without food. The body processes that are associated with acquiring oxygen are an immediate demand system. Breathing is an immediate demand because the body constantly requires oxygen. The body processes associated with water balance reflects a conservation system. Water must be conserved because water is a key element in the dynamic balance of concentrations and pressures that is responsible for the distribution of everything in the body. However, the body processes associated with acquiring food are a storage system. As humans, we don't have to eat constantly because our bodies are designed to store what we eat.

Food provides the types of molecules we call nutrients. Nutrients include sugars, fats, amino acids, vitamins and minerals. Some nutrients (sugars, fats, and amino acids) are the source of both organizational energy (ATP) and of building materials needed to make new cells or replace worn out structures and molecules within existing cells. Other nutrients (minerals and vitamins) participate in the regulation of many molecular reactions. Because regulation of nutrient flow in the body involves storage and controlled withdrawal from storage, the demand for nutrients from the environment is not acute. As long as adequate quantities of nutrients are available on an intermittent basis and as long as the system is functioning properly, acquiring nutrients from the environment is not an immediate problem faced by our bodies. However, if even small parts of the regulatory process malfunction, or if any of the nutrients are not periodically available, the consequences to the body are quite severe. Because of their importance, the storage and use of all nutrients is highly regulated.

Minerals: Minerals are elements. They are atoms (the fundamental units of matter) which we discussed in the second chapter. Biological systems simply use them "as is"; they don't need further processing (Fig. 7.1). Some of the minerals used in biological functions include sodium, potassium, chlorine, calcium, phosphorus, magnesium, iron, iodine, copper, manganese and zinc.

Figure 7.1 - Although not identical, the mineral composition of the water environment of the body is similar to sea water.

The common characteristic shared by the minerals is that by themselves in the water environment of the body they function as charged ions (Fig. 7.2). They lose electrons and therefore are positively charged, or else they gain electrons and are negatively charged. Because they are individual atoms, they represent relatively dense charge accumulations which generally move easily in water. Their charge density and mobility allow them to influence both the ordered structure of water and the structure of larger molecules such as proteins that are present in water. They also have a tendency to participate in reactions because of their charge. In addition, their charge density is used to impart certain reactive properties to larger biological structures such as membranes.

The simplest way to understand the role of minerals is to consider the kinds of processes in which they participate. For example, the ions that contain one charge sodium (Na^+), potassium (K^+), and chloride (Cl^-) participate in bioelectric processes. Many types of cells use metabolic energy to move these charges around, creating electrical potentials across membranes. These electrical potentials are like a battery with one side of the membrane being positive and one side being negative.

Like a battery, their voltage (electrical pressure) can be used as a driving force for many purposes. For example, hydrogen ion-based electrical potentials are used as a driving force for making ATP. Within cells, the controlled movement of hydrogen ions within compartments inside the mitochondria is involved in capturing the energy released from the breakdown of nutrients to form ATP. They can also be used for information transfer. The mechanism for the transfer of information along a nerve is based on the patterned release of electrical potentials across the nerve cell membrane (Fig. 7.3). These very common ions also participate in body water balance. Probably because they are so common, they are involved in some way or another in many biological reactions. In essence, these ions are an integral part of the water environment of the body. The body doesn't really store these atoms; it simply controls their acquisition through the mouth and their loss through the kidney.

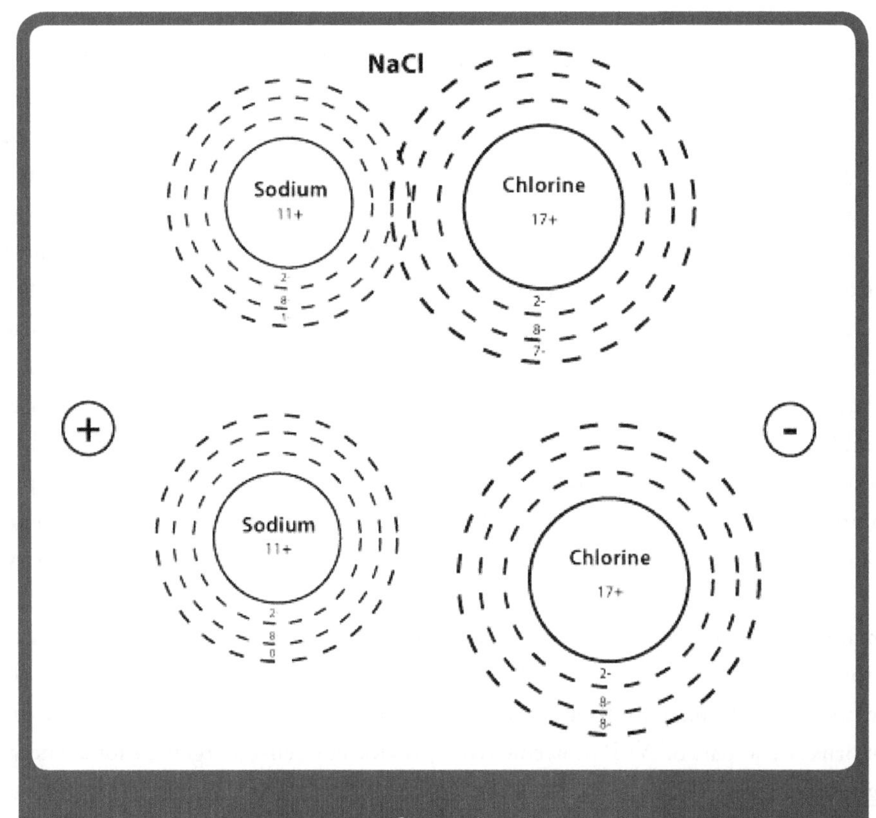

Figure 7.2 - Most minerals come from salts and exist as charged ions in the body.

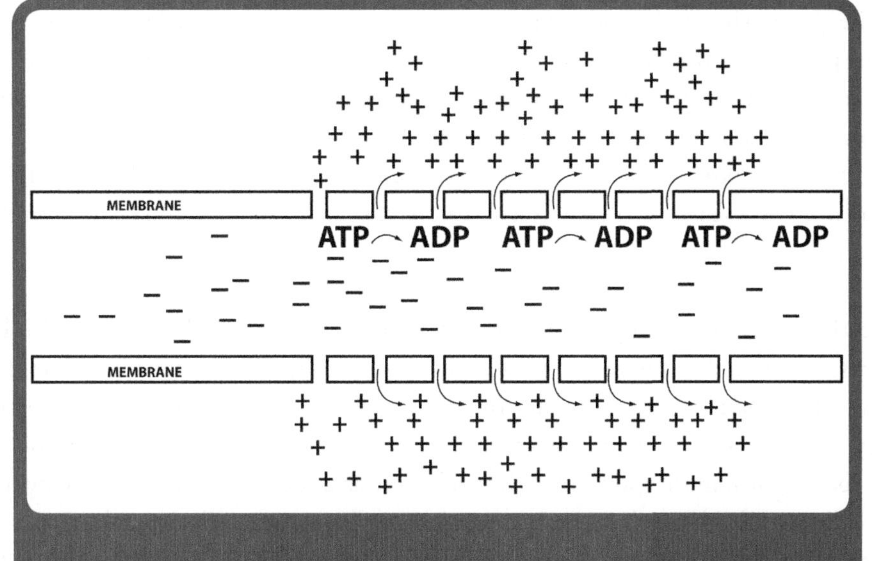

Figure 7.3 - Many cells expend ATP to create electrical potentials across their outer membranes.

In contrast, calcium (Ca^{++}), which participates in the regulation of many biological processes, is a stored mineral. Calcium is stored in bones. The calcium stored in the bones is available for release into the blood so that it can enter other cells and their environments. Free calcium ions in the water of the body participate in a large variety of essential reactions such as blood clotting, muscle contraction, and hormone secretion. If the calcium concentration of the blood is too low, a hormone is released into the bloodstream that causes calcium to be removed from the bones. If the calcium concentration in the blood is too high, another hormone is released that directs the removal of calcium from the blood and deposits it in the bones. From this point of view, the priority use of calcium is for biological regulation and not bone structure. If sufficient calcium is not available for long periods of time, calcium will be used for biological regulation and bone structure will be sacrificed. The problem of bone demineralization (osteoporosis) is not uncommon in older people. The origin of the problem is both insufficient calcium in the diet and lack of exercise. Like the build-up of muscle, exercise promotes the mineralization of bone. Similarly, calcium loss from bone can be a problem during pregnancy. If a pregnant woman has insufficient calcium in her diet, her bones will be demineralized in order to build the bones of the baby. This is the origin of the old saw "a woman looses one tooth for every child."

Similarly, phosphorus (P) is used for biological regulation and is stored as part of the structures of the bone. Among other things, phosphorus is used to help regulate the hydrogen ion concentration of the water environment of the body. In addition, phosphorus is part of the structure of DNA that forms the genetic material of cells. Phosphorus is also part of ATP molecules that provide condensed organizational energy that is used to drive biological reactions. Futhermore, the addition or removal of phosphorus from proteins is often a mechanism for turning on or off the activity of some proteins.

Iron, which can carry either two or three positive charges (Fe^{++}, Fe^{+++}) is also important to body function. Iron is stored by the body in the liver. Because of its ability to lose two or three electrons, iron is involved in processes that involve moving around electrons and the use of oxygen. For example it is an essential part of the oxygen-carrying protein hemoglobin. In addition, several molecules fundamental to the process of producing condensed organizational energy (ATP) with oxygen contain iron. When it comes to the control of iron, the body simply stores all that is available and releases it when it is needed. If a person has too little iron in his or her diet, he or she cannot produce enough red blood cells to carry oxygen around the body. This condition is known as anemia. If a person has too much iron in the diet he or she can form iron deposits in the liver which impacts function.

Another interesting ion is iodine. Thyroid hormone is an important chemical message which helps regulate body energy metabolism. Iodine functions in the body as part of thyroid hormone. If iodine is not available, then active thyroid hormone cannot be made. The consequence of a lack of iodine is a severe disturbance in body energy metabolism. In some places in the Midwest there is little iodine in the soil and therefore little iodine in foods grown in that soil. Until the problem was understood, many people living in that area developed a condition called goiter. With goiter, the thyroid gland excessively expands in size. The excessive growth of the thyroid gland is caused by the brain continually sending strong chemical messages to the thyroid gland directing it to make thyroid hormone. Because of the lack of iodine the task of making active thyroid hormone is impossible. The solution which solved this health problem was to add iodine to commercial table salt (iodized salt).

The importance of minerals as a group is that they participate in a myriad of essential reactions. Because of this their flow through the body is highly controlled.

In general, minerals enter the body through the digestive system and leave through the kidney (Fig. 7.4). Any condition that interferes with digestion threatens inflow, while any condition that interferes with kidney function can compromise controlled outflow.

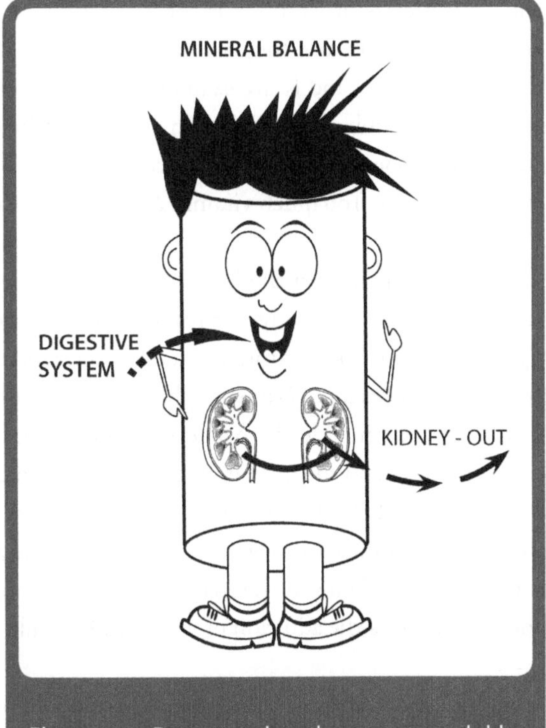

Figure 7.4 - Because minerals are water soluble, the amount in the body is a balance between ingestion and loss through the kidney.

Vitamins: Vitamins are a second class of nutrients that are essential for biological function. Vitamins are relatively small molecules. Like some of the trace minerals, only small amounts are required, but the availability of that necessary small amount is critical to the functioning of the body. In some respects vitamins, because of the low quantities necessary, are like natural drugs. They function in the body to regulate and facilitate essential reactions. Also in some cases, too much of a particular vitamin can have toxic effects. The roles of the various vitamins range from facilitating energy metabolism to serving as a hormone for the regulation of calcium. In all cases, vitamins are organic molecules. They are composed of carbon, hydrogen, and oxygen, plus in some cases nitrogen and other atoms.

There are two classes of vitamins. One class is the water soluble vitamins. Water soluble vitamins are charged molecules that move easily in the water of the blood. However, because of their charge, they can only enter cells through special channels in the membrane. The B-complex vitamins and vitamin C are examples of water-soluble vitamins. B vitamins are a group of eight organic molecules that participate in many aspects of cellular metabolism. A deficit of B vitamins results in a body that is incapable of producing sufficient condensed organizational energy for functioning. Too little of the B-complex vitamins has particularly distinct effects on the nervous system Another water-soluble vitamin, vitamin C, is essential to the structural integrity of the tissues of the body. Vitamin C is necessary for the formation of collagen. Collagen is a large structural protein that holds tissues together. Insufficient vitamin C in the body causes the disease called scurvy. In scurvy, the tissues of the body literally fall apart. For example, the gums degenerate and the teeth fall out. Control of the amount of water soluble vitamins is a function of a balance between inflow through the digestive system and outflow through the kidney.

The second class of vitamins is the fat-soluble vitamins. Fat-soluble vitamins are absorbed into the blood along with fats because they are not soluble in water. Fat-soluble vitamins are stored primarily in the liver. Vitamins A, D, E, and K are examples of fat-soluble vitamins. Vitamin A is necessary for vision, for healthy skin and for membrane linings such as those in the digestive system. Because vitamin A does not move about easily in the blood it can build up in certain areas. Toxicity results from the excessive accumulation in the liver and other places. Too much vitamin A is expressed as dry skin, swelling around the long bones and irritability.

Vitamin D functions in the body as a hormone. It is involved in the regulation of calcium. Vitamin D can be either made by the body as a result of the effect of sunlight on the skin, or it can be obtained through the digestive system. Like vitamin A, vitamin D is stored in the liver. Also like vitamin A, too much or too little of this vitamin is harmful. Not enough vitamin D leads to low body calcium and demineralization of the bones. Too much vitamin D leads to an excess of calcium in the body. The excess calcium causes vomiting, diarrhea, kidney damage and the formation of calcium deposits in many tissues. For a long time there was the problem of dark-skinned children living in northern climes developing weak bones (a condition called rickets). The problem was sufficiently severe that weight-bearing bones such as those of the legs were often bowed. The cause of the problem was that dark skin is a sun screen that reduced the production of vitamin D in the skin. The solution was the addition of vitamin D to milk.

Vitamin K participates in the process of blood clotting. Excess levels of natural vitamin K do not seem to be toxic. In fact, much of the vitamin K used in the body is made by "good" bacteria that live in the large intestine of all humans. The last of the fat-soluble vitamins is E. Vitamin E is an antioxidant. Antioxidants help prevent the build-up of certain very toxic highly reactive substances like hydrogen peroxide in the water environments of the body. Both aging and cancer have been linked to the build-up of these toxic substances that are destroyed by Vitamin E.

The control of the inflow, outflow and storage of essential vitamins and minerals is absolutely necessary to the functioning of the body. However, many drugs and a variety of conditions can disturb the various regulatory processes involved in the control of vitamin and mineral balance. For example, oral contraceptives and alcohol can influence the balance of several of the B vitamins by increasing their metabolism in the liver and decreasing their absorption in the small intestine. Vitamin K availability can be influenced by antibiotics that kill intestinal bacteria. Basically, any drug or condition that influences digestion, modifies absorption, alters kidney function, or alters liver metabolism has the potential of disturbing the regulation of one or more vitamins and minerals. The consequence of such disturbances is quite often an impact on the fundamental metabolic process associated with the use and distribution of the three major nutrients: sugars, fats and amino acids.

Sugars, fats and amino acids: First and foremost, sugars, fats and amino acids are sources of energy (Fig. 7.5).

The energy stored in the bonds of these molecules can be used to produce molecules such as ATP that can provide condensed metabolic energy. These high energy molecules are the source of immediate energy used to

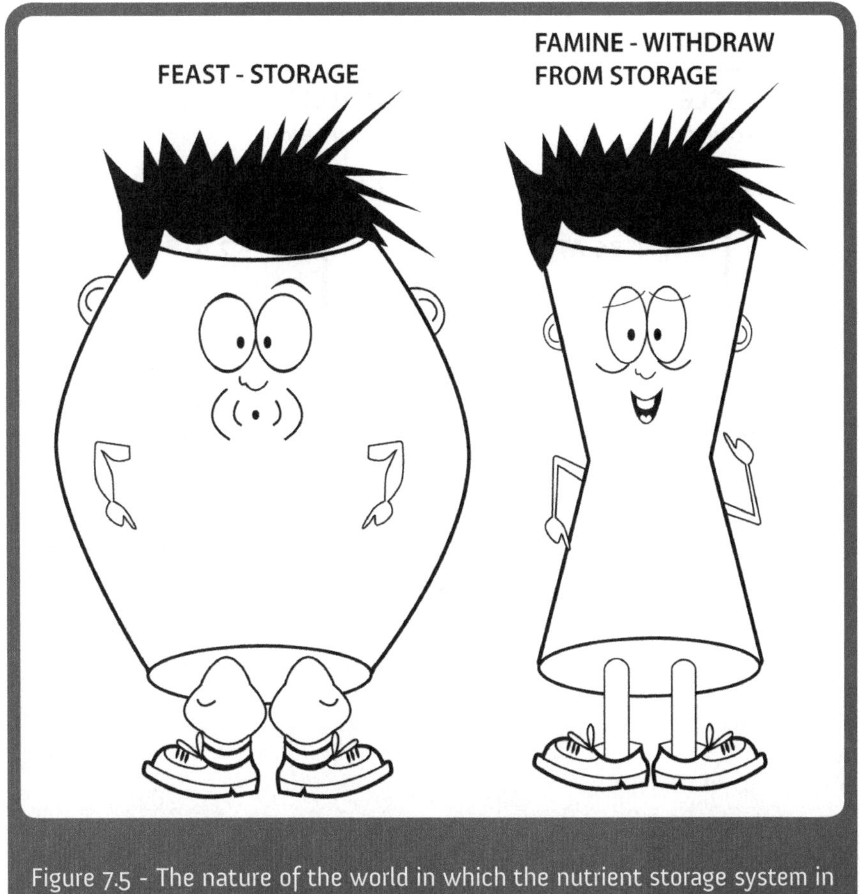

Figure 7.5 - The nature of the world in which the nutrient storage system in humans evolved was one of feast and famine. As a result, the human body will store all excess during times of plenty and withdraw from storage to survive during times of famine.

drive the various processes of the body. There are a variety of reaction sequences (much too complex to discuss in detail here) which can be used to make molecules like ATP out of sugar, amino acids and fats. However, sugar is special. Sugar is the only one of the three that can be used to produce molecules like ATP rapidly and directly without oxygen. Although the production of ATP without the use of oxygen only allows for a small fraction of the possible energy to be extracted from sugar, the ATP produced is quite important in some circumstances such as during exercise.

There are basically two types of exercise, aerobic and anaerobic (Fig. 7.6). Aerobic exercise occurs when an individual exercises at a level in which the lungs and cardiovascular system are able to bring in and deliver sufficient oxygen to sustain the exercise.

After a person finishes aerobic exercise, his or her breathing returns to normal quite rapidly. In contrast, during anaerobic exercise the individual is exercising beyond the capacity of the lungs and cardiovascular system

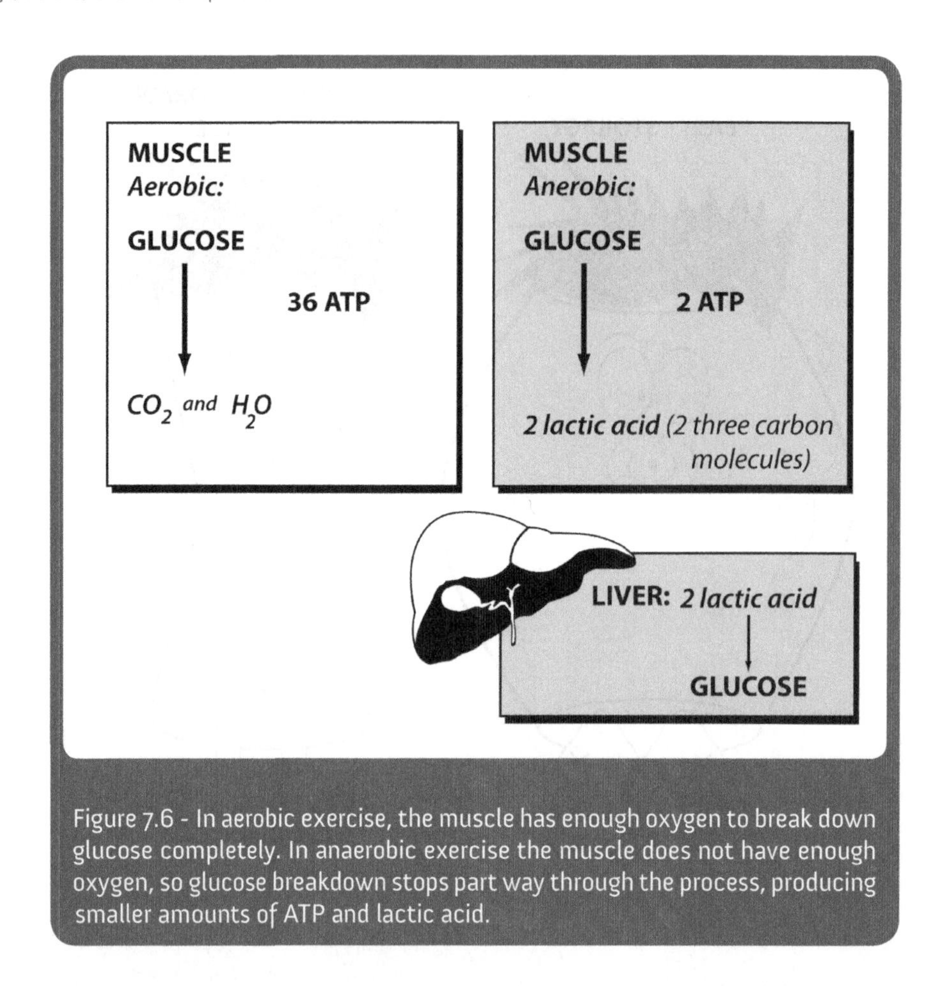

Figure 7.6 - In aerobic exercise, the muscle has enough oxygen to break down glucose completely. In anaerobic exercise the muscle does not have enough oxygen, so glucose breakdown stops part way through the process, producing smaller amounts of ATP and lactic acid.

to bring in and deliver sufficient oxygen to to sustain the mechanical work of skeletal muscle. In the case of anaerobic exercise, the excessive work of the musculature is sustained by producing ATP by extracting that small fraction of the energy from sugar that can be obtained without oxygen. The result is that the six-carbon glucose molecule is broken down to a three-carbon molecule called pyruvate. Under non-anaerobic conditions, pyruvate is used by mitochondria in an oxygen-requiring process to extract the remaining large amount of energy. In anaerobic exercise, the pyruvate is converted to another three-carbon molecule (lactic acid) which is released from the muscle. The lactic acid represents an "oxygen debt." Circulation carries the lactic acid from the muscle and delivers it to the liver. In the liver, the lactic acid is converted back to pyruvate and used to synthesize glucose, thus paying back the oxygen debt. Liver pays off the oxygen debt run up by the muscle because muscle is not capable of synthesizing glucose, a process known as gluconeogenesis. In fact, only the liver and to a much less extent the kidney are the only tissues that express the genes (make the proteins) that are necessary for gluconeogenesis. Fats and amino acids are also available for the production of ATP. However, they serve other purposes in addition to the production of usable energy (ATP). When the body requires them to be used for energy, mechanisms are available to shift their use for that purpose.

Energy availability and use in the body is in part a consideration of the demands of various individual types of cells, and in part a consideration of the flow of energy in the entire system. Some types of cells, like certain muscles, store large amounts of sugar as glycogen. This glycogen can be broken back down to free sugar, which can be used by those cells for the production of ATP. Most cells, like muscle, cannot release this glucose back out to the blood, so this storage of glucose as glycogen is for their own use only, and cannot be used by other cells of the body. However, the brain does not store sugar to any appreciable extent. The brain is dependent on the system as a whole to provide it with sugar on virtually a moment-to-moment basis. This fact, more than anything else, requires that the priority use of all three types of substrates be one of feeding the brain.

The body contains reservoirs for storing nutrients: As we have already mentioned, energy-containing nutrients are stored in various locations in our bodies so that we don't have to depend on constantly taking in these nutrients from the environment. These "cupboards," or nutrient reservoirs, allow us to store the food that we eat and use it at times when we are not eating. The body maintains reservoirs of sugar, fat and amino acids to be used to maintain a constant supply in circulation of molecules that can be used to produce ATP. The systemic reservoir for sugar is the liver, which stores this sugar as glycogen. The liver is the only tissue that can release stored sugar back into circulation. The reservoir for fats is the fat cells, which are distributed primarily underneath the skin and in people with excess, in the abdomen. Fat stored under the skin serves as both stored energy and insulation. Because fat is stored in the absence of water it is also a lightweight method for storing energy. The reservoir for amino acids is the bulk proteins of the musculature. These three reservoirs of stored energy together protect the brain from a lack of molecules that can be used for production of ATP. The mechanism by which this protection is accomplished is an interesting interplay of chemical messengers which provide for both storage and removal from storage under a wide variety of conditions.

Controlling the flow of sugars, fats and amino acids: When a person eats a meal that contains carbohydrates, large fats, and proteins that meal is digested yielding sugars, small fats and amino acids. The process of absorption from the intestines places the sugar and amino acids into the hepatic portal vein and therefore they are delivered directly to the liver. As a result, the blood going to the liver from the small intestine contains far more sugars and amino acids than is either necessary or advisable. Such excess, if allowed to continue, would strain the kidney and the general water and ion balance of the body. The fats drain through the lymphatics and in a short time they are in excess as well. However, such a circulatory feast is not allowed to continue. Excess nutrients act as a signal for a group of cells in the pancreas known as beta islet cells. The "excess signal" directs these cells to release a chemical messenger known as insulin.

Insulin is a protein. It is coded for in the DNA, but only beta islet cells make this protein. Insulin is released from the pancreas and travels through circulation and binds to specific "information ports" (called insulin receptors) on many types of cells. Of particular importance is the musculature. Muscles are responsible for about 75% of the insulin-directed removal of glucose from blood. Much of the remaining insulin-directed glucose removal is carried out by fat cells. Both muscle and fat can only take up glucose when insulin is in circulation.

The other important tissue for glucose disposal is the liver. However, uptake of glucose by the liver is controlled by the glucose concentration in the blood circulating to the liver, not insulin. Following eating, more than 70% of the blood entering the liver comes from the small intestine by way of the hepatic portal vein. This blood is rich in nutrients and the liver takes up as much glucose as it can as the blood transits on its way through the liver to general circulation by way of venous return. The liver uses this glucose to make a large sugar polymer called glycogen. Glycogen is the storage form of sugar. Glycogen consists of sugar molecules linked to sugar molecules linked to sugar molecules. The result is a reservoir composed of vast branching chains of joined sugar molecules (Fig. 7.7). The sugar that is not removed by the liver is taken up by other types of cells like the muscle and fat in response to insulin. Insulin directs muscles to not only take sugar (glucose) but also directs

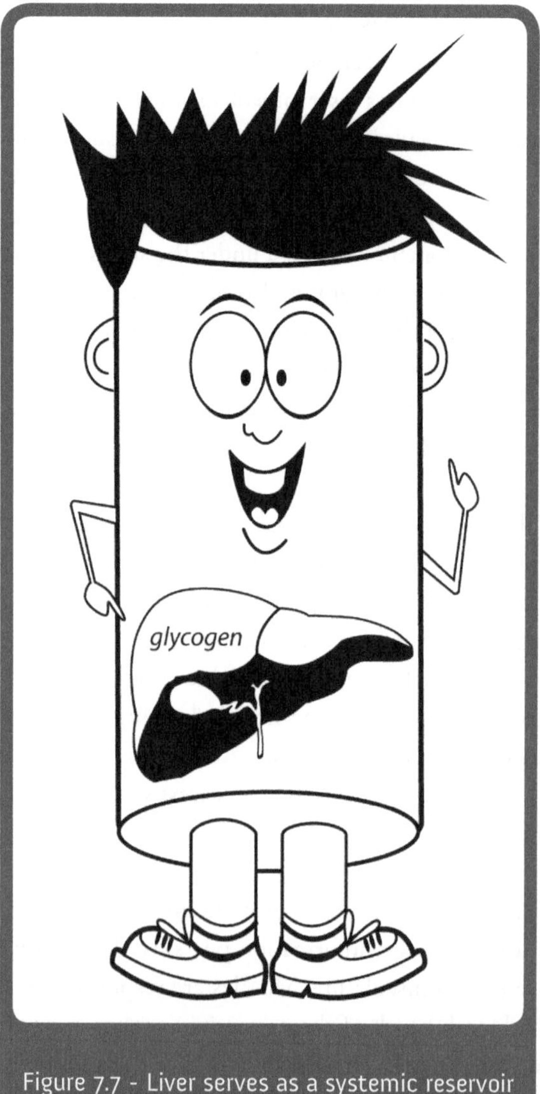

Figure 7.7 - Liver serves as a systemic reservoir for storing sugar in the form of glycogen for use by the brain.

Figure 7.8 - The muscle is the site of amino acid storage.

this tissue to make glycogen for storage. Glucose taken up by fat cells is used to synthesize fat for storage. The important difference between the stores of glycogen in the liver and the stores in the muscles is that the liver stores of glycogen are for the use of the body whereas the stores present in other types of cells are for their own use. Muscle, which stores huge amounts of sugar, uses its own glycogen for its own energy needs, whereas the liver glycogen is used to maintain proper levels of sugar in blood. Sufficient sugar in circulation is critical to brain function because the brain does not store nutrients.

Insulin also directs cells to take up amino acids and make proteins. Because of the large bulk represented by the musculature, most of the excess amino acids end up in muscles (Fig. 7.8).

The fat, which circulates by draining through the lymphatics, is picked up and stored in fat cells. All of these storage processes continue until the excess nutrients are removed from the circulating blood. However, if more amino acids or sugar are taken into the body than can be used, then the body stores the excess as fat (Fig. 7.9). When the excess is gone, the release of insulin is stopped. If one considers the large number of overweight people, it is clear that the ability of the body to store these nutrients is quite extraordinary.

Figure 7.9 - Excess fat molecules are stored in fat tissue in the body.

Within a short period of time after the excess nutrients are removed from circulation, the problem is reversed. The circulating concentration of especially sugar must be maintained by removing some from storage in the liver. Sugar is the preferable molecule for producing ATP by the brain. The full use of sugar, which requires oxygen, results in only one waste product, carbon dioxide. The sugar stored in the liver feeds primarily the brain. When not in excess in circulation, the brain is the only tissue that can take up sugar without being directed to do so by the insulin signal. When the concentration of sugar in circulation drops below a certain "set point" then another protein chemical messenger, glucagon, is released by another group of cells in the pancreas, the alpha islet cells. Glucagon acts as a "message" only for the liver.

The glucagon message directs the liver to break down glycogen and to release the resulting sugar into circulation. How long can the store of sugar in the liver last? If the liver stores are completely full, and a person is not physically active, the liver can provide sugar only for about 12 to 18 hours.

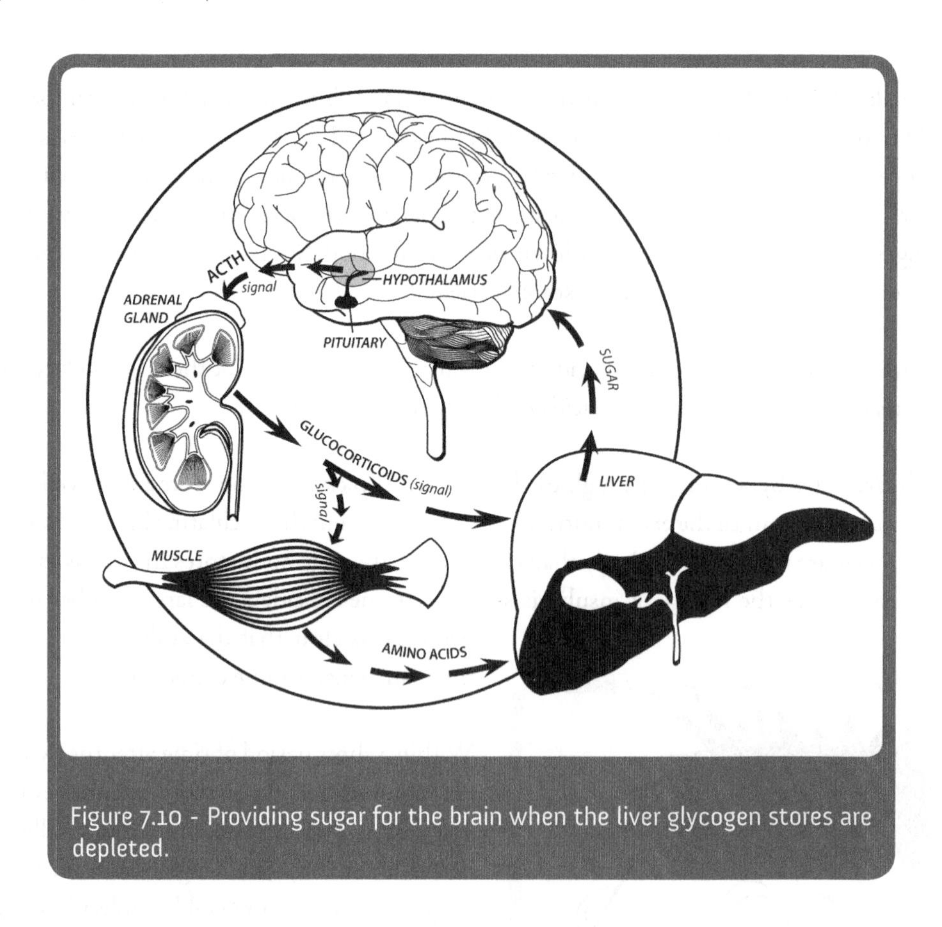

Figure 7.10 - Providing sugar for the brain when the liver glycogen stores are depleted.

However, there is another mechanism for providing sugar to the system that doesn't involve eating. This mechanism involves the relationship between the liver and the musculature (Fig. 7.10). The relationship between the liver and the musculature is mediated by another chemical messenger system. This chemical messenger system is controlled by the brain and involves the adrenal gland. A message from the brain directs the pituitary gland to direct the adrenal gland to release a steroid hormone called glucocorticoids. Glucocorticoids are always present in circulation, but the concentration varies in a pattern based on the light/dark cycle of the day. This circadian rhythm is designed such that glucocorticoids are high during the animals' active period when energy is needed and low during the animals' inactive sleeping period. In addition, demand (in the form of stress) will increase the amount of these hormones in circulation regardless of the pattern.

Glucocorticoids are a different type of chemical message than insulin and glucagon. Glucocorticoids are not proteins and therefore are not directly coded for in the DNA. Glucocorticoids belong to a class of molecules called steroids (estrogen and testosterone are other types of steroids that you may be familiar with). Specific enzymes direct the synthesis of steroids from cholesterol, which is a type of lipid. Like fat, steroids are not very soluble in water and must be carried around the body by proteins. However, steroids move very easily through the membrane of a cell. Unlike the receptors for insulin and glucagon which are proteins in the surface membrane of the cell, receptors for steroids like glucocorticoids are proteins in the water inside of the

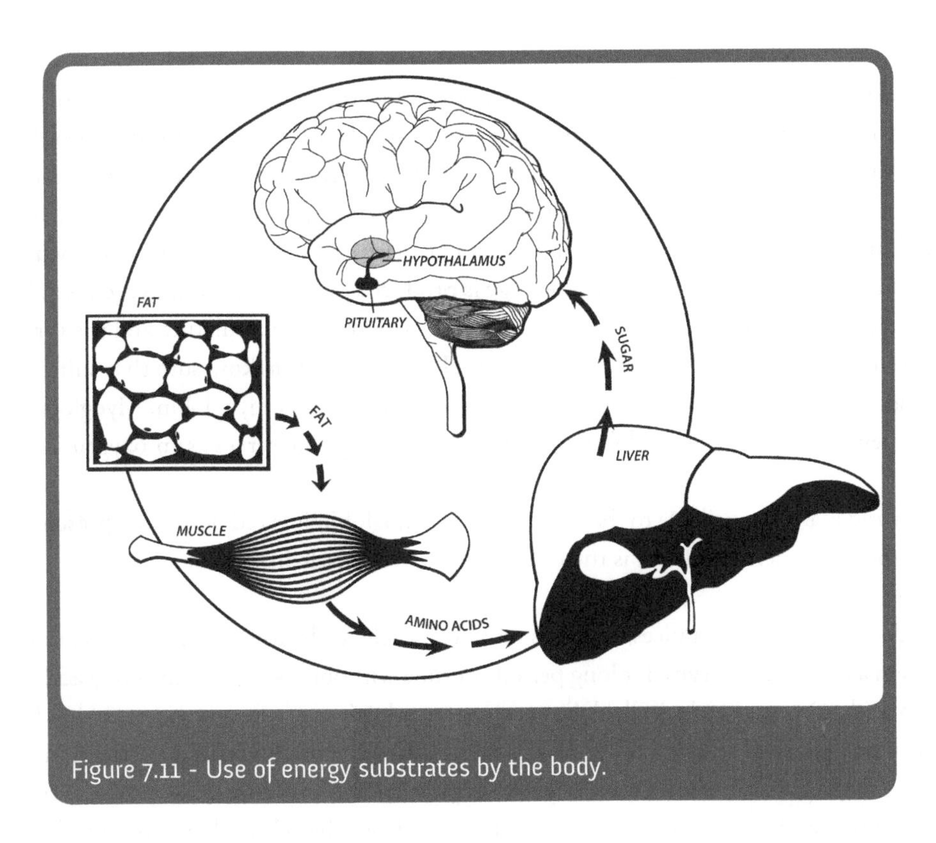

Figure 7.11 - Use of energy substrates by the body.

cell. When a steroid is delivered by circulation to a cell, it moves easily through the outer membrane and binds to its receptor inside the cell.

When a steroid binds to its receptor, the complex moves into the nucleus of the cell and directly binds to the DNA. The result is that the cell alters gene expression. Specifically, the binding of the hormone receptor complex to the DNA changes the amount of message (mRNA) being made for specific proteins. The change in the amount of mRNA for a protein ultimately results in a change in the amount of that protein. Since proteins are the functional molecules of the cell, steroids change the functional capabilities of the cell.

All cells of the body are constantly making proteins and breaking down proteins. The effect of glucocorticoids on the musculature is to tip the balance between making protein (synthesis) and breaking up proteins (degradation) such that degradation exceeds synthesis. The result is free amino acids to be released into circulation. Most of the released amino acids are picked up by the liver. Glucocorticoids, the same chemical message that directs the muscle to release amino acids, direct the liver to produce more of the protein enzymes necessary for making sugar out of amino acid carbon. The liver picks up the amino acids released by the musculature, cuts off the nitrogen, and makes sugar out of the remaining carbon skeleton. The nitrogen part of the amino acids is a waste product that the cell has no other use for. The nitrogen is used to make urea which is removed from the body in the urine produced by the kidney. The sugar made in the liver then becomes available to maintain the concentration of sugar in the blood.

The liver can also make sugar out of the glycerol part of fats but this is a more expensive process. Therefore, one may ask: what is the role of fat? Apart from serving as insulation to prevent excess heat loss from our bodies, most fat that is used for energy is used by the muscle. Exercise and the process of generating enough heat to maintain body temperature can deplete the store of glycogen in muscles quite rapidly. Muscle, because of its large bulk and the mechanical work that it does, is a primary user of energy in the body. Muscle produces and uses large amounts of energy. Much of this energy is produced by using fat in an oxygen requiring series of reactions. As long as sufficient oxygen is available, both the vast skeletal musculature and the very active heart muscle can produce large quantities of energy from fat. The process of making sure that sufficient oxygen is available is made easier because muscles express a unique protein called myoglobin. Myoglobin is a protein very much like hemoglobin. Like hemoglobin it binds oxygen. Myoglobin serves as an "oxygen sink" for muscle.

It binds oxygen and makes it available to the muscle. Like hemoglobin it is red due to the presence of oxidized iron as part of the molecule. Myoglobin is the protein that gives muscle its red color.

In essence, under normal circumstances, fat feeds muscle, muscle feeds liver and liver feeds brain (Fig. 7.11). With this system a person can survive for long periods of time without taking into the body additional sugars, fats and amino acids. One may technically define starvation as beginning after the last meal has been digested and stored. The first phase of starvation begins with supporting the brain with sugar stored in the liver. The second phase would begin as the liver stores begin to deplete and glucocorticoids activate the breaking down of the muscle protein and the synthesis of sugar in the liver from the released amino acids. This phase will continue for about three days with significant muscle wasting. However after three days the brain "decides" that times are really bad and it doesn't want to use up all of the musculature. It therefore reduces the signals telling the adrenal gland to make and release glucocorticoids. It then turns to another source of energy that involves primarily the liver and the fat. Large amounts of fatty acids are released into circulation. These fatty acids, in addition to providing energy for the musculature and the heart, are picked up by the liver. Fatty acids are used for energy by cutting off two carbon atoms at a time from the carbon chain and making a molecule called acetyl CoA. This molecule is then sent into the oxygen requiring energy producing pathways. What the liver does in this phase of starvation is produce far more acetyl CoA than it needs. The acetyl CoA and several conversion products are then dumped into circulation for use by the brain and other tissues. The group of molecules including acetyl CoA and its breakdown products are collectively known as "ketone bodies." Although this does not alleviate the need for some sugar derived from amino acid carbon, it does greatly reduce this demand. When this happens the individual is in what is called "ketosis." With starvation, the individual will remain in ketosis until all of the fat in the body is used up. Then the brain will again turn to the musculature for a short period of time before death. How long a person can tolerate starvation is directly proportional to his or her amount of body fat.

Apart from starvation, there are a number of reasonably common circumstances when this disturbance in nutrient flow happens to a person. One such situation is insulin dependent, or type 1, diabetes. In insulin dependent diabetes, the individual makes insufficient insulin to maintain the glycogen stores in the liver and muscle. Generally, such a diabetic person eats and then takes a shot of insulin. The shot of insulin allows the body to store the nutrients taken in by eating. However, if such medicinal remedies are not taken, then the first store that suffers from depletion is the small glycogen stores. When this happens, the body ends up running on fat and the brain ends up living off of ketone bodies and a small amount of sugar made by the liver from muscle-derived amino acids. The consequences to the body of chronic maintenance of such a situation are quite dire. In particular, ketone bodies strain the kidney and disturb the acid-base balance of the body. Such a condition is easily recognizable because the increased hydrogen ion concentration caused by the ketone bodies stimulates respiration through the hydrogen ion sensor in the carotid arteries. In addition, a small amount of acetone produced with the ketone bodies leaves the body as sweet smelling vapors through the lungs. The second relatively common situation that causes a person to go into ketosis is "fad" low carbohydrate diets. Very low carbohydrate diets deplete glycogen stores and force the body to initiate a starvation sequence. The initial phase, which lasts about three days, involves degradation of the musculature. Following the initial muscle degrading phase the body reverts to the use of fat through ketosis. The result of this ketotic state is a strain on the kidney, a disturbance of the acid-base balance of the body, and probably "acetone" breath.

Stress and metabolism: The above discussion describes the relatively normal flow of energy substrates in the body. However, there are situations that require the body to mobilize and use large amounts of energy substrates in a very short time. These situations involve a "stress override" of general energy metabolism. A good example of such a situation is sustained exercise. Exercise causes the release of a chemical messenger called adrenalin. (Another name for adrenalin is epinephrine.) Adrenalin provides for an increase in circulating substrates for making high-energy molecules. It causes the liver to release sugar, it causes the muscle to break down glycogen to provide for the increased ATP necessary for physical activity, and it causes the fat tissue to release fatty acid molecules for use by muscles and the heart. It also provides for increased blood flow to the muscles, the heart and the brain in order to deliver the substrates along with more oxygen. The body supports exercise from its nutrient stores. In most people the limit to exercise is oxygen delivery. However, as discussed above, even if enough oxygen is not delivered to the muscles, the muscle does not stop working. ATP is obtained by the muscles from breaking down sugar without oxygen. This leaves an intermediate molecule which builds up around the muscle as lactic acid. The buildup of this intermediate is called an "oxygen debt." Lactic acid buildup is an oxygen debt because oxygen must be available to either use the lactic acid to produce ATP or to build it back into sugar. Most of the lactic acid of the oxygen debt is released to the liver and is built back into sugar. A muscle will ache after exercise because some of the lactic acid is left in the muscle. In fact, creams and lotions used to relieve these aches are designed to increase blood flow in the muscle which facilitates sending this lactic acid to the liver.

Another situation where large amounts of substrates for producing ATP are mobilized occurs when an individual finds himself or herself in a situation that is either threatening or perceived as threatening. One of the immediate responses of the body to danger and anger is to mobilize substrates for production of energy in anticipation of exercise ("fight or flight"). Regardless of the nature of the situation, fear and anger cause the body to prepare to make an energetic response. The chemical messenger that mediates this type of stress response is also adrenalin. The first response to such an acute stress situation is for the brain to activate the body in the same way as it is activated during exercise. Adrenalin causes the liver to break down glycogen and release the sugar into circulation. Adrenalin also causes the muscles to breakdown glycogen for its own use in anticipation of mechanical function. In addition, adrenalin mobilizes fat from the fat cells to be used by skeletal and heart muscles. Finally, adrenalin opens up the airways to the lungs allowing them to bring in more oxygen to use these substrates. These metabolic responses, along with the circulatory responses discussed previously, prepare the body for "fight" or "flight." Adrenalin also sets up a response designed to replace sugar stores in the liver. Adrenalin acts on the brain, which acts on the pituitary, which directs the adrenal to increase the output of glucocorticoids. Glucocorticoids cause the muscle to release amino acids and direct the liver to make glucose from these released amino acids. Thus, human bodies are designed to respond to threatening situations with exercise.

It is very clear that the body has evolved a highly sophisticated set of responses designed to provide for a constant supply of energy in a large variety of situations. However, what has not evolved is a very efficient mechanism for putting mobilized substrates back into storage if they are not used. Anger, fright and other stressful situations cause the mobilization of substrates. If the individual does not respond to such situations by using more energy, then everything must be put back into storage. Sugar and amino acids are relatively easy to put back in storage. However, fat is more difficult to put back into storage. Because fat has low water solubility, it does not travel in circulation by itself as a free molecule. Fat travels in circulation in two forms. One form is as free fatty acids bound to proteins in the blood. The other form is called low density lipoproteins (LDLs). Low density lipoproteins are globs of fat surrounded by a little protein. The little bit of protein around the glob of fat gives the structure charge. This charged surface allows for solubility in the water of the blood. These globs of fat are delivered to many areas of the body, including the heart, in response to the adrenalin message. When they arrive at their destination, much of the fat is released to the tissue. The package is now a high density lipoprotein (HDL), because now the ratio of fat to protein in it has changed. It appears that these HDLs can be recycled as carriers of fat. The problem is that in order for the unused fat to be returned to storage, it must be repackaged for transport. To the degree that the unused fat is not returned it stays where it was delivered. At present there is a great interest in understanding the mechanisms that control the movement and use of fat in the body. The reason is that years of unreturned fat mobilized in anger or stress appears to be a major contributor to cardiovascular disease. The fat clogs the arteries. In addition, many drugs like amphetamines, cocaine, nicotine, alcohol and caffeine either directly cause or indirectly promote the mobilization of fat.

From the above discussion it should be clear that given the interconnected balances involved there are many situations that can cause disturbances in aspects of body metabolism. Some of these situations can be easily treated, such as too little of one of the B-complex vitamins. Others are more complicated such as too little or too much thyroid hormone. Perhaps the most prevalent is the types of diabetes that are often called "sugar diabetes." In this type of diabetes, the person does not store glucose (sugar) following eating. The immediate consequernce is "high" blood sugar. A derivative consequence is the lack of stored glycogen. There are actually two types of "sugar diabetes." One type, referred to as type 1 or juvenile diabetes, is an autoimmune disease. For some reason which is currently not understood, the body's immune system attacks and destroys the beta islet cells of the pancreas which are the only cells of the body that express the gene for insulin. The approach to treating type 1 diabetes is to try and balance the introduction of human insulin from outside the body with the ingestion of food. The most common way this is done is by injection. However, the use of insulin pumps, which are able to sense blood sugar levels and replace the "right" amount of insulin, are an interesting developing technology. The other far more common type of diabetes is referred to as type 2 or adult onset diabetes. This type of diabetes develops from a complex interaction of genetic predisposition and lifestyle. In type 2 diabetes, the body tissues are unresponsive to insulin. Genetic disposition is generally not simply a "bad" gene. Rather there are small variations in the structure of genes called alleles that can make them a bit more or less active. Each individual has two alleles for each gene, one from his or her mother and one from his or her father. A person can be born with a balance of alleles of the many different genes associated with insulin responsiveness that can make him or her susceptible to developing type 2 diabetes in response to certain life style choices. In general, lack of exercise and excessive high fat, high carbohydrate diets promote diabetes. The exercise aspect is very important, because the musculature is responsible for about 75% of the insulin-directed glucose disposal and exercise regulates the sensitivity of muscle to insulin. The more exercise, the more sensitive the muscle is to insulin. In general obesity and type 2 diabetes are associated. That does not mean that all obese people will develop type 2 diabetes or that all type 2 diabetics are obese. It just means that obesity is a risk factor that increases the probability of developing type 2 diabetes. As discussed above, the liver receives blood from the hepatic portal vein which is rich in nutrients taken in by the small intestine. The first disposal of glucose is insulin independent by the liver because the high concentration of glucose allows the liver to take up glucose through its non-insulin dependent glucose transporter. The nutrients then enter general circulation where the musculature and the fat take up glucose through their insulin dependent glucose transporter. If this doesn't happen efficiently because those tissues are insulin resistant the remaining glucose is either lost through the kidney or in the next pass through the liver more glucose is taken into the liver. In response to insulin, the liver is supposed to synthesize glycogen from the glucose taken in from circulation. If the liver is also insulin resistant then the sugar taken up by the liver is used to synthesize fat instead. It is important to understand that insulin resistance is not an "all or nothing" phenomena; it is a process that starts small and becomes worse with time. One common consequence of insulin resistant diabetes is the excessive accumulation of fat in the liver. Another very common consequence of type 2 diabetes is failure of the beta cells in the pancreas. The unresponsiveness of the system to insulin causes the beta islet cells to put out more and more insulin, which eventually causes the beta cells to "burn out" so now the person must take insulin medicinally. Because much of the sugar that was not taken into the tissues in response to insulin is used to synthesize fat, people with diabetes often develop the type of dysregulation of lipid metabolism that causes athrosclerosis. Similarly, chronic hyperglycemia causes abnormal additions of sugar molecules on to serum, vascular and kidney proteins which interfere with their function. In fact, glysosylated hemoglobin (hemoglobin with abnormal sugar substitutions) is used to determine glycemic control in diabetic patients. This is the "HbA1c" that doctors measure periodically along with blood glucose in diabetic patients. It is thought that the abnormal substitution of sugar molecules on to vasculature proteins contributes to the vascular degeneration that occurs with time in diabetic patients leading to blindness and amputations. Similarly, the combination of strain on the kidney due to the excessive glucose along with the abnormal substitution of sugar onto kidney proteins begins to cause the kidneys to fail. Finally, failure of the kidney along with the arthrosclerosis causes hypertension (high blood pressure). Type 2 diabetes is an excellent example of how failure of one part of the system cascades throughout the entire system.

chapter eight

the brain in control of it all

Adjusting the body: The body can be viewed as a society of specialized groups of cells. Cells are "specialized," which means that they don't all do everything that is done in the whole organism. Rather, different types of cells do a limited number of things, allowing for more complicated functions to be carried out by the whole organism. Each group of specialized cells in this society carries out its limited set of functions based on the proteins that it makes.

Figure 8.1 - Stresses on the body may be small or large.

For example, fat cells store and release fat molecules to be used for energy, but they don't produce insulin. Conversely, the beta islet cells of the pancreas produce the hormone insulin, but don't store and release fat molecules.

The body can also be viewed as a collection of adjustable functions. For example, blood vessels open and close, channeling more or less blood to different parts of the body. The cells in the heart cause it to pump faster or slower. Glucose may be taken up and stored by the liver, or released to the bloodstream for other cells to use. Functions are constantly adjusted because the internal and external circumstances of the body are always

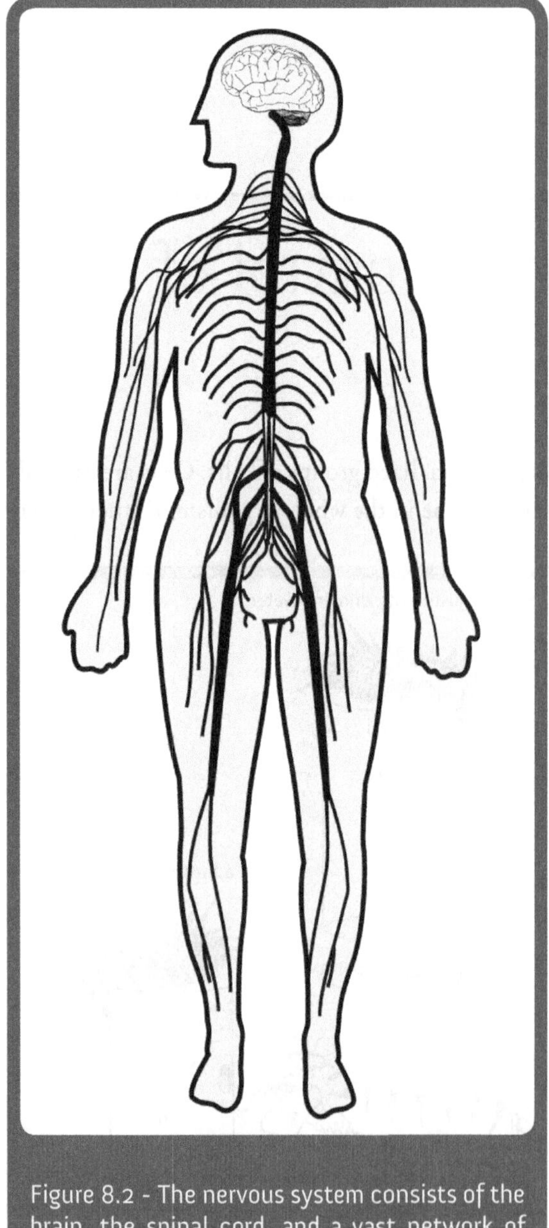

Figure 8.2 - The nervous system consists of the brain, the spinal cord, and a vast network of nerve cells which are distributed throughout the body.

changing. If an individual simply drinks a glass of water or sits down, a vast number of functions must be adjusted in order to accommodate even these small changes. In a stress situation, much larger adjustments in body functions occur in order to be ready to deal with that stress (Fig. 8.1). However, each adjustment in any particular function must be coordinated with all other functions.

Coordinating the vast number of adjustments that are constantly occurring in the body is primarily the responsibility of the nervous system. Such coordination requires that a massive amount of information be constantly collected and continuously evaluated so that appropriate directions can be given to all of the various specialized types of cells. Everything that occurs in the body or outside the body serves as information to the nervous system. This is the information on which adjustments in the many functions of the body are made.

The flow of information: The nervous system consists of the brain, the spinal cord and a large network of nerves that are distributed throughout the body (Fig. 8.2).

The nervous system is a large group of diverse cells whose entire purpose is to acquire information, to process information, and to direct the functioning of the body based on that information (Fig. 8.3).

In addition, many of the glands which make and release hormones (chemical messages that travel in circulation) are also part of the communications network that is controlled by the brain. Information constantly flows into and out of this communications network. The most fundamental purpose of this flow of information is simply the maintenance of the quality of the internal environment around the brain. In order to preserve its own environment it must collect information and control the environment of the entire body. Over and above accomplishing this basic purpose of keeping the workings of our body in balance, there is the flow of information that is associated with every action, behavior, feeling, thought and memory of the individual.

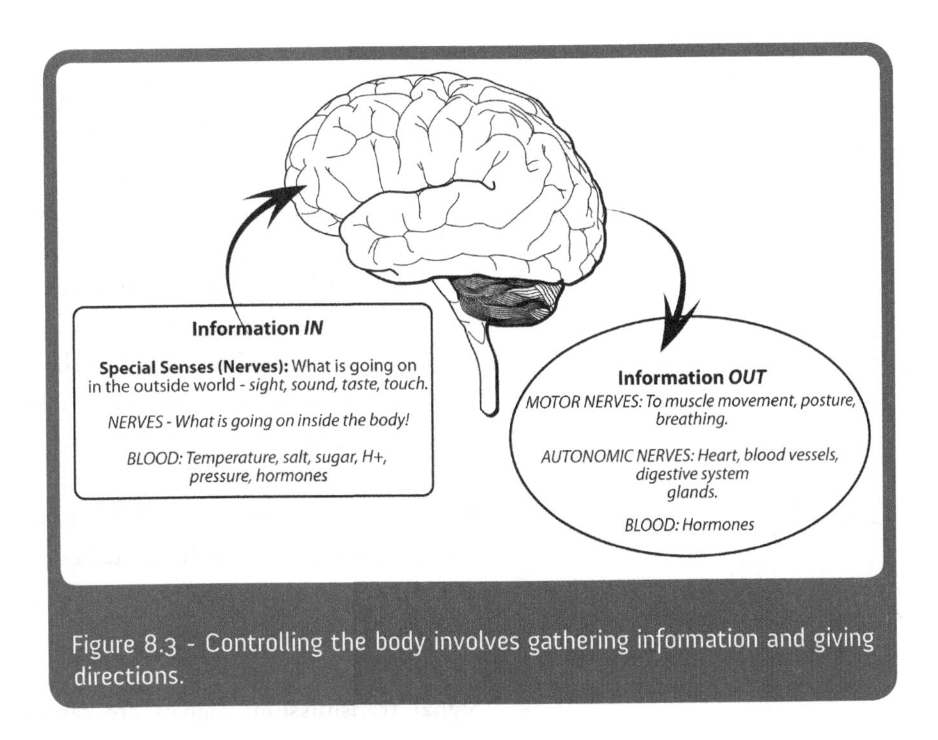

Figure 8.3 - Controlling the body involves gathering information and giving directions.

These higher brain functions have evolved in concert with the evolution of the brain. If you look at brain structures in lower vertebrates, the underlying brain structures are very similar to those in higher vertebrates. What has evolved and become larger and more complex are regions of the brain that are involved with carrying out these more complex brain functions (Fig. 8.4).

Nerve cells and signals: The nerve cell, or neuron, is the basic unit of this information flow. In general, nerve cells are polar, that is they have two distinctly different ends which are highly organized (Fig. 8.5). Different parts of the nerve cell are responsible for different aspects of the nerve cell function. The nature of this polarity is that information flows in at one end of the cell and out at the other. At one end they have a branching set of projections called dendrites that receive signals: this is the "listening" end of the nerve cell. Dendrites lead into the cell body. Like all cells, nerves must carry out basic life processes such as making proteins and ATP. The cell body contains the nucleus and DNA of the nerve. The cell body also contains the protein manufacturing capacity of the nerve cell. Leaving the cell body at the other side is another branching projection called an axon. The axon is the "talking" end of the nerve cell. Axons transmit signals to dendrites on other nerve cells. Signals flow from dendrites to cell bodies to axons.

A nerve cell receives a signal from another nerve cell in the form of a type of chemical message called a neurotransmitter. There are many different neurotransmitter molecules which are used by different kinds of nerve cells. All nerve cells communicate by releasing some kind of neurotransmitter. The tip of the axon of a signaling nerve cell releases a large amount of one specific type of neurotransmitter into the very small space between the tip of the axon and the next dendrite (Fig. 8.6). Because there are many neurotransmitter

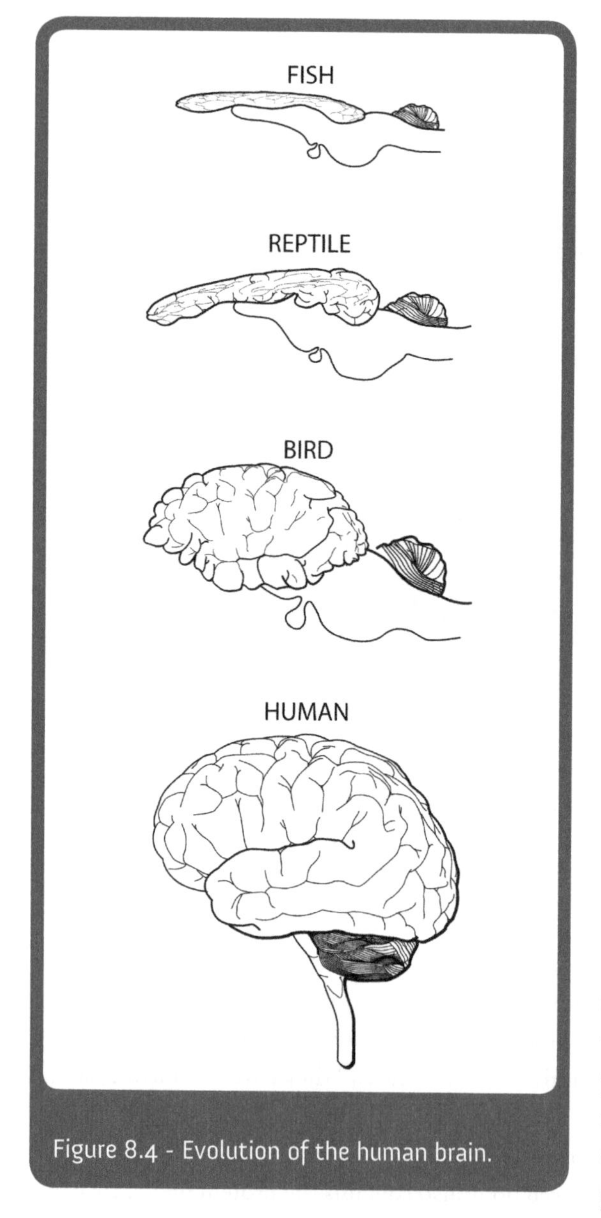

Figure 8.4 - Evolution of the human brain.

molecules released near the tip of the axon and few near the dendrite, neurotransmitter molecules move by diffusion across the small space and some bind to receptors on the dendrite. These receptors are the "information ports" in the membrane of the dendrite. Each individual neurotransmitter molecule binding to a single receptor changes the electrical activity of the membrane of the dendrite a tiny bit. The accumulation of the individual tiny changes on each dendrite branch flows towards the cell body of the nerve. At any instant in time the sum of the changes from the dendrites determine whether or not a signal is passed down the axon of the nerve cell to the dendrites of other nerve cells. At any instant in time a nerve cell is either "sending a signal" or "waiting" to send a signal.

Signal transmission: Signals are carried from one place in the neuron to another by patterned changes in ion-based electrical potentials. The neuron expends metabolic organizational energy to move ion charges from one side of its membrane to another.

In this way nerve cells create a difference in the distribution of ions across their outside membrane. Such differences form an electrical potential (Fig. 8.7). It is a potential because of the natural pressure to make the ions on both sides of the membrane the same. The potential is electrical because it is based on the differential distribution of charged ions such as sodium (Na^+) or chloride (Cl^-) across the membrane. These potentials can be created and released in a local area of the membrane at an extremely rapid rate. By using a mechanism of creating and rapidly releasing and recreating the electrical potential in local areas of the membrane, signals are carried from one end of a dendrite to the cell body or from cell body to the end of an axon or nerve terminal. The signal simply moves along the cell as a cascading pattern of changing electrical potentials. When such a cascade of electrical potentials reaches the end of the axon it causes neurotransmitter molecules to be released. The neurotransmitter molecules diffuse "downhill" from the axon terminal across the small space to a branch of the dendrite of the next nerve. Some bind to the receptors in the membrane of the dendrite branch and the distribution of electrical potentials on the next neuron is changed. The entire transfer of signal is extremely fast: it only lasts a few thousandths of a second and then it is terminated. The

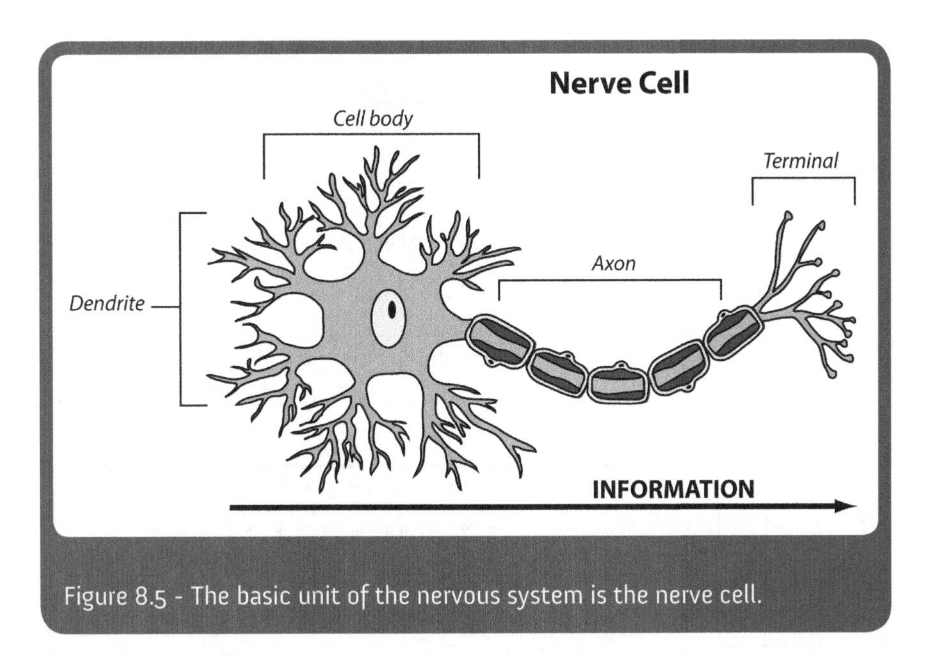

Figure 8.5 - The basic unit of the nervous system is the nerve cell.

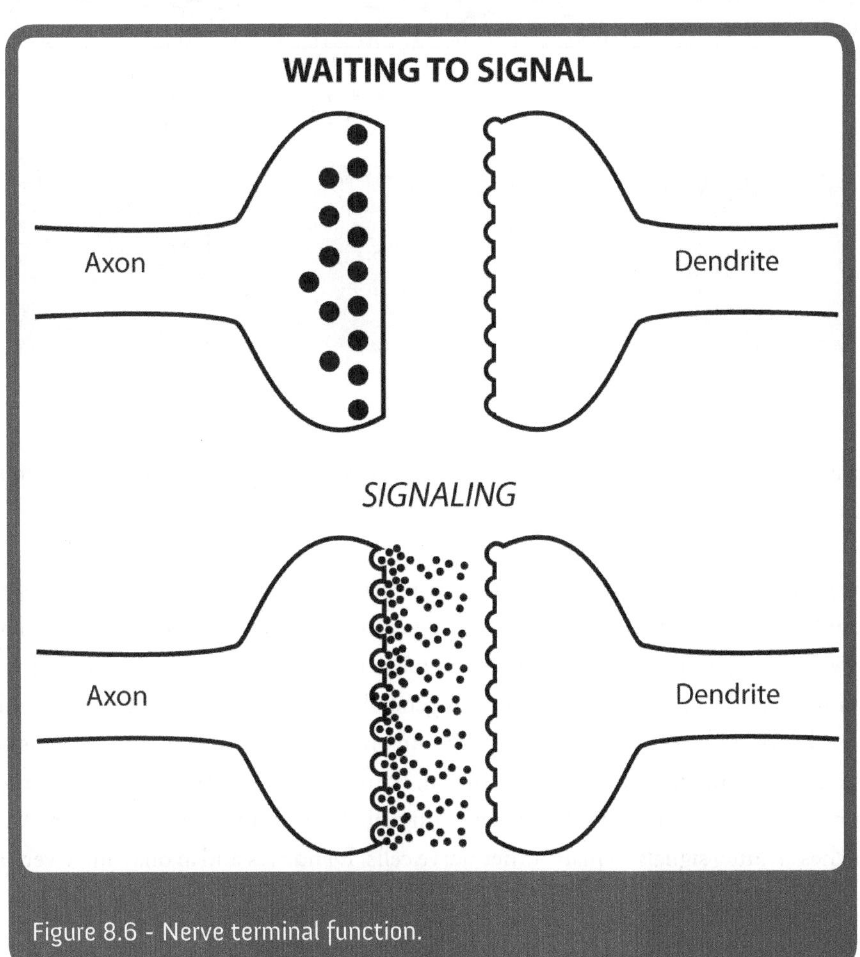

Figure 8.6 - Nerve terminal function.

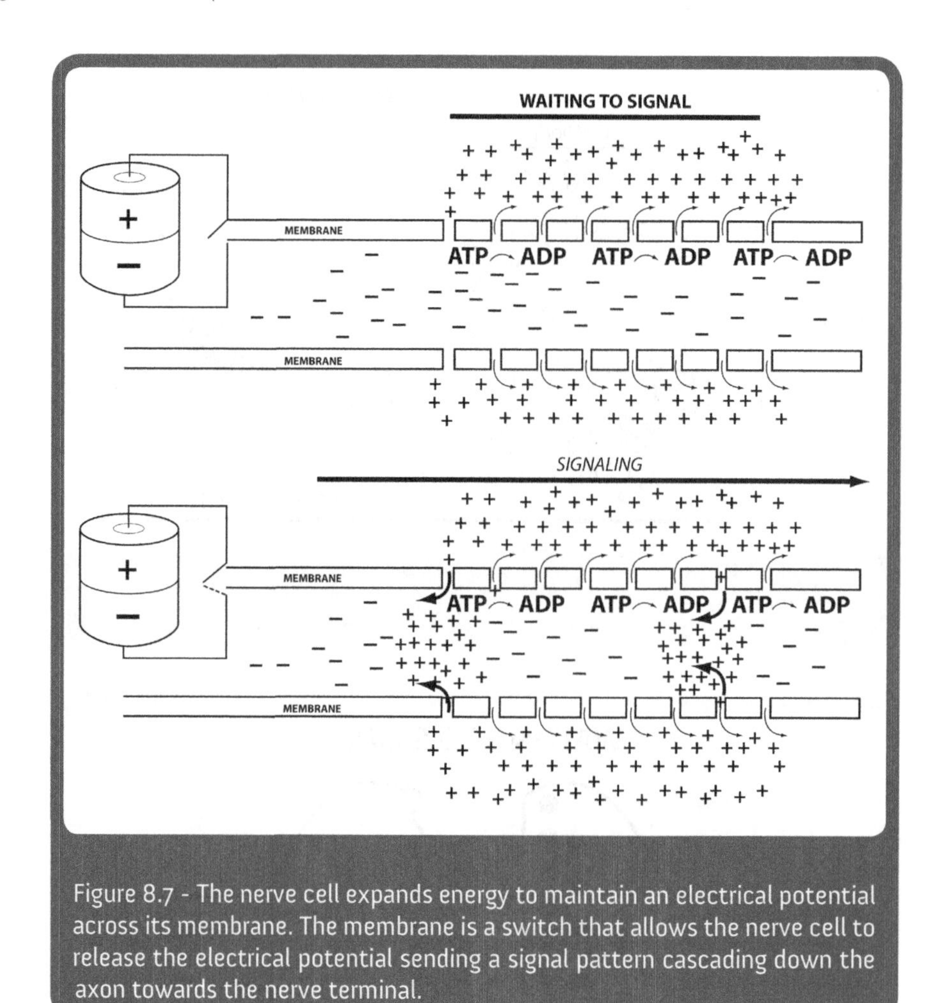

Figure 8.7 - The nerve cell expands energy to maintain an electrical potential across its membrane. The membrane is a switch that allows the nerve cell to release the electrical potential sending a signal pattern cascading down the axon towards the nerve terminal.

signal is terminated by a combination of destruction and recovery (reuptake) of neurotransmitter molecules back into the axon tip.

There are many different molecular structures used as neurotransmitters in the nervous system. Each neurotransmitter binds to a unique receptor molecule. A cell can only receive a particular neurotransmitter signal if it has the corresponding or "right" receptor in its membrane for that neurotransmitter. This becomes very important when we consider drugs and the brain. In general, a neuron makes and releases only one type of neurotransmitter. In contrast, a neuron generally contains receptors for and receives signals in the form of many different types of neurotransmitters. In addition, there are many variations in the structure of nerve cells. A cell can have many dendrite branches receiving signals from many other nerve cells. An axon can also have many branches sending signals to many other nerve cells. Dendrites and axons can be very short or many feet long.

Besides neurons, the nervous system contains another type of cell called glia. Glial cells are the "helper cells" of the nervous system. Glial cells do not conduct electrical activity; rather, they support the activity of neurons. For example, the long dendrites and axons are covered by a sheath constructed of special glial cells called myelin. The myelin sheath speeds up the rate of signal transmission in the nerve. The myelin sheath also serves as a type of insulation to prevent "cross talk" between different nerves running together in a bundle. The importance of this myelin sheath can be seen in the disease multiple sclerosis (MS). In MS, because of the "incorrect" actions of the immune system, the myelin sheath is attacked by immune cells and degenerates causing many problems in brain function.

Both nerve cells and glial cells derive from the same stem cell population. However, in the adult, for some reason currently not understood, the stem cells are able to reproduce new glial cells but not able to reproduce new nerve cells in the living animal. Interestingly, if the stem cells from the adult brain are removed and placed in culture, they can be "teased" by manipulating the environment to differentiate into nerve cells. Because of the ability of stem cells to differentiate into glial cells, gliomas are the only type of cancers that develop in the adult brain. The only type of nerve cell cancer is a neuroblastoma which is found in very young children and had its origin during fetal life when the brain was forming.

Nerve cells make decisions: In a sense, each nerve cell is like a decision-making microcircuit connected with many other decision-making microcircuits. Because of the branched structure of the dendrite end of most neurons, many signals from many other nerves are received simultaneously. These signals are processed and a signal may or may not be transmitted through the cell body to the axon and the other end of the nerve cell. Whether or not the sometimes large number of signals received at the dendrite end of the cell results in neurotransmitter release out the axonal end depends on the combined nature of all of the different signals being received by the cell at that instant in time.

In the most simplistic sense, the signals entering a nerve cell can be divided into two categories. One category of signals can be viewed as positive. Positive signals are excitatory: they promote depolarization of the dendrite membrane. These depolarizing signals reduce the electrical potential in a local area in the membrane. The function of a depolarizing signal is to direct the nerve cell to pass along the signal. The other category of signal can be viewed as negative. Negative signals are inhibitory. They promote hyperpolarization of the nerve cell membrane. Hyperpolarizing signals increase the electrical potential in the local area of the membrane.

The function of a hyperpolarizing signal is to direct the nerve cell not to pass along the signal. The resolution of the excitatory and inhibitory signals entering the cell body from the dendrite branches of the nerve determines at any instant in time if the cell is releasing neurotransmitter at the end of its axonal branches. The resolution of the many signals into one response, whether it be to signal or not to signal, is a basic integrative mechanism of the nervous system. To either signal or not signal is a "decision" by the nerve. Such decisions are integrative mechanisms involved in information processing. Both the signal and the lack of a signal are part of the balance of the informational mechanism (Fig. 8.8).

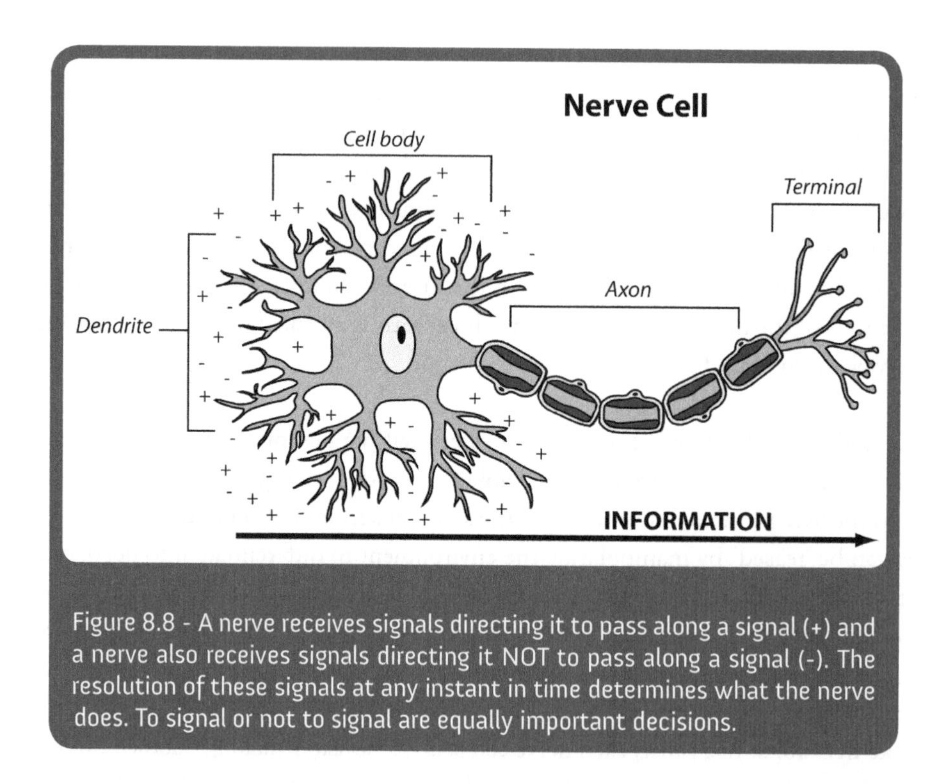

Figure 8.8 - A nerve receives signals directing it to pass along a signal (+) and a nerve also receives signals directing it NOT to pass along a signal (-). The resolution of these signals at any instant in time determines what the nerve does. To signal or not to signal are equally important decisions.

In an informational sense, the nervous system carries out its function by balancing and resolving large numbers of signals. In many respects, one can view a nerve cell as an individual element in a vast computing system. Similar to a computer that operates on binary language, at any instant in time each element is set to either "+" or "0." Just as a computer can store and process vast quantities of information using a simple binary code, so a brain with its trillions of nerve cell elements continuously operates the body on a binary system of each individual cell at every instant in time either signaling or not signaling. If the balance between excitation and inhibition is disturbed the consequence can be quite dire. For example, the poison strychnine disturbs the signaling balance. Strychnine blocks one of the inhibitory neurotransmitters. The result is that excitation dominates in some nerve pathways and uncontrolled muscle contraction results.

Throughout the nervous system the range in the complexity of such signal resolutions is vast. Some neurons receive a few connections, while others receive hundreds. Some neurons connect to a few other neurons and some connect to hundreds. In addition, axon-dendrite contacts are not static. At least in part, brain function involves a continuous change in these microscopic relationships between neurons.

The brain—collecting information and giving directions: At the center of the information flow in the nervous system is the brain. The brain is not a single group of identical cells, but rather it is organized into many smaller clusters of cells. The brain is composed of groups of organized cells within groups of organized cells within groups of organized cells.

Signals flow up and down the hierarchy of organization, bringing many types of coded information together in pathways concerned with the regulation of many different processes in the body. Because each bit of coded information is widely distributed in the nervous system, directions controlling many different processes in the body can be coordinated.

Signals are constantly flowing into and out of the brain. The spinal cord, the cranial nerves (large nerves which enter directly at the brain) and the peripheral nerves carry signals to and from the brain. In addition, the blood entering the arterial circulation going to the brain brings in information. There is also a direct connection between the brain and pituitary gland (a small gland that sits at the base of the brain, and is sometimes called "the master gland" because it controls many physiological functions). This direct connection gives the brain control over the endocrine system and the many different chemical messengers of the body. Because of

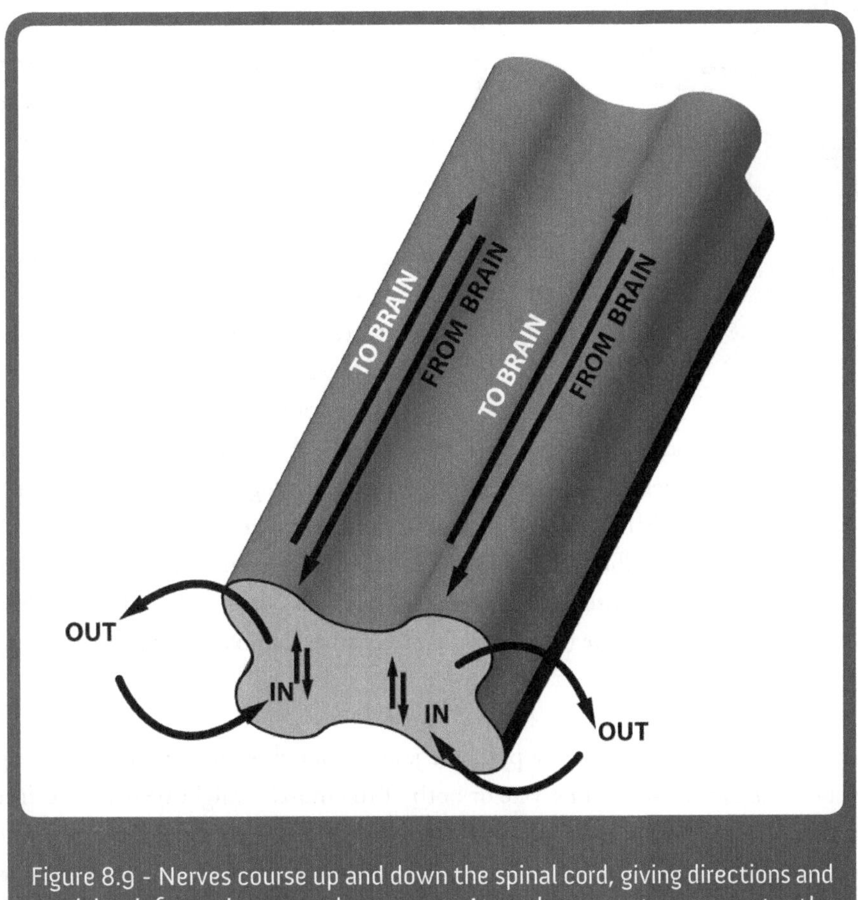

Figure 8.9 - Nerves course up and down the spinal cord, giving directions and receiving information at each segment. At each segment nerves enter the back of the spinal cord bringing in information and leave the front of the spinal cord taking instructions to the body.

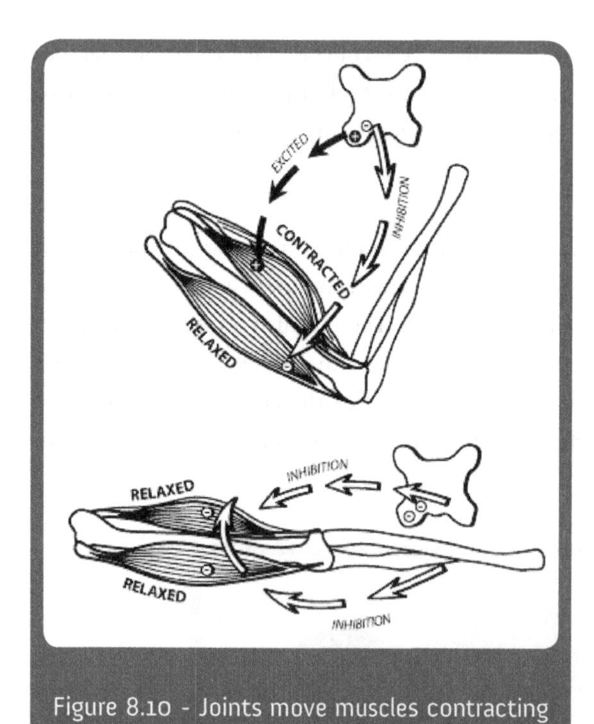

Figure 8.10 - Joints move muscles contracting and pulling them in one direction. When nerves direct a muscle to contract and pull the joint in one direction, the nerves to the opposing muscle receive strong inhibiting signals.

the connection to the pituitary, the blood leaving the circulation of the brain carries out directing signals to the body in the form of chemical messages.

The spinal cord—gathering information: The spinal cord is the main trunk of nerve fibers entering and leaving the brain (Fig. 8.9). Large bundles of nerve fibers course up and down the spinal cord carrying information. Nerves in this bundle connect with other nerves at each segment of the spinal cord. In addition, at each segment along the spinal cord, nerves called afferents enter the spinal cord, bringing information in from the body to be transmitted up to the brain. Some of these afferents carry signals from sensory endings associated with the surface of the body. These nerve fibers carry information concerning such things as pain, touch and skin temperature. Other of these nerve fibers carry signals in from sensory organs associated with muscles. These nerve fibers carry information concerning such things as the stretch of muscles or the orientation of joints. Other nerve fibers carry signals from sensory structures associated with the internal organs of the body. They carry information concerning such things as the stretch of the lungs or food in the digestive system. The incoming nerve fibers connect with other neurons in the spinal cord and signals are sent up to the brain as well as being passed to the other major groups of neurons in each segment of the spinal cord, the outgoing nerves, efferents. Efferent nerve fibers carry signals away from the spinal cord out to the body. Some of these efferents carry signals to muscles telling them to contract. These outgoing signals are coordinated with the incoming sensory signals from the sense organs associated with various muscles. This coordination provides a good example of integration of positive (excitatory) and negative (inhibitory) signals by the nervous system (Fig. 8.10).

Muscles work in opposing pairs. One muscle pulls the joint in one direction and the opposing muscle pulls it back. If both contracted at the same time, one or both of the muscles might be detached from the joint. If an incoming signal indicates that a muscle is contracted, then a signal to contract the opposing muscle will be blocked by strong inhibitory signals. In some cases an afferent signal will pass through only one connecting neuron to an efferent nerve, bypassing the brain. This simple, rapid "automatic" information flow is called a reflex. Reflex responses are rapid because the information processing occurs at the level of the spinal cord, before the information even reaches the brain. For example, one does not "think" about removing a hand from a hot stove.

Another type of efferent leaving the spinal cord carries signals to smooth muscles, to various glands, and to the heart. These nerve paths are collectively referred to as the autonomic nervous system. The autonomic nervous system controls many physiological processes in the body that just adjust as necessary: we don't have to "think about" controlling them. These nerves carry information for directing processes such as the opening and closing of blood vessels, the release of digestive enzymes and the speeding up or slowing down of the heart. Nerves that leave the spinal cord carrying signals to smooth muscles and glands always connect with a small cluster of nerve cells outside of the spinal cord. The signal is passed on to the cells in these clusters of nerves. The cells in the cluster integrate the signal from the spinal efferents with other signals. Depending on the resolution of all of the incoming signals, the cells of the cluster may send directing signals to the particular smooth muscles or glands. Signals to smooth muscles, to the heart, and to glands through these efferent paths play a major role in controlling the priorities of the body. Because of the involvement of the autonomic nervous system in controlling priorities, there are often two components to the innervation (Fig. 8.11).

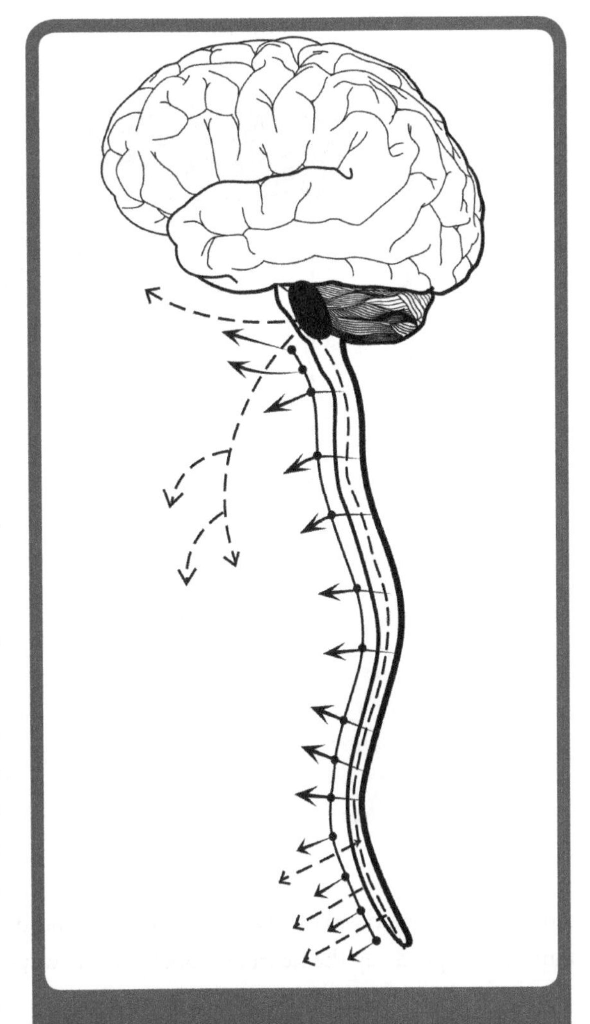

Figure 8.11 - The autonomic nervous system is composed of both parasympathetic nerves (dashed lines) and sympathetic nerves (solid lines).

One component of autonomic innervation, often referred to as the parasympathetic, controls the coordination of the heart, the particular blood vessel or the particular gland with the more relaxed processes of the body like digestion. When the parasympathetics signal a target, the target is adjusted to coordinate with the calmer, more relaxed housekeeping functions of the body, like digestion. The other autonomic component often referred to as sympathetic, coordinates the heart or the particular smooth muscle or gland with the acute stress "fight or flight" response of the body. When the sympathetics signal, the target is adjusted to coordinate with the danger responses of the body. In many respects parasympathetic and sympathetic nerve connections can be looked at as two parallel control systems (Fig. 8.12). The parasympathetic system controls organs and blood vessels during calm times. It directs more blood to the digestive system. It directs more blood to the kidney for filtration. It slows down the heart. In contrast, the sympathetic system controls organs and blood vessels during emergency situations. It reduces

Figure 8.12 - The parasympathetic part of the autonomic nervous system governs basic functions, and the sympathetic part governs functions when change is required.

blood flow to the digestive system and kidney and directs the blood to the heart, the brain, and the skeletal muscles. It speeds up the heart and opens the airways. It prepares the body for activity.

The special nerves of the head: There are three other modes of communication to the brain. The first is the cranial nerves. These are nerves that enter and leave the brain itself. The function of these nerves are quite varied, but all with the exception of one deal with processes localized in the head and neck region. Some of these nerves are purely sensory afferents such as those that bring in signals associated with smell, vision, and hearing. Others contain both sensory afferent and motor efferent branches. These nerves send out and receive information related to facial sensation, the movement of facial muscles and the movement of the muscles of the eyes.

The single exception is the vagus nerve. The vagus nerve carries major parasympathetic direction to many groups of cells in the thoracic (chest) and abdominal cavities. An experiment with the vagus nerve can be used to illustrate a very important point about sympathetic and parasympathetic control.

Both calming and activating the body are active processes. For example, the parasympathetic signaling actively slows down the heart. If one applied a strong electrical charge to the vagus nerve one could virtually stop the heart. In contrast, if one cuts the vagus nerve the heart usually speeds up.

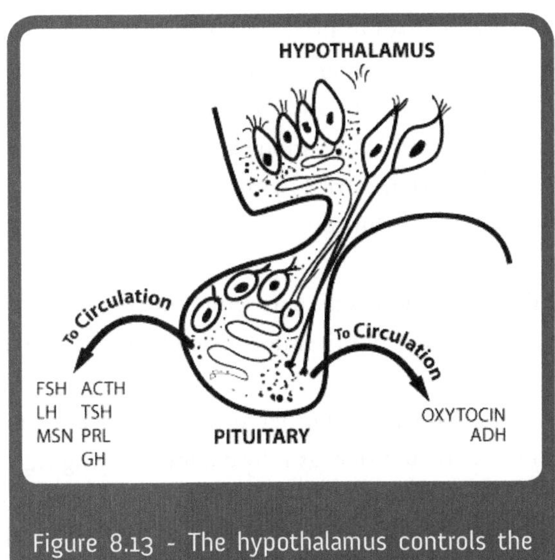

Figure 8.13 - The hypothalamus controls the pituitary gland which releases many hormones.

The brain controls hormones: The pituitary gland provides another major vehicle by which the brain can communicate with the body. The pituitary gland sits below the brain and just above the roof of the mouth. The pituitary gland consists of two parts. The front part of the pituitary is not developmentally a part of the brain. It is a gland which is closely associated with and controlled by the hypothalamus of the brain. The mechanism by which the hypothalamus controls the front part of the pituitary involves a special type of blood vessel called a portal system. Like the hepatic portal vein that runs from the capillary beds around the small intestine to the capillary beds in the liver, the hypothalamic pituitary portal vessels are designed to specifically deliver, by means of blood flow, something from point "A" to point "B." In the case of the hypothalamic pituitary portal system, special chemical messages are delivered from groups of cells in the hypothalamus of the brain to the cells of the front part of the pituitary. The cells of the front part of the pituitary make and release a variety of chemical messages (hormones) which direct many aspects of body function. The chemical messages traveling in the hypothalamic pituitary portal vessel are called releasing factors because they direct the synthesis and release of the chemical messages from the front part of the pituitary. Hormones from the front part of the pituitary participate in regulating body growth, stress responses, general body metabolism, reproduction, and milk production in the female. The back part of the pituitary gland is part of the brain. Nerves run directly from the hypothalamus of the brain to hormone releasing cells in the back part of the pituitary gland. Hormones from the back part of the pituitary participate in regulating water flow in the kidney, the birth process, and milk release by the breasts. In short, the eight or nine chemical messages released by the both parts of the pituitary under direction of the brain play a major role in controlling many of the major life processes of the individual (Fig. 8.13).

Blood carries many types of information: The last vehicle for communication between the body and the brain is the blood that enters the circulatory paths of the brain. Specific groups of nerve cells, primarily in the hypothalamus of the brain, monitor various aspects of the quality of the blood such as temperature, ion content, and sugar content. In addition, the blood entering the circulatory paths of the brain carries chemical messengers produced by various glands of the body under direction from the brain. These chemical messengers such as glucocorticoids, androgens, estrogens, thyroid hormone, and adrenalin feed back information to the brain. These hormones, in essence, inform the brain of the consequences of its own actions. This allows the directions from the brain to be adjusted based on results.

The mechanism of information flow—a summary: All of the various molecular and cellular processes are simply mechanisms of transmitting signals. Nerves carry signals in and nerves carry signals out. Blood carries chemical signals out, and blood carries chemical signals into the brain. However, these are purposeful signals which transmit information from one place to another. The real function of the brain is to decode and evaluate the information carried by the various types of signals coming in. It then uses this information to direct the myriad of body functions.

Information flow is the nature and the purpose of the nervous system's function. The signals are the molecular processes used, but the information flow is carried by the changing patterns of the molecular processes.

Actually, understanding nervous system function involves decoding the information which is contained in the patterns. In addition, as information-carrying signals move through the nervous system, they are translated from form to form. Ripples of depolarizations moving along a nerve are translated at the nerve ending to patterns of release of a neurotransmitter, which then may be integrated with other signals and translated into another ripple of depolarization in another nerve or a patterned release of a releasing factor from the hypothalamus, which may eventually be translated into a temporal pattern of a particular hormone moving through the circulation of the body. In this sense, the information flow in the body is simply a cascade of patterns being translated from form to form. Understanding the information flow involves decoding the language of the patterns and following the cascades of patterns from structure to structure and analyzing changes in function.

One person talking to another is a good illustration of pattern translation. When a person speaks he or she pushes air over his or her vocal cords. Patterns of motor nerve depolarization cause patterns of release of neurotransmitters at the individual muscle fibers of the vocal cords. The vocal cords vibrate and organize the air leaving the mouth. The air is organized much like the ripples one sees when you toss a pebble into a pond. The organized air moves out and into the ear of the other person. This causes a membrane in the ear to vibrate in concert with the organized air. The vibrating membrane causes a ripple to move through a fluid in a curved cylindrically shaped structure in the ear. The ripple deflects hair-like structures projecting into the fluid. The deflection of the hair-like projections causes a pattern of depolarization to occur in the auditory nerve. The pattern of depolarization in the auditory nerve is simply a translation of the pattern of vibration in the other person's vocal cords.

Organization of the nervous system: The nervous system is composed of the peripheral nerves, the spinal cord and the brain. The peripheral nerves, which are connected to the spinal cord, run throughout the body collecting information and providing direction to various structures around the body. The spinal cord runs down the center of the trunk of the body and serves as the site of entrance and exit for the peripheral nerves.

An important structural characteristic of the spinal cord is that it is segmented. The spinal cord itself is protected by bone (our "backbone"). Underneath each of those "bony bumps" along the center of the back is a segment of the spinal cord. At each segment nerves enter and leave the spinal cord. Specific nerves to and from particular structures of the body enter and leave at specific segments. A nerve bringing information into the spinal cord from one side of the body will connect to several other nerves at that segment. It will connect to nerves that will provide the information to the nerves that leave the spinal cord at that segment. In this way, the information coming in is coordinated in a very immediate sense with the directions going out of the spinal cord. A nerve bringing information into the spinal cord also connects to other nerves that carry the information up the spinal cord to the brain. Finally, a nerve bringing information into the spinal cord at a particular segment will connect to nerves that provide that information to the other side of the spinal cord at that segment.

The crossing of the information allows for bilateral coordination of the body. In other words, the left hand is immediately told what the right hand is doing. Such bilateral distribution of information in the nervous system occurs both in the spinal cord and in the brain. However, in the brain, the bilateral distribution and coordination of information takes a rather interesting twist. In the spinal cord, the left side controls the left side of the body and the right side controls the right side of the body. In the brain, this becomes reversed. Especially at the level of the cerebral cortex (the "higher brain centers"), the left side controls the right side of the body and the right side controls the left side of the body. For example, damage to the left cerebral cortex will impair the right side of the body. This bilateral twist includes both motor outputs from the cerebral cortex and sensory inputs to the cerebral cortex. For example, the left visual field is perceived in the right cerebral cortex. Interestingly, the center of the field is perceived on both sides. In essence even though all information coming into the brain from either side is distributed to both sides, as it moves towards the cerebral cortex, opposite side specialization occurs. This specialization reaches its epitome at the cortical surface. The surface of the left cerebral cortex specializes in the right side of the body and the surface of the right cerebral cortex specializes in the left side of the body. Specialization of the cerebral cortex also appears to extend to the higher intellectual functions of the brain. In this regard, it appears that in humans, the left cerebral cortex specializes in language while the right side seems to specialize in spatial activities.

Organization of the brain: Leading in from the spinal cord at the center of the brain is the evolutionarily oldest part of the brain. At the base of the brain is the medulla oblongata. In the middle is the hypothalamus, and at the front is the limbic system. It is these structures that have not changed a lot throughout the course of evolution. These three sets of structures, and the paths and structures in between, constitute a route for the flow and evaluation of information associated with the most fundamental processes and motivations of life.

The medulla oblongata: The medulla contains many "centers" vital to controlling internal body functions (Fig. 8.14). Injury to the medulla results in death.

Many of the final directions for both sympathetic and parasympathetic responses are initiated in the medulla. The final central direction for changing such factors as blood flow, respiration, heart function, and digestion are initiated in the medulla. From a simple reactive point of view, sensors such as those for blood hydrogen ion concentration and blood pressure send their signals directly into the medulla. This direct flow of information to the medulla allows for very rapid graded adjustments to maintain balance in all kinds of functions.

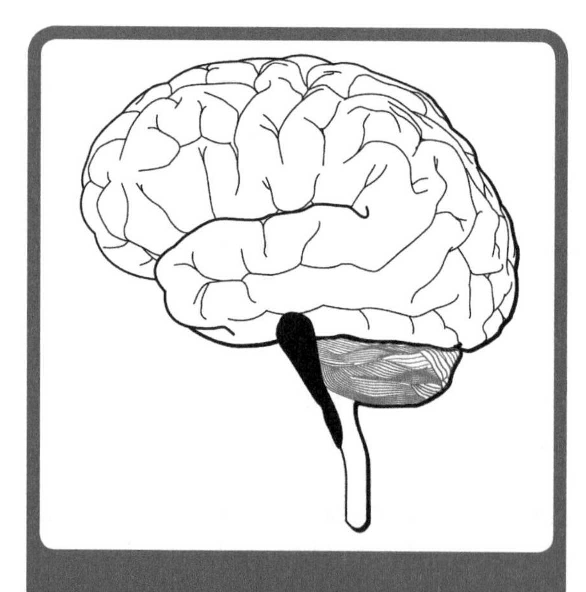

Figure 8.14 - The medulla controls the autonomic nervous system and many other basic functions such as breathing.

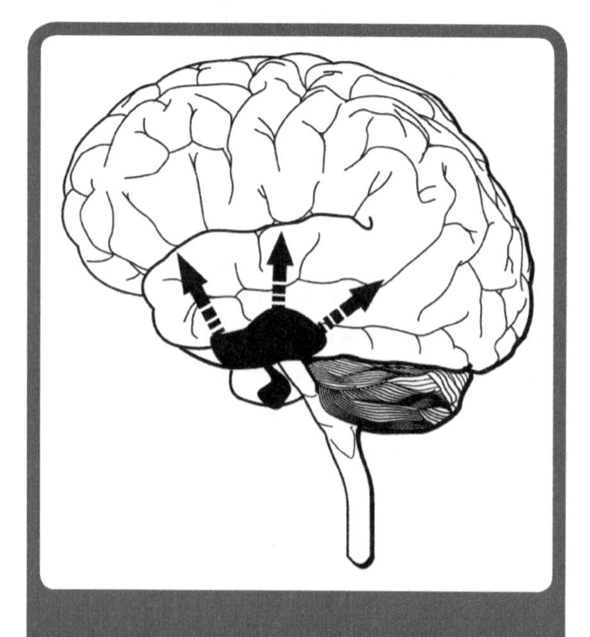

Figure 8.15 - The hypothalamus has substantial interactions with the medulla, the limbic system and higher brain centers; it also controls the pituitary gland.

For example, if the hydrogen ion sensors in the carotid artery indicate that hydrogen ion content in the blood on the way to the brain is too high, signals from the medulla initiate a signal which speeds up respiration through the parts of the brain that control muscles. The result is that the individual breathes, blows off carbon dioxide and reduces the hydrogen ion content of the blood. (See the carbonic anhydrase equation in Chapter 6.)

The hypothalamus and the pituitary gland: Similar to the medulla, the hypothalamus collects signals and information from incoming signals (Fig. 8.15). However, in this case many of the signals are carried by the blood. Different groups of cells in the hypothalamus are sensitive to a large variety of different hormones such as glucocorticoids, thyroid hormone, estrogen, progesterone, and testosterone. In addition, specific groups of cells in the hypothalamus are sensitive to such characteristics as blood ion content, sugar content, and temperature. Based on such information, the hypothalamus can direct signals to the medulla, which feeds into the "decision-making" processes in the medulla. In addition, the hypothalamus has another way of regulating the body. This other way is through its control of the pituitary gland. Like the medulla, the hypothalamus reacts to information.

Many of the reactive processes of the hypothalamus are based on the hormonal signals carried in from the body by the blood. The "feedback" of information by these hormones plays a major role in the hypothalamic control of the release of pituitary hormones. In general, the chemical messages coming in provide "negative feedback" on the release of pituitary hormones. Such "negative feedback" is the basis of the mechanism by which birth control pills work (as we shall see later). A good illustration of how negative feedback works involves the condition of goiter. People who lack the essential mineral iodine in their diet develop goiter.

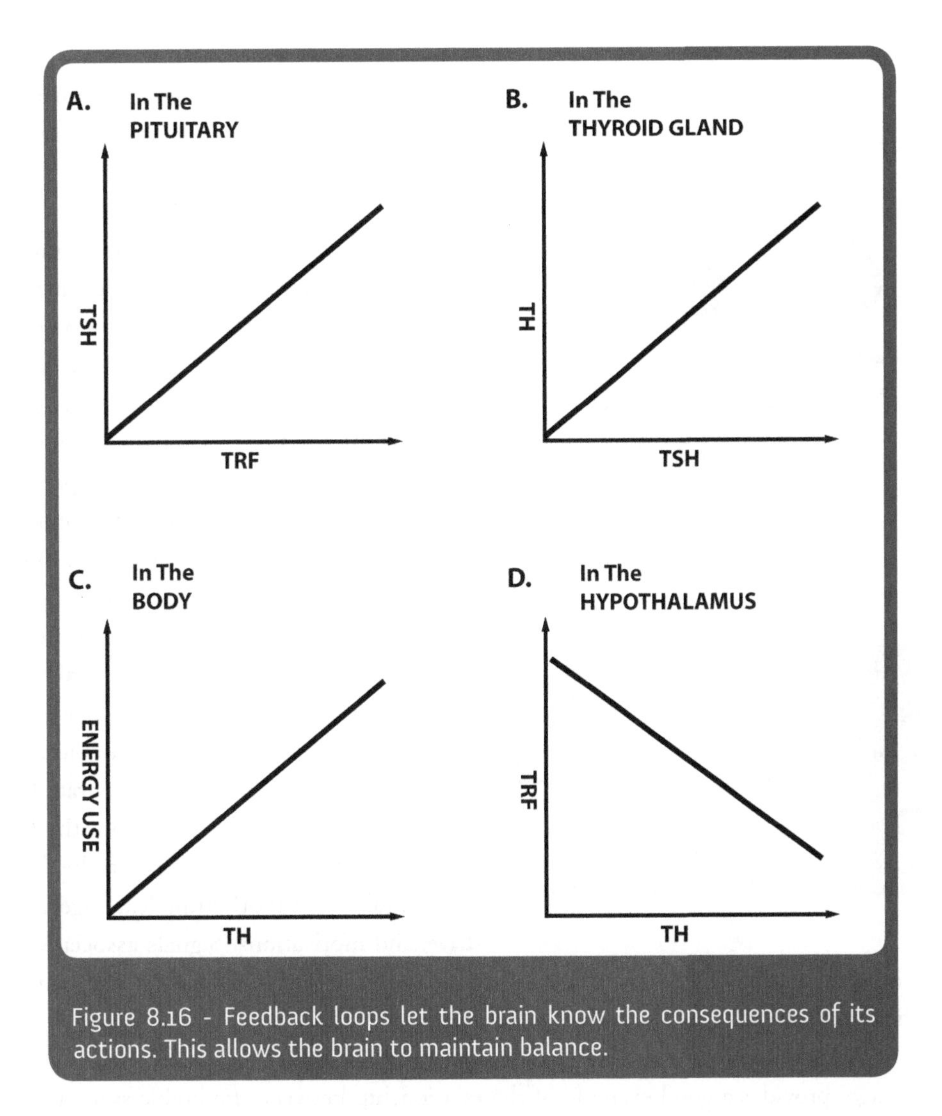

Figure 8.16 - Feedback loops let the brain know the consequences of its actions. This allows the brain to maintain balance.

Goiter is an enlargement of the thyroid gland. If the condition persists, the thyroid gland can increase to many times its normal size. Iodine is required by the thyroid gland in order to make active thyroid hormone. The thyroid gland produces and releases thyroid hormone under direction from a chemical message released by the pituitary gland called thyroid stimulating hormone. Thyroid stimulating hormone is released from the pituitary under the direction of a chemical message released into the portal blood system by the hypothalamus. This chemical message is called thyroid stimulating hormone releasing factor. Balance of the amount of thyroid hormone in circulation is accomplished by the negative feedback of thyroid hormone on the hypothalamic release of thyroid stimulating hormone releasing factor. As the level of thyroid hormone in circulation increases, it feeds back on the hypothalamus and reduces the output of the releasing factor, which in turn reduces the output of thyroid stimulating hormone. Thus, balance is maintained (Fig. 8.16). However, if iodine is absent, then stimulation of the thyroid does not result in the presence in circulation of active thyroid hormone. The result is that the negative feedback is not present, and the hypothalamus continues to

Figure 8.17 - The limbic system is the origin of basic drives, motivations, fear, anger, and pleasure. The limbic system wraps around the hypothalamus. It projects both down to the hypothalamus and medulla and up to the frontal cortex.

tell the pituitary to continue to stimulate the thyroid gland. Such continual stimulation causes the enlargement of the thyroid gland.

The limbic system—the primitive emotional brain: The reactive adjustments by the hypothalamus and the medulla to the incoming information from the body represent only a small part of the information controlling outputs to the body from these two areas of the brain. At the front of this very basic route of information flow controlling fundamental processes of the body is the limbic system and the limbic cortex (Fig. 8.17). The limbic cortex is the most primitive part of the cerebral cortex. The limbic cortex is quite distinct in an informational sense from the more "intellectual" part of the cerebral cortex, which is called the neocortex. It is the neocortex which has evolved so much in humans. In many respects, the limbic cortex and its underlying structures are quite separate from the rest of the cerebral cortex. Signals from the limbic cortex tone the regulatory processes of the hypothalamus and medulla with biorhythm, fear, rage, hunger, sexual drive, and motivations. Signals associated with primitive drives descend out of the limbic cortex and influence the more simplistic reactive adjustment processes occurring in the hypothalamus and pituitary.

The drive of hunger provides a good example of the relationship between the limbic system and the hypothalamus. Cells in the hypothalamus control the drive of hunger from the limbic system. Animal experiments demonstrate that if one destroys a small number of cells in one part of the hypothalamus one can create an animal that is bulimic: it will eat itself to death.

Conversely, if one destroys a small number of other cells in the hypothalamus one can create an animal that is anorexic: it will starve itself to death. The hypothalamus balances the expression of hunger. Understanding this relationship has very important implications for human conditions such as anorexia, bulimia, and obesity. When an individual eats, nutrients are in excess in the blood. This nutrient excess in circulation causes insulin to be released from the beta islet cells of the pancreas. Both insulin and the nutrient excess are information that cells in the hypothalamus respond to by suppressing the drive of hunger from the limbic system. Drives and emotions exist in the brain for a valuable purpose: they motivate and protect. The limbic system of a human is not very different than that of a shark, a fish or a reptile. What has evolved in humans is the neocortex. In the front part of the neocortex is the area of the brain where decisions are made. This area has an extensive relationship with the limbic system. This allows drives and motivations to influence our decisions. If the connections between the limbic system and the front part of the neocortex are destroyed the result is a human with no drive, no motivation, and no interest is anything. A frontal lobotomy is the type of psychosurgery described in "One Flew over the Cuckoo's Nest": it simply disconnects the part of the brain responsible for drives and motivations from the part of the brain that makes decisions. This type of surgery is no longer practiced, at least in this country. More than any other animal, humans learn. One of the most important facets of human learning is the connection between drives and motivations and decisions. The experiences of the child will determine how pain, pleasure, and fear are connected to decisions.

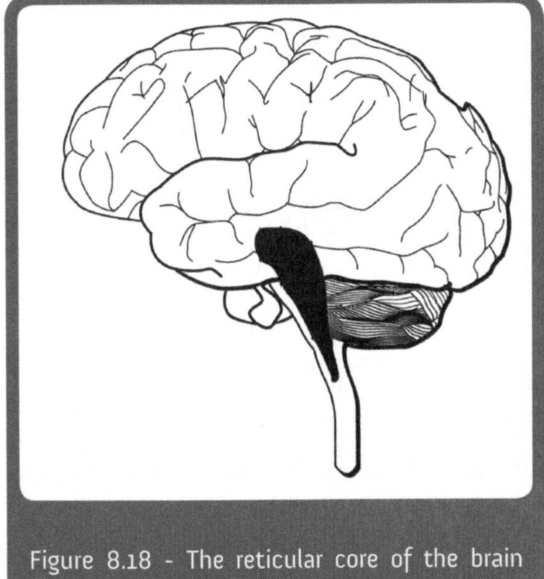

Figure 8.18 - The reticular core of the brain allows the brain to focus attention on incoming information.

The reticular core of the brain: Information flowing into the brain is not just involved with temperature control, ion balance, and other such basic housekeeping functions. It is also the basis for consciousness, learning, memory, and purposeful movement of the body by skeletal muscles. Incoming signals carry all kinds of sensory information which support consciousness. The massive amount of incoming information is fed into the reticular core or "reticular activating system" of the brain (Fig. 8.18).

Activity in this central core of interconnecting pathways determines the consciousness or unconsciousness of the individual. When the individual is awake, signal activity in the reticular core is high and variable. When a person is asleep, signal activity is low and more constant. In some respects, signals seem to lose their identities in the reticular core. Incoming signals of any nature increase activity in the reticular core, but it is difficult to sort out the contributions of the different types of signals. However, different signals appear to be more or less activating at different times. For example, if attention is focused on looking at an object, then visual signals are promoted and produce more activity in the reticular core. How this occurs is unclear. What is clear is that damage to the reticular core generally leads to prolonged unconsciousness. In addition, suppression of activity in the reticular core by anesthetics appears to cause unconsciousness. Conversely adrenalin, which is associated with the "fight or flight" response, raises the level of activity in the reticular core and focuses attention.

The cerebral cortex: If there is one group of specialized cells that distinguishes humans from other animals it is the cerebral cortex. Interestingly, the distinction is not really apparent in either structure or organization. The distinction is mainly important in function. Contained within the organization of the cells of the cerebral cortex is the capacity to learn, think, and reason. Learning, thinking, and reasoning are "what humans do best."

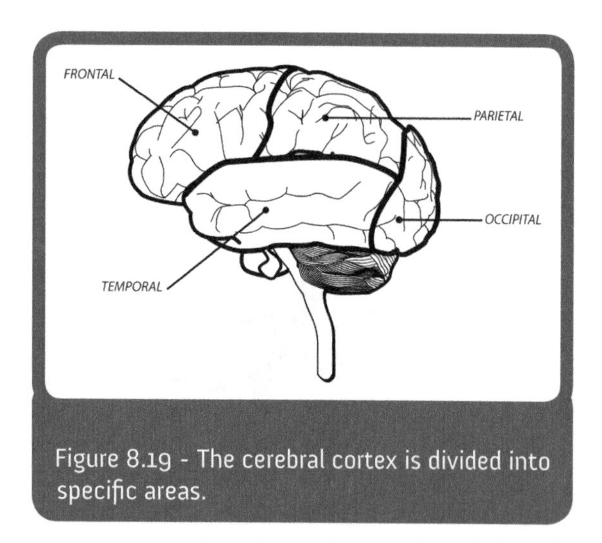

Figure 8.19 - The cerebral cortex is divided into specific areas.

The cerebral cortex is composed of two hemispheres: one on the right side and one on the left side. These hemispsheres are further distinguished by four areas, or lobes (Fig. 8.19). In many respects, the two hemispheres are two distinct intellects that "talk" to each other through a series of pathways in between called the corpus callosum. As discussed previously, there is a twist in specialization in the cerebral cortex. The left hemisphere specializes in the right side of the body and the right hemisphere specializes in the left side of the body. However, this cross-specialization is not really simple. In actual fact, most information is projected to both hemispheres regardless of the side of the body of origin. The cross-specialization reflects the fact that the right hemisphere dominates left side body processes and vice versa. Coordination of the two sides involves hemispheric communication through the corpus callosum. However, hemispheric specialization has expanded beyond the simple control of the right and left side of the body.

Much of the capacity of the human intellect involves language. Language involves sound and sight. Speech involves the fine control of the muscles of the vocal cords. Writing involves fine control of the muscles of the hand. In the human brain the processes of language appear to have been specialized into one hemisphere. Although still a matter of some controversy, it appears that at least in the vast majority of humans, the left hemisphere specializes in language processes. In many respects such specialization is most apparent when damage occurs. For example, if a normal right-handed individual has a stroke that damages the left hemisphere of the brain, it is not uncommon for the individual to be unable to either speak or write.

Although such a stroke may paralyze parts of the right side of the body including the right writing hand, the inability resides in the left hemisphere, in the capacity of formulating and recognizing language itself.

Just as the left hemisphere has its specializations, so does the right hemisphere. In many respects, right hemisphere specialization is more abstract and difficult to understand. The right hemisphere specializes in spatial, mathematical relationships. Humans live and manipulate objects in a three dimensional space. We order our world and describe it in mathematical relationships. Such conceptual manipulations appear to be the activity of the right hemisphere.

Specialization also occurs within each hemisphere. For example, there are areas in the back two-thirds of each hemisphere that specialize in receiving coded information concerning sight, sound, and touch. In these areas the information is represented spatially. Sensory information such as touch and the visual field can actually be mapped on the cortex. Because of the cross-specialization of the hemispheres the mapping is crossed. For example it is possible to map a picture of the right side of the body on the touch area of the left cerebral hemisphere. Interestingly, such a mapping yields a distorted picture of the body. The distortion occurs because the map reflects the density of sensory input from the body, not the size of the area of the body. Because of this the hand will map onto the cortex as much larger than the entire back. Similarly, the thumb will be disproportionately larger than the little finger. When one looks at the pattern of specialization of the human cerebral hemisphere what is apparent is that the areas specializing in representational sensory information are quite small. These representational areas, called primary sensory areas, are not really special in humans. What is unique to humans are the areas in between.

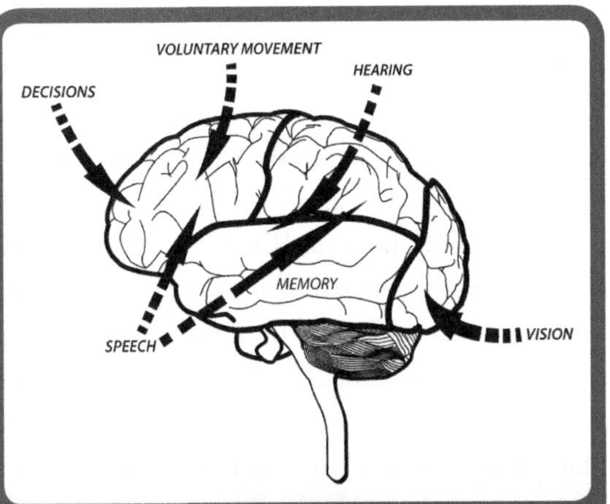

Figure 8.20 - In between the primary sensory areas are areas of the cerebral cortex that process complex associative functions such as speech.

The areas in between are association or interpretive areas (Fig 8.20). In these very large areas the different types of primary sensory information are brought together. Not only is such information as visual and sound patterns stored in these areas but also the relationship between such patterns is stored in and evaluated in these interpretive areas. By far language is the best example of the focus of interpretive information in an area of specialization. Humans hear, see, write, and speak language. At the center, where interpretive areas for sight, sound and touch meet is an area called Wernicke's Area that specializes in language. In this area of the brain humans conceptualize and interpret language. Before we speak, we decide what to say in this area of the brain. However, humans neither speak nor write from this area of the cerebral cortex. Speaking and writing requires the control of muscles.

Muscles are the sole way that the brain can act on the world. In front of the sensory areas of the cerebral cortex is the motor control area. All voluntary muscles are controlled in this area. Like the sensory areas, the motor area is representational and controlled by the hemisphere opposite to the side of the body. Also like with the sensory areas, the representation is distorted to reflect the importance of the muscle. For example a disproportionately large area of the motor cortex called Broca's area controls the vocal cords that control the muscles of speech. Returning to the example of speech, humans conceptualize words in Wernicke's area and speak the words from Broca's motor area, but do not make the decision to speak in either place.

Decisions originate in front of the motor areas in the prefrontal cortex. Humans make decisions on when to use those muscles—that ability to act on the world—with the front part of the cerebral cortex. The actions of the brain on the world are initiated in the front part of the brain. However, much of the motivation for these actions is supplied by the primitive limbic system which sits beneath the prefrontal cortex. Drives such as hunger, thirst, sex, anger, pleasure, and aggression bubble up into our decisions from the limbic cortex. Conversely, the prefrontal cortex exercises control over the primitive drives of the limbic system. In a prosaic sense, the "Jekyll and Hyde" relationship of Robert Louis Stevenson resides in the balance between these two parts of the brain.

Activity and attention in the cerebral cortex: In addition to the direct projection of information to both the primary sensory area and the association areas of the cerebral cortex, there is a special relationship with the reticular core of the brain. The level of activity in the reticular core appears to be critical, determining the level of activity in the cerebral cortex. Actually, signals move in both directions. Signals move from the reticular core to the cerebral cortex and from the cerebral cortex to the reticular core. Somehow in the exchange of signals sensory attention is focused. For example, if the cerebral cortex is "paying attention" to sound, the threshold for auditory signals is reduced allowing this information to pass more easily up to the cortex. In essence it appears that different types of information must signal through the reticular core for attention in the cerebral cortex. In some way, the cerebral cortex can adjust the relative ease of passage of signals to it through the reticular core.

Learning and memory: A 100-year-old man may remember in great detail an occurrence that happened 90 years ago but not be able to recollect what he did ten minutes earlier. Much of what we label as intelligence is a function of learning. A Down's child with an extra piece of chromosome 21 is less able to learn than a normal person with only two copies of those genes. Some people can see and remember while others must struggle to learn. Sometimes we cannot remember something that we "know." Such is the nature of human memory. However, learning is a process that can be carried out by even the most simple of nervous systems.

In the most simplistic sense, learning and memory can be considered as an input or event that alters the response to subsequent inputs or events. How such occurrences happen in their various forms is still a subject of much study. Perhaps one aspect that is agreed on is that memory does not reside in single cells but rather involves sequences or circuits of cells. The establishment of such sequences of cells for signal flow seems to involve a number of different short- and long-term mechanisms. For example, if one sends a signal through a sequence of neurons, then for a short period of time, signals move more readily through the same sequence of neurons. Such a facilitation of signal transmission is probably involved in the more immediate aspects of learning and memory. In contrast, long-term memory processes seem to involve the establishment of new anatomical connections between neurons. For example, if one denies visual input to a young animal and then later examines the organization of nerve cells in the visual cortex, one finds that the ordered relationships between the nerves in this area of the brain are not present. In some way sending information to the visual cortex induces orders in the structural relationships between the cells. When this input of information is blocked the structural relationship between the cells remains in disorder. Such mechanisms allow one to speculate on the nature of memory. Perhaps in a young animal, novel information begins to establish pathways of relationships between cells. As the animal grows older and gains more experiences, less and less information is novel. Those parts of the incoming information that are similar to previous experience flows through previously established circuits while that part of the incoming information that is novel establishes new pathways.

The cerebellum—more control of movement: Movement occurs because a muscle is directed to contract. A ripple of changing electrical potentials moves down an axon and causes a neurotransmitter to be released onto the fibers of a muscle. This causes a ripple of changing electrical potentials to move along the membranes of the muscle. The result is that calcium ions flow into the muscle, ATP is expended and the muscle shortens thereby moving an arm or leg. Such a sequence can originate as a simple reflex arc to sensory input at the level of the spinal cord, or it can originate in the cerebral cortex. However, the fine coordination associated with purposeful movements such as picking up a pencil is not controlled by the cerebral cortex. Coordination is provided by the fine guidance control system of the cerebellum. The cerebellum sits behind and below the cerebral cortex above the medulla (Fig. 8.21).

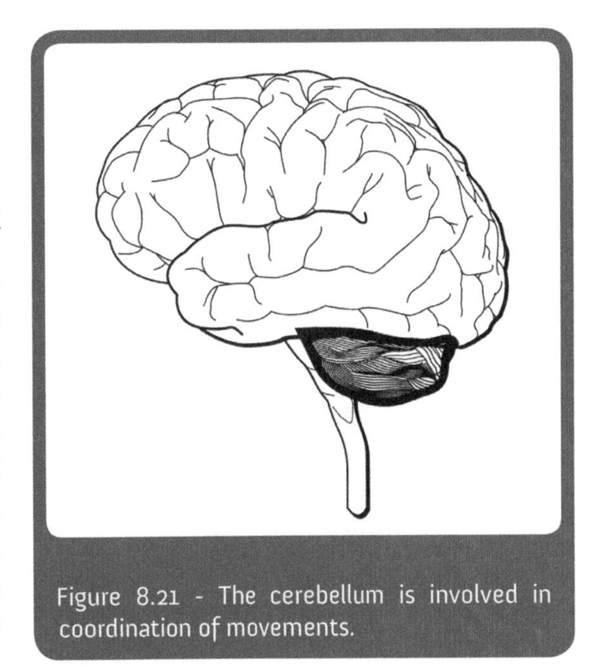

Figure 8.21 - The cerebellum is involved in coordination of movements.

Although it is representational like the cerebrum, the representational projection is nowhere near as fine. Incoming information relating to the movement of the body in space is sent to the cerebellum. In some way, the cerebral cortex and the cerebellum integrate information. This integration allows for directed and highly coordinated activity.

It is all a matter of information: Blood vessels open and close, more or less blood goes somewhere in the body. Perhaps the body temperature is being regulated. Perhaps the individual is standing up. Perhaps the individual is sleeping, or eating, or running, or making love. Nutrients are moved from one place to another and are converted from one form to another. Perhaps the individual is awake or asleep, happy or sad, frightened or calm, or just thinking about buying a new car. Digestive enzymes are released, nutrients are absorbed, sodium is filtered and recovered, a leg moves, blood pressure increases, more or less oxygen is brought into the body. Something changes in the body and many functions must adjust. All of the changes and adjustments occur for some purpose. However, the consequences of all of these actions and reactions are information to the brain. The skin gets cold, the concentration of an ion changes, more or less sugar is available, gravity pulls on the blood, there is the smell of food or the sight of a tiger. The brain collects information and acts and reacts. The individual takes a drug that speeds up the heart, or changes blood flow, or changes the concentration of an ion, or alters absorption, or does anything and the brain acts and reacts to that information. Directly or indirectly, every little change in the body is information to the brain. The brain acts and reacts to change by causing more change based on the constant flow of information.

chapter nine

reproduction

Sex and reproduction go together (at least in organisms like us): From a biological perspective, the term sex refers to trading genetic material between two different individuals, while reproduction involves making new organisms. Not all organisms couple mixing genetic information with reproduction. However, in humans and many other types of organisms, reproduction is coupled to sex: reproduction in humans involves sex. Two individuals—one male and one female—collaborate in making a new organism (Fig. 9.1). In doing so, they both contribute genetic information necessary for forming a new individual who is a genetic mix of both parents. This coupling is both recreational (sex) and procreational (making a new individual with a mixing of genetic material). Mother Nature designed a compelling system when she coupled sex and reproduction.

Figure 9.1 - Reproduction: survival of the species.

Built into the primitive part of the brain (the limbic system) is an interest in sex. Survival of humans depends on an interest in sex.

Controlling the quality and quantity of sex has been for centuries of interest to humans. Promoting or blocking fertility has also been an important objective of many humans. Many people wish children and cannot have them. Others conceive and do not wish them. Some wish for a child of a particular sex. All people wish for a healthy child. From aphrodisiacs to predicting genetic defects in an unborn child, from sexual preference to birth defects, humans are interested in studying and manipulating sexual functions.

Reproduction—survival of the species: The reproductive system is an oddity in the human body. It is unusual because it has no direct survival value for the individual. All other systems in the body such as the digestive system, the circulatory system, or the respiratory system have a specific task to perform within the complex functioning of the individual. If any one of these other systems fails in their specific task, the human body fails. This is not true of the reproductive system. An individual, in terms of health and physical well-being, can function and survive quite well without reproducing. In fact, sex and reproduction are generally a stress. Why do salmon waste their bodies to death swimming upstream simply to reproduce? They do so because of a compulsive drive from their limbic system. A similar compulsion exists within the human brain: the difference is that in humans the expression of the compulsion resides to a great extent in learning and perhaps culture rather than simple instinct. In females, pregnancy clearly constitutes a major stress on the body. The physical needs of the developing infant take precedence over the physical needs of the mother's body. If nutrition is not adequate during pregnancy then the mother's body is "cannibalized" to build the baby's body. The question of whether or not sexual activity or the lack thereof comprises a stress on the individual is at minimum a debated subject. So what is the value of sex? The sole purpose of sex from a biological perspective is reproduction, which is necessary for the survival of the species. Any species that loses an interest in reproducing ceases to exist. Any other value of sex is probably nature's way of seducing animals into participation.

Males and females—it's in their genes: Male and female human beings are different. Physically, males and females differ in both the primary sex characteristics (males and females have different reproductive structures), and in the secondary sex characteristic (such things as "breasts and beards"). It is also a much debated subject as to whether or not males and females differ in some brain functions such as personality traits, emotions, and behaviors. Regardless of the answer to such questions it is the genes—the DNA molecules—that contain the biological information that initially determine the answer to the big question: male or female. The genes provide the potential for what the individual might be. However, how this potential develops is highly dependent on circumstances. Events which occur after conception can drastically alter the results.

Autosomes—they are the same in males and females: Ultimately, all biological information is contained within the genes. A gene is a stretch of DNA which contains the information that allows the cell to produce a single kind of protein. There are in the order of 30,000 individual genes in the human genome. Genes are organized into larger structures called chromosomes. At certain times in the cell cycle, if stained properly with the right dyes, chromosomes are large enough to see using a microscope. The chromosomes reside in

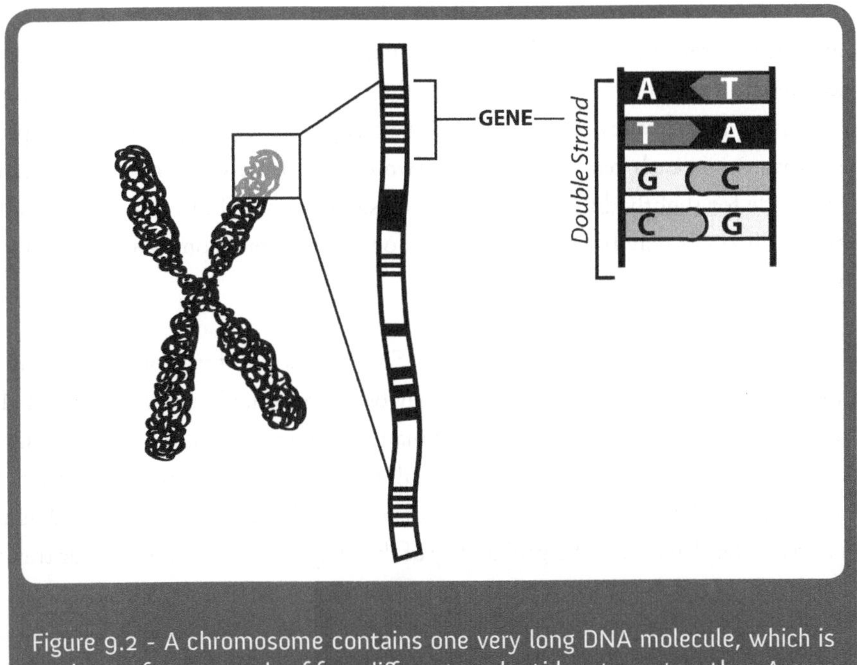

Figure 9.2 - A chromosome contains one very long DNA molecule, which is made up of two strands of four different nucleotides strung together. A gene is a small region along this DNA molecule: genes are arranged linearly along the chromosomes, with regions on non-coding nucleotides in between. The sequence of the four nucleotides within a gene contains the information for making a specific protein.

the nucleus of the cell. Chromosomes are very long DNA molecules, each containing thousands of individual genes. You will remember that DNA is a large molecule made up of individual smaller molecules called nucleotides that are linked together (Fig. 9.2). There are four different nucleotides (which we abbreviate as A, T, G, and C) . It is the sequence of nucleotides within a gene that provides the information for making a protein. A chromosome is a very long molecule of DNA containing many individual genes, but with extra stretches of "non-coding" nucleotides interspersed. Also, there are many proteins which control the activity of the various genes that are associated with the DNA of the chromosomes. These proteins control which individual genes are expressed in a particular type of cell at a particular time. As discussed in Chapter 2, not all genes are expressed at a given time in any given type of cell. Although all cells have the same genes, or genetic information, only some genes are actively expressed as the synthesis of specific proteins by a particular cell. Which genes are expressed as proteins is what makes a liver cell different from a skin cell.

Within all of the different types of cells of the human body, there are 46 of these chromosomes. The 46 chromosomes can be divided into 23 pairs. Twenty-two of these pairs are "matched sets." They are "matched sets" in that both members of a pair are the same size and shape, and they contain genes that code for the same protein at the same location on the chromosome. These chromosome pairs are called "autosomes." By convention, these chromosomes are numbered from one to 22. Each cell therefore has two copies of chromosome number one, two copies of chromosome number two and so on.

The autosomes are not different in male and female humans. Thus, 22 of the 23 pairs of chromosomes in males and females are the same in the genes they contain. As we said, both chromosomes in each pair of autosomes carry the same set of genes. Thus, there are two copies of every gene for every protein which is coded for on the autosomes in each cell. However, the two copies of the genes may not be exactly identical. The two copies may code for slightly different forms of the same protein, but basically they still code for the same protein. Some individuals may have identical genes on both chromosomes while others may have two different forms of a particular gene or genes.

The expression of a gene is the making of the protein. Sometimes one form of a gene may be selectively expressed while in other cases both variants of a gene may be expressed. Genes code for a single protein, but a single protein may influence a few or many traits, and many traits are influenced by more than a single gene. Some traits, especially many genetic diseases, are the result of a single gene. Some examples of this are sickle cell anemia or Huntington's chorea (a nervous system disorder). Traits that are inherited as the result of a single gene are said to be Mendelian, and the probability of the offspring having a particular trait is predictable.

In recent times, with the advent of new technology, the origins of more and more genetic diseases are being located in the human genome. For example, the genetic defect that causes sickle cell anemia has been located to one arm of chromosome 11 (Fig. 9.3).

This disease occurs because the gene coding for hemoglobin is changed so that the hemoglobin protein does not carry oxygen as efficiently. Sickle cell anemia is inherited in a Mendelian fashion. In this case, the individual

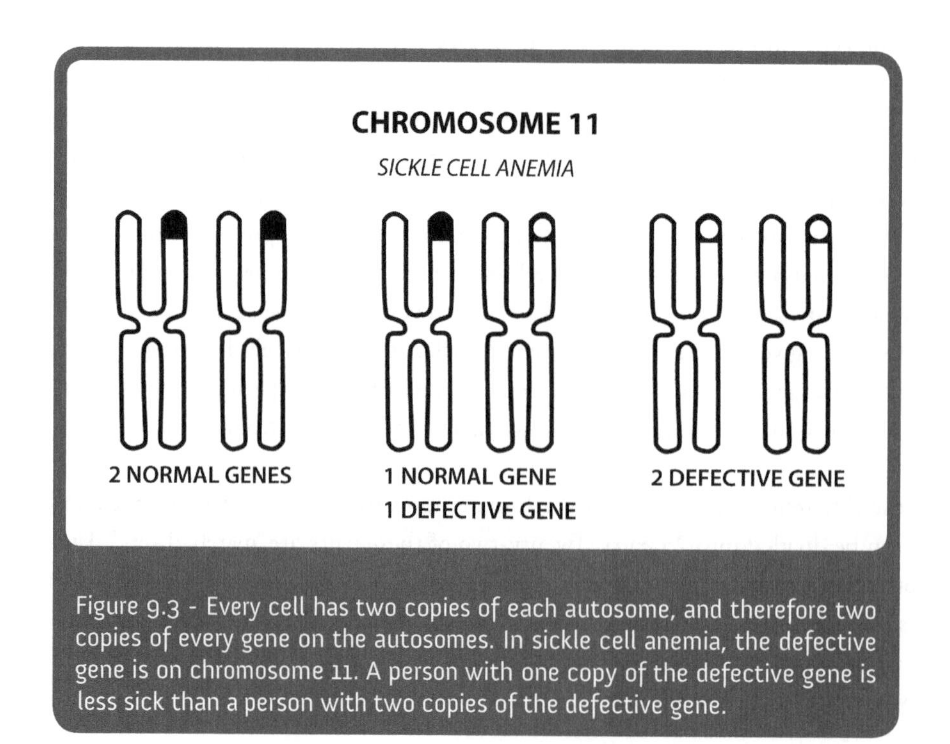

Figure 9.3 - Every cell has two copies of each autosome, and therefore two copies of every gene on the autosomes. In sickle cell anemia, the defective gene is on chromosome 11. A person with one copy of the defective gene is less sick than a person with two copies of the defective gene.

with one copy of the defective gene is less sick than the person with two copies because neither gene dominates the other. In other instances a person can have one defective gene and one normal gene and his or her health will be totally unaffected. The genetic disease PKU (phenylketonuria) is a good example of this type of genetic disease. As long as an individual has one normal copy of the gene her or she has no disease. In order to inherit this disease both parents must carry a copy of the defective gene. When this situation occurs, the probability of those parents having a child with phenylketonuria is one in four. In addition, the probability that any child born of these parents will be normal but carry one copy of the defective gene is two in four (one half). Only one in four of the children will be both normal and not carry a defective gene. . This type of trait is called recessive, because two copies of the gene must be present to show the trait or disease. In other cases, one "bad" copy of the gene can result in malfunction, such as in the neurological disorder called Huntington's chorea. In these types of traits, the "bad" or non-functioning protein produced from the altered gene overrides the function of the normal protein produced from the normal gene. Such a trait is said to be "dominant."

However, many traits such as eye color or height are generally a function of a mixture of many genes. Although each gene that contributes to the trait is still inherited in a Mendelian manner, it is hard to predict exactly what mixture of these genes an offspring will get. That is why two quite tall parents can have a child much shorter than either parent. Traits such as this are said to be polygenic: the apprearance of the trait results from many different genes which act to produce the trait. Polygenic traits are usually graded, rather than being there or not there. A person is not either tall or short: he or she can have heights that range in a continuum from very short to very tall. In general, polygenic traits also have a strong environmental component which affects exactly how that trait will be expressed in the person.

The importance of also having the appropriate amount of DNA is well illustrated by the genetic disease known as Down's syndrome. In this case, no single gene is defective. In Down's syndrome, the individual has three copies of chromosome number 21, instead of the normal two copies. The existence in every cell of three good copies of every gene on chromosome 21 in some way causes mental retardation, heart malformations and a variety of other changes in normally expressed characteristics. Having an extra copy of part or all of a chromosome (called a chromosome abberation) has devastating consequences for normal development. Why this is so is not currently understood.

Sex chromosomes—the difference: The last pair of chromosomes are referred to as "the sex chromosomes." These are only in part different in males and females. There are two totally different sex chromosomes. One of these two chromosomes has been named "X," and the other "Y." Under a microscope they look very different in size, in shape and in the way that they stain with certain dyes. Unlike the autosomes, the X and Y chromosomes carry very different genes.

A female human being carries two copies of the X chromosome in the cells of her body; a male has one X and one Y chromosome in his cells (Fig. 9.4). It is this complement of sex chromosomes that determine the potential for "maleness" or "femaleness." That is not to say that all of the genes coding for all of the proteins that determine female characteristics are on the X chromosome, or that only genes coding for female traits

Figure 9.4 - Females have two X chromosomes, and males have one X chromosome and a smaller Y chromosome. All of the other pairs of chromosomes are the same in males and females.

are on the X. This is certainly not the case. There are clearly many genes coding for many proteins on the X chromosome that have nothing at all to do with sexual differences. The Y chromosome is quite different. Only a very few genes besides those associated with being male have been located on the Y chromosome. Since a normal male has only one Y chromosome, every gene is expressed at the appropriate time and passed from father to son. For example, a gene that determines a rather rare trait referred to as "hairy ears" is found on the Y chromosome. If you are a male and your father has "hairy ears," then you too will have "hairy ears." Somewhat more controversial, a gene which may contribute to alcoholism has also been reported on the Y chromosome. One perspective is that it makes sense that nothing too terribly important could be carried by the Y chromosome because females seem to get along fine without one.

When cells of our bodies reproduce (such as stem cells that form new skin or liver cells, or as a new individual develops from a single cell into a whole organism), they duplicate their DNA in each of their 46 chromosomes, and then the cell splits apart forming two daughter cells, each with 46 chromosomes. Each cell that is produced is an exact genetic duplicate of the cell that it came from. If human beings were to reproduce new individuals in this way, it would be rather genetically boring. Each new individual would be an exact genetic duplicate of its parent. This process is called "cloning" or asexual reproduction. Recently, cloning has been accomplished with sheep and many other lower animals such as dogs. The only genetic variation that would arise in such a population would come from random mutation. A mutation is a mistake in the process of exact duplication of the chromosomes. Mutations occur naturally with a very low probability. However, human beings reproduce in a much more interesting way. A male and female are involved in the process. This process is called sexual reproduction.

Germ cells—the exception to the rule of 46 chromosomes: Females have ovaries and males have testes. Contained within the ovaries and testes are germ cells. Germ cells are made into the sex cells involved in reproduction. The male individual produces a sex cell called a sperm. A female produces a sex cell called an egg. These sex cells are different from the normal body cells of that individual. The sex cells have only 23 chromosomes, not 46. The sex cells contain only one of each of the pairs of autosomes, and one of the pair of sex chromosomes. This occurs because the cells from which the sperm or egg developed duplicated in a different manner. In this case, after the DNA in the chromosomes has duplicated, it divides twice, yielding

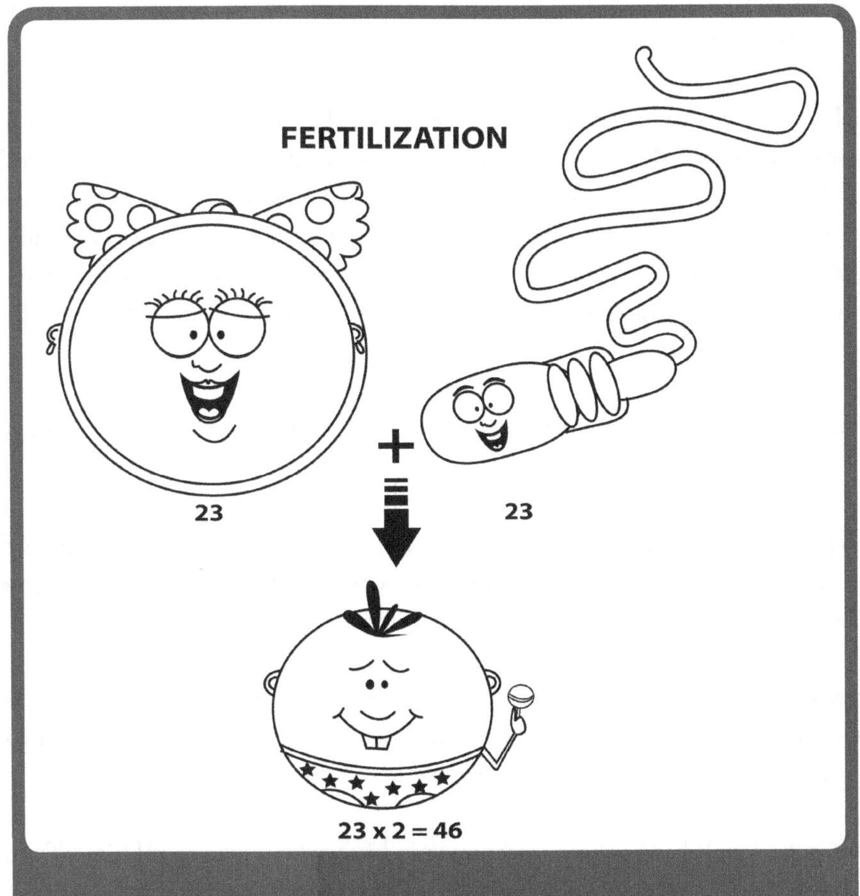

Figure 9.5 - Both the sperm and egg contain only 23 choromosomes (one of each pair). When the sperm fertilizes the egg, the resulting cell contains the full complement of 46 chromosomes.

four cells. The 46 chromosomes were divided in two, with one of each pair going to a new cell. In females, the beginning stage of producing every egg she will ever have is thought to be complete before birth. Each egg waits in a dormant state until the correct series of signals causes it to complete development later in life. An egg, because the cell from which it arose had two X chromosomes, will always have an X as its sex chromosome. In males the situation is quite different. Males produce new sperm throughout their reproductive life. However, a sperm may either have an X or a Y as its sex chromosome, because the cell from which it came had one X and one Y chromosome. Thus, 50% of sperm will carry an X chromosome, and 50% a Y.

A new cell capable of forming a new human individual is produced when a sperm and an egg meet in the right place at the right time. The sperm penetrates and fuses with the egg. This fusion of sperm and egg now brings the complement of chromosomes back up to 46 (Fig. 9.5). The result is a totally unique new cell with two copies of each gene except for those of the X and Y chromosomes of males. Even in the case of the X chromosome in females, where two copies exist, only one of the two copies is active. Very early in female development, in each cell one X chromosome is inactivated. The result is that only the genes on one of the X chromosomes are expressed as proteins. Since the inactivation is random in each cell, half the cells of a female may express one form of the gene while the other half may express the other form.

The fact that males have only one X chromosome and each cell of a female only expresses one X chromosome has some interesting and medically important implications. First, there are a substantial number of genetic diseases that are linked to the X chromosome and therefore expressed severely in only males but can be expressed moderately in females. Duchenne muscular dystrophy is one such disease. Secondly, the fact that females are mosaics, containing cells which differ in the expression of genes on the X chromosome, results in some interesting expressed characteristics. A calico cat is the classic example. Virtually all calico cats are female.

The black and orange patches on the white background are the result of a mosaic of cells due to the inactivation of one of the two sex chromosomes in the female.

The process of reproduction "shuffles the genetic deck." Half of a person's genes come from its father's sperm, and half from its mother's egg. With each generation, the whole deck of genes is shuffled in every possible combination. The result is that genetic diversity is propagated and promoted.

A person inherits a complexity of traits derived from the mixture of genes from its male and female parents. These traits are mixed and matched and shuffled about in new combinations, providing in part (or at least allowing for) the uniqueness that is every new human being.

Daddy determines the sex: It has not been uncommon in history for a man to rid himself of a wife because she did not give him a son. This is biologically ridiculous. The egg always contains an X as its sex chromosome because the cell from which it derived contained two X chromosomes. In contrast, the sperm may carry either an X or a Y because the cell from which it derived contained one X and one Y chromosome. If a sperm containing an X chromosome fertilizes an egg, the new individual will be female. If the sperm happens to carry a Y chromosome, a male will develop. Thus, ultimately it is the sperm of the father that determines the genetic sex of the new individual.

There are mistakes: Sometimes, abnormalities occur in the number of sex chromosomes in the single cell that starts an individual. The result is that every cell of the individual will carry an abnormal genetic compliment. Instead of being XX or XY, a person's cells could contain XXY, XYY, XO, or XXX. This happens because something did not quite go right during the division creating the egg or the sperm. Instead of each of the new cells getting only one each of the sex chromosomes, one got both and the other got none. When this egg (or sperm) carrying an abnormal sex chromosome number is fertilized by a normal sperm (or fertilizes a normal egg), the cells of the resulting offspring have an abnormal number of sex chromosomes.

What does this abnormal number of sex chromosomes mean to the new individual? Like with Down's syndrome, the unusual compliment of genes causes developmental abnormalities that influence the individual throughout their life. An XO individual develops as a female, but she develops abnormally. This condition is called "Turner's syndrome." The individual is very short in stature, has some structural abnormalities such as webbing of the neck and a very small jaw, is often mentally retarded and usually does not mature sexually. The XXY condition is referred to as "Klinefelter's syndrome." This individual is born with male structures, but has the genetic compliment of both a male and a female. Abnormalities include immature genitals, some breast development, usually extreme tallness and elongated bones, some mental retardation and of course, sexual immaturity. The XXX syndrome has the rather strange name of "superfemale." Although female, this individual is hardly "super." (The name superfemale comes from the same condition in fruit fly genetics) These individuals are usually (although not always) retarded and usually (although not always) infertile. Other "versions" of the so-called superfemale exist with chromosome complements of XXXX. In general, the more X chromosomes, the more infertile and the more retarded the individual. It was not until the mid-1960's that the XYY condition was recognized, and its existence as an "abnormality" has been highly debated. Although no striking developmental abnormalities occur as with the other aberrations in sex chromosome number, these individuals are thought to be taller and more aggressive than normal XY males. In fact, at least one study has demonstrated a much higher proportion of such individuals than should occur by chance amongst males in maximum security prisons. The XYY syndrome was even used, albeit unsuccessfully, in the defense trial of a famous case involving the murder of eight student nurses in Chicago in the late 1960's. One possible condition has not been mentioned, that of the YO. This is because, although theoretically possible, no individuals with this chromosome complement are known to exist. Undoubtedly, this is due to the absence of any X chromosome. The absence of all of the important genes on the X chromosome is lethal.

Steroid hormones determine further sexual development: After the egg is fertilized by the sperm, the new cell that is formed by this fusion begins to divide. The new cell divides in two, these two cells into four, and so on. About seven days after fertilization, the small ball of cells implants into the wall of the mother's uterus. After implantation, the cells continue to multiply throughout the course of the nine-month human pregnancy. Cell division is more rapid at first. Very early in this process, cells begin to differentiate into the various cell types that make up the structures that will become a new viable human being. Most of the structures and organs that are formed actually develop during the first three months (the first trimester) of pregnancy. The remainder of the pregnancy is mainly devoted to growth and maturation of the organ systems of the fetus. During the sixth to seventh week, either ovaries in the female or testes in the male begin to develop. This is determined by the presence or absence of the Y chromosome. In a male, once the testes are formed they begin to put out, for a relatively brief period of time, a hormone called testosterone. Testosterone is a "male sex hormone." Testosterone is an androgen and belongs to a class of hormones called steroids. This output of testosterone by the testes of the male fetus at about seven weeks of gestation "casts the dye" as to the further sexual development of the individual. By week 12 of development, the remainder of the structures (ducts and external genitalia) of the reproductive system is formed. Male structures will begin to develop if testosterone is present during this period. However, female structures will develop if it is not. If you took a normal genetic male (XY) and stopped the output of testosterone at this point in development, the fetus would develop as a "phenotypic" female. Outwardly, the individual would appear as a normal female at birth. The external genitalia would be female. However, the embryonic testes would exist, no ovaries would be present, and genetically "she" would still be male. Such an individual is referred to medically as a "male pseudohermaphrodite." Similarly, if you introduce testosterone during this period to a genetically normal female the fetus would develop into a phenotypic male, a "female pseudohermaphrodite." Therefore, it is the output of testosterone at this particular stage of development that sets things going down a particular track with respect to sexual development. This is a good example of where the environment (in this case, the presence of testosterone) influences what happens to the genetic potential. A female is, in a sense, the "default function" in terms of development. If testosterone is not present at the proper time during development, then the fetus develops as a female.

Development—forks in the road: Development is the process of forming the body in all of its intricate detail. This is followed by a lifetime of maturing. Often in development, once one thing happens other things automatically occur. Stages in development are often "forks in the road" on the pathway of further development. Sexual development is a good example of such a process. There is even some evidence in animals that the output of testosterone at critical times affects the brain. For example, evidence exists that certain male and female sexual behaviors are imprinted into the brain by the presence or absence of testosterone at this or other stages of development. Testosterone may not only directly affect the events at this stage of development, but also its presence in fetal development may have some effect that influences behavior throughout the lifespan of the individual.

Puberty—more sex steroids: For a period of time after the fetal stage of development until a period when the child is approximately seven years old, the story of sex is fundamentally dormant. There is a very high degree of development occurring during the first three months of pregnancy, and then development levels off until shortly before puberty.

At about the age of seven years old, the brain starts "turning on" with regard to the production of hormones that are involved in sexual reproduction. This process takes several years to fully develop. Although incompletely understood, the activation of reproductive capability in some way involves a part of the brain called the pineal gland. The pineal gland is often called "the third eye." It receives direct input from the visual system, and has many connections to the hypothalamus. In some vertebrates that live in areas where the length of the day varies considerably between summer and winter, animals become incapable of reproduction during the winter months when days are short. Animal experiments demonstrate that the turning on and off of reproductive capability in these animals involves changing levels of a neurochemical called melatonin in the pineal gland. One factor controlling the level of melatonin is the light cycle. In the dark of night the level of melatonin increases and in the light of day it decreases. When days are short, the melatonin levels remain higher which appears to suppress the output from the hypothalamus of the brain hormones involved in reproduction. It has been demonstrated in humans that a condition known as precocious puberty involves abnormally low levels of melatonin in the pineal gland. Conversely, profound blindness in which light input to the pineal does not occur often is associated with late onset of puberty. It is also thought that jet lag and the stress which often accompanies shift work are associated with disruption of light cycle input to the pineal gland. The onset of puberty is associated with a drop in melatonin in the pineal gland. It is thought that this drop in melatonin is one factor allowing the hypothalamus to initiate puberty.

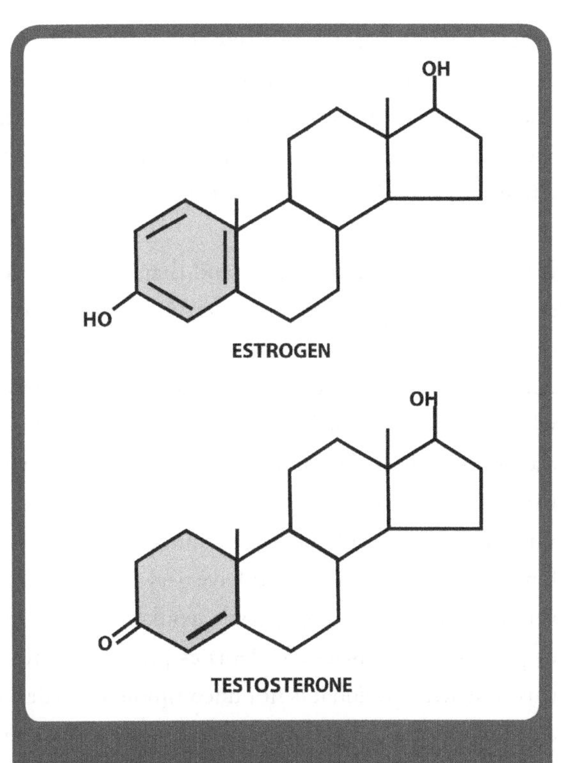

Figure 9.6 - The difference between the female hormone estrogen and the male hormone testosterone.

For probably a number of reasons (besides just those associated with the pineal gland) at about seven years of age the hypothalamus starts putting out chemical messages that will act on the pituitary gland. In both males and females, the messages from the hypothalamus occur in the form of two "releasing factors" FSHRF (follicle stimulating hormone releasing factor) and LHRF (luteinizing hormone releasing factor). These releasing factors act on the pituitary, causing it to release two other chemical messages. LHRF causes the release of a hormone called LH (luteinizing hormone) from the pituitary, and FSHRF causes the pituitary to release FSH (follicle stimulating hormone). These two hormonal messages in turn act on the gonads, initiating maturation and instructing them to put out the appropriate sex steroid hormone (testosterone from the testes in males and estrogens from the ovaries in females). At puberty, there is a rise in the output of these hormones, and it is the presence of either testosterone or estrogen that leads to the development and maintenance of the secondary sex characteristics (such as "beards and breasts"). At this stage, production of functional sperm or eggs occurs due to the presence of the appropriate hormones, and in females menstruation begins. Reproduction is now physically possible.

From a chemical structure point of view, the differences between testosterone and estrogen are extraordinarily little. Both belong to a class of molecules called steroids. Both are made from another steroid called progesterone. They differ from each other by only a single hydrogen atom, and it is very easy for the body to convert one into the other (Fig. 9.6). As will be discussed later, their similarity and inter-convertability is very important when such steroids are used as drugs.

It is also important to understand that the difference between males and females with respect to the type of sex hormones produced (testosterone or estrogen) is not absolute. It is NOT the case that males have only testosterone and no estrogen, and females have estrogen and no testosterone. Rather, the differences that occur are in the relative amounts of these hormones. Males have higher testosterone and lower estrogen levels, while females have higher estrogen and lower testosterone levels. It should also be noted that "high and low" are not absolutes. Hormone levels vary from individual to individual and within an individual from time to time. If one measured the level of these hormones in the blood of a large population of individuals of the same sex, you would find a wide range in the relative amount of either of these hormones in both males and females. Some people would have very low levels compared to the rest of the population, others would have very high levels, but most people would fall in the middle of the two extremes. In fact, during the early stages of puberty, both males and females produce significant amounts of both testosterone and estrogen. During this transition period it is not uncommon for males to develop pseudo-breasts due to the presence of estrogen and for females to develop acne and increased muscularity due to the presence of testosterone.

As discussed above, both the releasing factors produced by the hypothalamus (LHRF and FSHRF), and the hormones produced by the pituitary (LH and FSH) are the same in males and females. What differs in males and females is the presence of either testes or ovaries and the relative amount of the steroids produced by each. Another difference between males and females is that female sex hormones are produced in a monthly pattern, while in males this is not the case. In both males and females, the sex steroids act as a negative feedback on the hypothalamus. Like the example of thyroid hormone, the feedback loop participates in regulating the balance of the hormone.

Hormone patterns in males: Reproductive capability is controlled by hormones. The male role in reproduction is to make and deliver sperm. In many species there are distinct hormonal patterns that turn on and off the reproductive capability of the male. These are primarily yearly patterns based on the amount of daylight. The degree to which such yearly patterns in sex hormones exist in the male of a species varies greatly. The regulation of such patterns is very complex and not completely understood. However, like the onset of puberty, it involves the pineal gland. Although in some species male reproduction is totally dormant in seasons of short days, in human males this cycling is at most vestigial and at least poetic. ("In spring a young man's fancy turns to…"). In general, human males are always capable of producing sperm and are always ready to reproduce.

Male reproduction: The hypothalamus releases two hormonal factors: FSHRF and LHRF. FSHRF directs the pituitary to release FSH, and LHRF directs it to release LH. LH and FSH travel through the blood to the testes. These hormones together prepare the testes to produce sperm (FSH), and direct the testes to

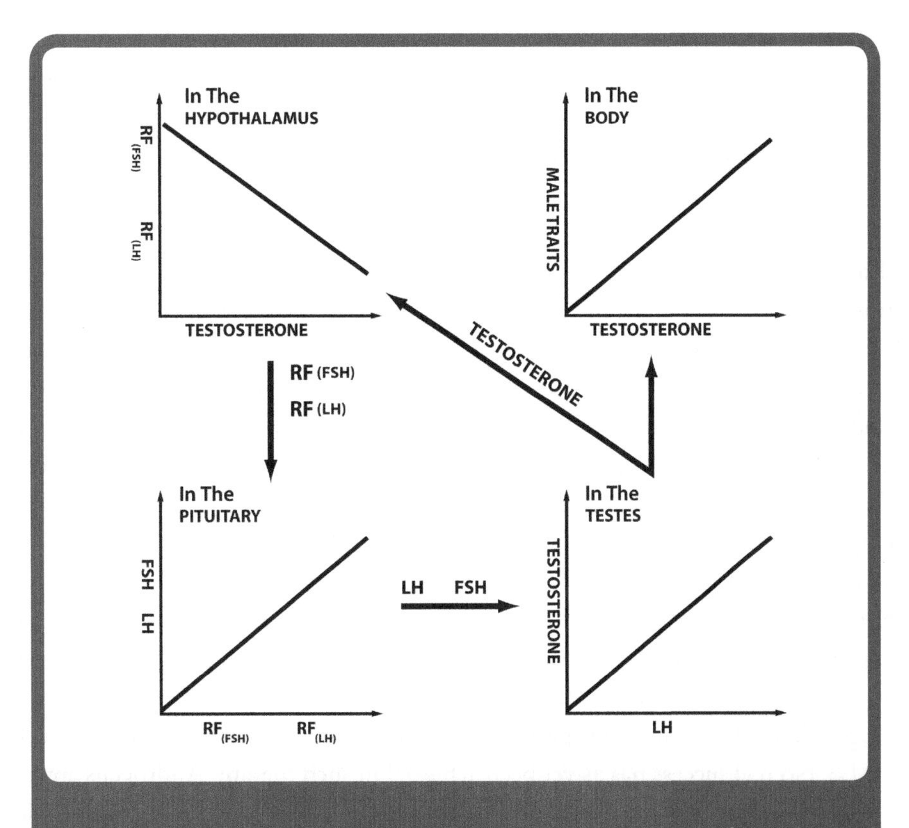

Figure 9.7 - Nerve cells in the hypothalamus secrete releasing factors (RFs) that travel through capillaries in a portal system that go directly to the pituitary gland. The RFs leave the capillaries and bind to receptors on cells in the pituitary gland. The RFs direct pituitary cells to release FSH and LH into general circulation. FSH and LH act on the testes. FSH prepares the testes to produce sperm, while LH directs the testes to produce testosterone and release it into general circulation. Testosterone acts on many tissues of the body causing the development of male secondary sex characteristics. Testosterone also acts as a negative feedback on the hypothalamus and tells them to stop producing RFs. The negative feedback maintains a balance of testosterone in circulation.

produce testosterone (LH). Testosterone is a chemical message that affects a number of things in the body. It is necessary for the production of sperm and sexual activity. It is responsible for facial hair growth and baldness. It appears to promote aggressive behavior, and has some not at all well understood effects on the body musculature. Testosterone also provides negative feedback on the hypothalamus, and directs it to reduce production of LHRF and FSHRF. Like the example of the negative feedback control of thyroid hormone, the negative feedback of testosterone on the hypothalamus maintains balance in the amount of testosterone produced (Fig. 9.7). There has been much interest in attempting to produce antibodies to these releasing factors for the purpose of male birth control. If you could knock out the right releasing factor, you would knock out sperm production. However, there are problems in the development of this approach. The testes and maturity of male structures are maintained by the presence of LH and FSH. Therefore, when you stop the production of either, you not only turn off sperm production but you also cause testicular atrophy and impotency. Although this is an effective method of birth control, not many people would find its side effects worth its efficacy.

Steroids and bodybuilding: Testosterone is an androgen. Androgens are also known as "anabolic steroids" and are sometimes taken by people for the purpose of bodybuilding. Interestingly, anabolic steroids appear to have no effect on the musculature in the absence of exercise. Some evidence suggests that the way that anabolic steroids influence the musculature is by making exercise more useful in building muscle. The use of high doses of androgens by athletes has become a subject of great controversy in recent years. Again, because of the negative feedback aspect of androgen action, a side effect of the use of androgens for this purpose can be testicular atrophy and impotency. An athlete using anabolic steroids is a "He Man" in frame only. Another side effect of taking androgens may be the development of female secondary sex characteristics (such as breast tissue development). This occurs because some of the androgens may be converted to estrogens by the removal of one hydrogen atom from the molecule. Both of these side effects in the use of steroids for bodybuilding are reversible. However, other side effects such as cardiovascular disease, liver damage and kidney damage are permanent. All of these effects will depend upon the dose and length of time the androgens are taken. Androgens have also been used experimentally as a drug for some muscle diseases such as muscular dystrophy and ALS. No real success has as yet been achieved in such therapy. Androgens appear to have less effect on "sick" muscles than on healthy ones, and of course the side effects of these drugs would be the same as in bodybuilders.

Hormone patterns in females: The female role in reproduction is more complex than the male role. The female role is to produce eggs, receive sperm, and to house and nurture the new individual if pregnancy should occur. Unlike males, female humans are not always ready to reproduce. They are not always ready because the female role requires a coordinated sequence of events. The egg must be there when the sperm arrives and the uterus must be prepared to accept the fertilized egg. The optimum situation in human females is for one egg to be produced every 28 days. The uterus is prepared for a short period of time to accept and nurture the small ball of cells that developed from the fertilized egg. Usually, one egg is produced each cycle, normally alternating back and forth between the left and right ovary. Often a woman can tell which ovary is ovulating that month because of an ache on one side of her lower back. The release of the egg is a timed event. It is timed to coordinate with the preparation of the uterus for implantation.

Each of the various chemical messages involved in the reproductive process in females appear to have their own unique pattern (Fig 9.8). However, in reality, all are part of the same interconnected pattern based on chemical messages controlling each other. The entire process involves six chemical messages. Two chemical messages are from the hypothalamus: FSHRF and LHRF. Two others are from the pituitary: FSH and LH. Two chemical messages are produced by reproductive structures associated with the ovaries: estrogen and progesterone. The messages from the hypothalamus control the messages from the pituitary. The messages

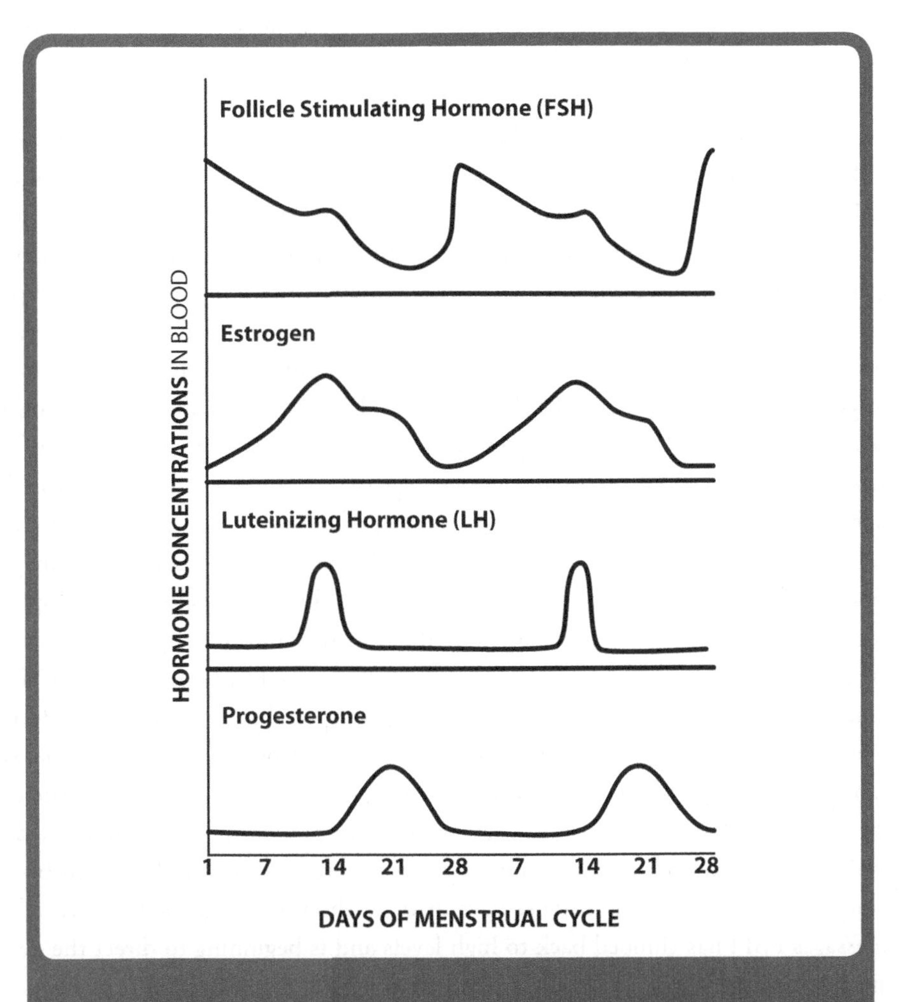

Figure 9.8 - Menstruation begins on the first day of the cycle. FSH is high, which causes development of an egg to begin in one ovary. FSH also causes cells in the ovaries to begin to produce and release estrogen. Estrogen levels continue to increase, causing a build-up in the uterine lining. Estrogen also provides a negative feedback on the hypothalamus, driving down the level of FSH. Estrogen levels peak around day 14 of the cycle. The surge of LH causes ovulation. The cells of the ovaries begin to put out progesterone at well as estrogen. Together, estrogen and progesterone provide negative feedback on the hypothalamus to further drive down the level of FSH. As FSH declines, the cells of the ovaries stop putting out estrogen and progesterone, withdrawing support for the lining of the uterus. The cells that have built up in the uterine lining begin to die. On day one of the next cycle, menstruation begins, washing away the dead cells. The entire cycle repeats.

from the pituitary control the messages from the ovaries, and the messages from the ovaries in turn control the messages from the hypothalamus (more feedback loops).

The two messages from the hypothalamus (FSHRF and LHRF) direct the pituitary to make and release FSH and LH. The more FSHRF released from the hypothalamus the more FSH released from the pituitary. The more LHRF released from the hypothalamus the more LH released from the pituitary. FSH from the pituitary directs the ovary to both begin to develop an egg and to develop cells around the maturing egg that put out estrogen. Estrogen has three effects. First, estrogen directs the body to prepare for pregnancy. Of particular importance is the expansion of the lining of the uterus in preparation for implantation of a fertilized egg. Second, estrogen directs the hypothalamus to reduce the output of FSHRF. As a consequence, as the amount of estrogen increases the amount of FSH decreases. Finally, the high levels of estrogen reached in mid-cycle causes a release of LHRF from the hypothalamus, which in turn causes a release of LH from the pituitary. LH has two effects. First it causes the developed egg to be released. This is called ovulation. Second, it directs the specialized cells developed in the ovary as part of the process of maturing the egg to release progesterone. Progesterone has three effects. First, it expands the preparations for pregnancy. Secondly, it directs the hypothalamus to stop releasing LHRF, which in turn directs the pituitary to stop releasing LH. Finally, progesterone adds to the estrogen signal telling the hypothalamus to stop putting out FSHRF. The result is that the FSH message is driven to very low levels. Without the FSH signal, the specialized cells that developed around the maturing egg begin to die. These are the specialized cells that first developed as the source of estrogen, and then after ovulation as the source of both estrogen and progesterone. As these cells began to die, estrogen and progesterone levels decline, removing support for the lining of the uterus. When this happens the cells of the lining of the uterus begin to die and are shed (menstruation). By the end of one cycle and beginning of the next cycle estrogen and progesterone are both back to low levels because the cells that made the large amounts of these hormones have died. Without the negative feedback of estrogen and progesterone messages FSH has climbed back to high levels and is beginning to direct the development of another egg and new estrogen-producing cells around that egg. Estrogen begins to increase again driving down FSH. By day 14 of the cycle, estrogen is high enough again to cause the LH release that causes ovulation. If fertilization and implantation do not occur within a few days, the high estrogen and progesterone drive down the FSH signal necessary to support the cells making the estrogen and progesterone. Again the cycle ends because of negative feedback. The "average" length of the menstrual cycle in humans is 28 days, but individuals can have regular cycles which are longer or shorter. This is not considered abnormal and does not generally impact reproductive ability.

The menstrual cycle is an exquisitely well timed series of events whose specific purpose is to allow fertilization of a single egg and preparation of the woman's body for the proper nurturing and growth of the ball of cells derived from that fertilized egg. Early in the cycle, under the influence of high FSH levels, the egg develops and matures. In the meantime, the uterus is being prepared for acceptance of the few hundred cells that will become the placenta and a new individual. At mid-cycle, the egg is released under the direction of an LH spike: it is at this time (that is, within a window of a few days) that fertilization must occur. At roughly the same time the uterus is further prepared, under the influence of estrogens and progesterone, for implantation.

The uterus is a muscular sac that lies within the pelvis behind the urinary bladder, and connects with the vagina (in the lower portion of the body) and the Fallopian tubes (the tubes lying very close to the ovaries) at the top. When an egg is released from the ovaries, it enters one of the Fallopian tubes. It is in the Fallopian tubes that fertilization takes place if it is to occur. The fertilized egg divides to form two daughter cells. These cells divide, and their daughter cells divide and their daughter cells divide forming a ball of cells traveling down the Fallopian tubes to the uterus. Under the influence of estrogens, the cells lining the inside of the uterus have thickened and become engorged with many blood vessels. It is within this layer of cells that the ball of cells will implant. Implantation occurs about one week after fertilization, or around day 21 of the cycle.

If implantation does not occur, the levels of estrogen and progesterone drop. When the levels of estrogen and progesterone decrease, the blood vessels in the lining of the uterus constrict, cutting off "life support" to this layer of cells. These cells then die. Briefly, these blood vessels open up again, and the flow of blood washes away the dead cells. This process is called menstruation, and occurs around "day one" of the cycle.

Pregnancy: If the egg is fertilized in the Fallopian tubes around day 14 of the cycle it takes it about a week for the dividing ball of cells to wander its way down to the uterus. If it reaches a prepared uterus about day 21 of the cycle and implants, a different sequence of events occurs. When the ball of a few hundred cells implants in the wall of the uterus, a structure called the placenta starts to form. The placenta will become the interface between the mother and the developing baby. Early in pregnancy, the placenta acts mainly as an endocrine gland, secreting hormones. Later in pregnancy the placenta is very important in bringing nutrients to and removing wastes from the developing baby. The placenta is the structure that joins mother and baby during pregnancy, and contains the umbilical cord. Early in pregnancy, the placenta not only secretes estrogens, but it also releases another very important hormone called HCG (human chorionic gonadotropin). HCG acts on that special group of cells that developed with the egg, telling them to keep putting out estrogens and progesterone. HCG has taken the place of FSH in supporting these cells that produce estrogen and progesterone. The level of HCG reaches its peak at about the third month, and then slowly declines. From about five months on, the level of HCG has reached its lowest point and remains there throughout the rest of the pregnancy. Since HCG is a hormone produced only during pregnancy, its presence in the urine is the basis for "home pregnancy tests" sold over-the-counter.

During pregnancy there are many changes in the woman's body. It takes approximately 38 weeks from the time of conception for a new individual to grow and mature to the point where it is optimally ready for survival outside of the mother's body. That does not mean that a child born in less than 38 weeks will not survive. With modern technology it is now possible for a baby to survive after only 18 to 20 weeks. The limit of viability is greatly dependant on the maturation of the lungs and their ability to exchange oxygen and carbon dioxide. Much of the actual development of the fetus occurs during the first three months of pregnancy. During this "first trimester," all of the physical structures and organ systems develop. By around 12 weeks, a miniature fully-formed human exists. It has only to grow and mature from this point on.

Because a fetus represents a state of very rapid development, it is extremely susceptible to environmental factors. It is a general principle that the earlier the stage of development of the organism, the more sensitive it is to external factors. These external factors range from drugs and infections to maternal nutrition. Thus the developing fetus is more sensitive than the newborn, which is more sensitive than the young child, which is more sensitive than the adolescent, which in turn is more sensitive than the fully-formed adult. Another important thing to remember is that virtually all drugs are able to cross the placental barrier from the mother's body into the fetus. Thus, it is highly recommended that a pregnant woman takes no unnecessary drug, prescription or non-prescription. Of course this includes the use of two common drugs: alcohol and cigarettes. Studies have shown that the use of alcohol by the mother during pregnancy causes many problems with development. These range from the severe damage of fetal alcohol syndrome to relatively more difficult to define problems such as a higher percentage of miscarriages and premature births, low birthweight babies, and increased fetal abnormalities. Smoking also leads to a higher increase in low birthweight babies and other problems after birth. Since such effects are undoubtedly not an "all or nothing" phenomena, it is often recommended that even moderate use of such drugs be avoided. In fact, it can be argued that since there is so much we do not know concerning the myriad of intricate events that comprise development of the fetus, it is prudent to avoid the use of any drug at any level during pregnancy unless absolutely necessary. Of course one must, as with any situation in which drugs are involved, weigh the risk/benefit ratio of taking versus not taking a particular drug. Will not taking the drug be more harmful to the fetus and/or mother than taking the drug?

Pregnancy causes a large number of physical changes in the body of a woman. There is no question that the priorities of the body change during pregnancy. The welfare of the mother's body is no longer the main priority of the system. Rather, survival and proper nurturing of the developing fetus becomes the prime focus of the mother's body. For example, if the intake of nutrients is not adequate, the mother's body will "sacrifice itself" in order to maintain and nurture the growing fetus. Approximately 300 extra calories a day are needed in pregnancy. In addition, much higher levels of iron as well as higher levels of almost all vitamins and minerals are required. Proper nutrition is extremely important in pregnancy for this reason.

The most obvious change in the body during pregnancy is maternal weight gain. A healthy weight gain for the entire pregnancy is now considered to be approximately 25 pounds. Of that, 7–8 pounds can be accounted for in baby, four pounds in placenta and amniotic fluid, two pounds in increased weight of the uterus, two pounds in increased weight of the breasts, four pounds in water and approximately two pounds in fat. Other physical changes in the mother include a 1525% increase in metabolic rate, and 3040% increases in heart rate, respiration rate and blood volume.

There is no question that pregnancy represents a physical stress on virtually all systems in the mother's body. Much of the stress on the mother's body is caused not only just by changes in the body but also by very substantial increases in the circulating level of the chronic stress hormone, glucocorticoids. Much of the origin of this increase in glucocorticoids comes from the increase in progesterone. Glucocorticoids, like testosterone and estrogen, are made from progesterone (Fig. 9.9).

Glucocorticoids have many effects on the body that can facilitate the development of the baby. They demineralize bones and provide calcium. Such demineralization is the origins of the old saw "a woman looses one tooth for each child." It is also the origins of osteoporosis in many older women. Glucocorticoids also make the mother's body insulin resistant. Insulin resistance raises the circulating nutrient levels in the body by reducing uptake into the mother's tissues. The increased nutrient levels assure support for the developing baby. If a woman is going to develop insulin resistant diabetes, also known as adult-onset diabetes, it is likely that she will first do so during a pregnancy.

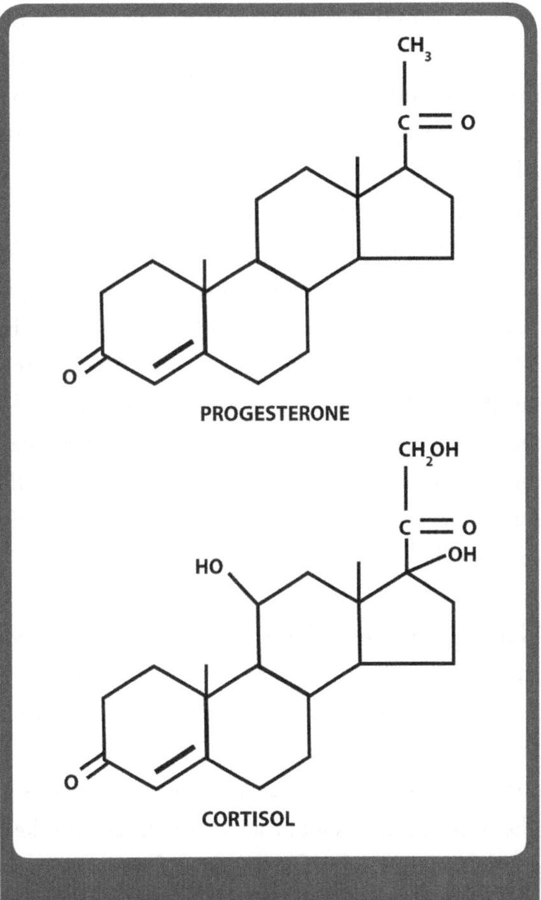

Figure 9.9 - The difference in structure between progesterone and cortisol.

Birth: No one completely understands all of the signals that are involved in the birth process. At any rate, it appears that some signal comes from the baby indicating to the mother's body that it is "ready to be born." This signal may come from the baby's adrenal gland, in the form of glucocorticoids. These hormones influence some important developmental changes in the baby that prepare it for birth, including lung maturation. The signal also crosses the placenta into the body of the mother and tells her that the baby is "ready." The amount of estrogen and progesterone in the mother's blood then sharply declines. Another hormone called oxytocin is released from the pituitary. Oxytocin causes the uterus to contract and birth to occur. However, oxytocin appears to be only one chemical signal involved in the birth process. It also appears that at least two other chemical mediators (a substance called relaxin and a class of compounds called prostaglandins) are also necessary for this process. For example, both of these compounds are necessary for the mouth of the uterus to dilate (open up) and provide enough room for the baby to get out. Several years ago, it was "in vogue" for physicians to induce labor for convenience through the use of oxytocin.

However, this required many more "forceps deliveries" and Caesarean sections than should be necessary because of the lack of other appropriate signals for labor. Today, even though more of the signals are known, this practice is carried out less routinely, and labor is generally induced only when deemed medically necessary.

Lactation: Lactation is the process of milk production and release by the breasts. As we have said, the breasts develop during puberty under the influence of estrogens. However, the changes that occur with puberty do not lead to milk production. During pregnancy the breasts enlarge even more. These changes are due to the very high levels of estrogen and progesterone during pregnancy, as well as another hormone called prolactin. Estrogens and progesterone cause the pituitary to make and release a protein hormone called prolactin. Milk production begins to be turned on about the fifth month of pregnancy due to the high levels of prolactin. Although prolactin is the primary signal for milk production, significant levels of milk are not produced during pregnancy itself. This is due to an inhibitory effect on milk production caused by the very high levels of estrogen and progesterone. At birth there is a decrease in the levels of estrogen and progesterone, and milk production shifts into "high gear." After birth, high levels of prolactin generally last as long as the mother, in "consultation" with the baby, wishes them to last. Every time the mother nurses, there is a tenfold surge in the release of prolactin. Thus, milk production continues for as long as the mother continues to nurse. However, the actual release of the milk to the baby is caused by another hormone, oxytocin. The sensory stimulation of the baby's mouth on the breast causes the back part of the pituitary to release oxytocin. The oxytocin causes the release and ejection of the milk. Another effect of oxytocin which occurs along with milk ejection is contractions in the uterus. The value of such uterine contractions is that it helps return the uterus to normal size after birth.

Manipulating reproduction: For a whole variety of reasons humans are interested in manipulating reproduction. For the most part these manipulations stem from an interest in either preventing or facilitating pregnancy. To that end we have developed a variety of methods to intervene in the reproductive processes. The simplest, albeit not the most effective, method for preventing or promoting pregnancy is for a woman to know her own body. There are three to four days in a woman's cycle during which she can become pregnant. Theoretically this should allow a woman to take advantage of or avoid this "window of opportunity." The problem is that many factors can influence the regularity and length of a woman's cycle. If a cycle is longer or shorter than normal, then the "window" becomes less predictable.

There are a number of reasons why a male may not be able to participate in the conception of a child (other than no woman wishes to participate). These physical reasons relate to the male role in the process: production and delivery of sperm. Each time a normal male ejaculates, about 500 million sperm are released to "chase" the single egg. It takes a lot of sperm to assure that one sperm "makes it." Males that release less than about 150 million sperm are generally sterile. This is because the probability of successful fertilization is too low. The solution to this problem is to collect multiple ejaculates, concentrate them and artificially introduce them into the female vagina at the appropriate time in the woman's cycle. This process is called artificial insemination. Other problems causing male infertility include no sperm production or structurally abnormal sperm. The only solution to those problems is artificial insemination with donor sperm. In addition, certain disease processes such as diabetes can interfere with the male's ability to deliver sperm. These are generally diseases that influence the vasculature in a way that interferes with the attainment of an erection. Depending on the cause, in some cases these problems can be resolved. Recently, drugs such as Viagra have become available, which increase blood flow to the penis and allow an erection to occur.

A common problem that causes infertility in women is blockage of the Fallopian tubes. A common reason for such blockage is scar tissue resulting from an infection. The solution to this problem is in-vitro fertilization, originally called a "test tube baby." For this procedure, the woman is treated with gonadotropin to stimulate the production of eggs. There are currently two such "fertility drug" preparations that can be used for this purpose. The first is chorionic gonadotropin, originally purified from the urine of pregnant women. The active compound in the preparation is HCG. The second is a mixture of FSH and LH, originally purified from the urine of post-menopausal women. In both cases the treatment usually causes the development of multiple eggs. When the eggs have matured they are surgically removed from the woman and mixed with sperm in a laboratory dish for fertilization. After only a small number of cell divisions of the fertilized eggs, one or two of the small balls of cells are introduced into the uterus of the women for implantation. The remaining fertilized eggs can be stored at ultra-low temperatures for further attempts.

A second problem causing infertility in women is under-stimulation of the ovaries by pituitary hormones. The solution here is to use the same fertility drug preparations to stimulate the egg production. In this case, a male is usually "on call" to carry out the fertilization using natural procedures. One side effect of using fertility drugs is that multiple eggs are usually produced and the result is often multiple births. Generally when you read of a woman giving birth to five to eight babies at one time, fertility drugs were involved.

The other side of the coin is intervention to prevent pregnancy. At present there is only one effective procedure to prevent males from delivering sperm in an ejaculate. This method is called a vasectomy. Sperm are produced in the testes and travel through a tube called the vas deferens to be mixed with seminal fluid before entering the urethra for ejaculation. A vasectomy is a surgical procedure in which a small section of the vas deferens is removed so that sperm cannot mix with the seminal fluid. The result is that the ejaculate does not contain sperm.

Safe and moderately successful types of birth control are also provided by a variety of devices which work on the principle of "blockage." If the sperm never reaches the egg, then fertilization cannot occur. This type of approach includes physical barrier devices such as condoms and diaphragms. It also often involves "chemical warfare agents" used against the sperm such as spermicidal creams, jellies, and foams. These drugs are designed to kill sperm, or at least inhibit their ability to swim, thereby preventing them from ever reaching the egg. In recent times, the use of condoms has gained additional popularity for another reason: the prevention of AIDS and other sexually transmitted diseases.

There are a variety of approaches to birth control that involve only the woman. One approach is the equivalent to the vasectomy, a tubal ligation. This is a surgical procedure in which a section of each Fallopian tube is removed and the ends are closed. Like with blocked Fallopian tubes, the sperm are unable to reach the egg. Another type of birth control device is the IUD (intrauterine device). An IUD is a smallish structure (often a coil of brass) that is implanted into a woman's uterus. Although not completely understood, the implanted devise seems to cause a chronic low-level inflammatory response in the uterus which inhibits implantation of a fertilized egg. This method at one time was rather popular. However, after it was found that the "string" attached to the device could serve as a conduit for the entrance of bacteria into the uterus causing Toxic Shock syndrome most IUDs were removed from the market.

Drugs are also used to prevent pregnancy. The oldest drug method, which dates to the early 60s, is the "birth control pill." These drugs are a relatively high dosage combination of estrogens and progesterone The high dose of estrogen and progesterone act as a negative feedback on the hypothalamus. The result is that the hypothalamus does not direct the pituitary to release FSH. Since FSH is necessary for the development of an egg, the result is no egg and therefore no pregnancy. In a sense, the pill "tricks" the body into thinking it is already pregnant. This approach has several side effects and is not indicated under several conditions. The side effects are directly related to the fact that the woman's body is subjected to hormone levels similar to those that occur during pregnancy. The result is chronic intermittent large hormonal adjustments in the body leading to water retention, weight gain, over-stimulation of the uterus and breast, and insulin resistance. Because of the water retention a woman who smokes or has a family history of high blood pressure should not take "the pill." Because of the high levels of hormone present, women with a family history of hormone sensitive tumors such as those of the breast should not take "the pill." Because of the insulin resistance, women who have a family history of adult-onset diabetes should not take "the pill." In fact "the pill" is recommended against if a woman has almost any health problem because, like pregnancy, almost all organ systems such as the liver and kidney are subjected to increased functional stress.

A different type of birth control pill was developed in France. This drug, RU486, sometimes referred to as "the abortion pill" was originally developed as a progesterone antagonist. RU486 blocks the action of progesterone. As a result, the fertilized egg cannot be maintained, and in a sense, a "mini-abortion" or miscarriage occurs. The recommended approach to its use is that if the woman has any reason to think that it is possible that she is pregnant (i.e., she had intercourse around day 14 of her cycle) she takes the pill once at the time her normal cycle would end. The result is that, for a short period of time, support for the uterine lining is blocked and close to normal menstruation occurs. The advantage over "the pill" is that the body is subjected to an abnormal hormonal environment for only a short period of time. Interestingly, as the drug has continued to be studied, it has been found to block glucocorticoids as well. What other effects the drug may have is a subject for current research. Herein lies a reason for caution. At the present time, perhaps the only significant reservation that can be raised about this drug from a medical point of view is that it has been around for such a short time. The significance of this concern is apparent when one considers the drug diethylstilbestrol (DES). This was a drug given to women in the late 1940's and early 1950's to prevent miscarriages. Although the drug was ultimately found to be useless for that purpose, what has been discovered many years later is that children born of mothers who took the drug have a higher incidence of certain kinds of cancer. An even newer approach to preventing pregnancy after unprotected sex is the regimen of drug treatment referred to as Emergency Contraceptive Pills (ECPs) which are sometime referred to as "the morning after pill." As the name implies these ECPs must be taken within a relatively short time following unprotected sex. The pills themselves are rather high doses of either progesterone alone or estrogen and progesterone taken once or twice within 72 hours after sex. There is still some debate as to whether the dosing regimen prevents ovulation or prevents implantation following fertilization.

Some perspectives on sex and drugs: Who would have thought that 40 years later society would be dealing with a higher incidence of cancer in a population of individuals whose mothers were given a drug to prevent miscarriage. How can a drug have a side effect so many years later? Who would have thought that a mild tranquilizer (thalidomide) given to pregnant women in the early 1960's to control "morning sickness" would cause such massive birth defects in children born of those mothers. Infertile women can now have children conceived in a "test tube." Choices can be made concerning when and how many children a woman might have. Drugs are the potential for manipulating the quantity and quality of life including reproduction. How we benefit from that potential and understanding the risks of that potential is the central question. It is the central question because ultimately it is evaluated at the individual level. Statistics concerning risks and benefits apply to the effects of drugs on populations of people. However, the consequences always apply to individuals.

chapter ten

the immune system

The body must control births and deaths of individuals: The society of cells that we call the human body is made up of many types of cells carrying out different activities necessary to the functioning of the organism. The most important goal of this society is to maintain the quality of the environment. Nutrients are brought in and used, waste products are removed, temperature is maintained just right, and the proper balance between salts and water is maintained. A very important factor in maintaining "social balance" is control of the demands on the system. Resources are limited, as is the ability to eliminate waste. As we have discussed, the brain controls the priority distribution of the resources by controlling how much blood flows where in the body. The brain can also limit demands on the system through chemical messages such as thyroid hormone which regulates the rate at which cells consume nutrients and produce waste. In addition, there are local controls on populations of cells. For example, skin cells reproduce only fast enough to replace those that die. If there is damage to the skin then the rate of reproduction increases until repair is accomplished. When the damage is repaired then the rate of reproduction returns to one of replacement. Uncontrolled, promiscuous reproduction would cause havoc in the society of cells. If individuals were allowed to randomly reproduce, control of nutrient availability would be lost, control of waste removal would be lost, and space and resources that should be devoted to some function might be consumed by the wrong individuals. Competition for resources and death of the "society" would ensue. This does not happen because all normal individuals in the society pay attention to the signals controlling their reproduction.

What is an infection? For our purposes, an infection is an invasion of "society" by individuals that do not pay attention to control signals. It is an invasion of outlaws that do not follow the reproductive rules of the society. Such individuals can take many forms. There are invaders such as viruses, bacteria, fungi, protozoans and parasitic worms. These are invaders from the outside world that enter our bodies and take up residence. There are also cancers. Cancers are "invaders" from within our own bodies. Cancers are cells of our own body that become altered and no longer respond to the controlling signals in the environment. All invaders share in common the characteristic that they take resources from the society, produce waste, and contribute nothing. In addition, cancers can grow large enough to damage organs or body structures by their sheer size. Such an invasion can occur in many different places in the body. If the body did not have some way to protect itself against unwanted "guests," the person would soon be overrun by these invaders and die. But the body does have a special group of cells whose function is to protect it from both invasions by other organisms and from cancer.

The immune system is designed to defend against foreign structures: The immune system is composed of a number of different populations of cells that work together to protect the body. The immune cells seek out and destroy invading cells that don't belong. The key to the immune system is that it "tolerates" a limited set of molecular structures that it recognizes as belonging to the body and attempts to destroy all others. Those structures that are tolerated are viewed as "self" and all others are viewed as "non-self." The body is mainly composed of large molecules. From the perspective of immune recognition, large molecules contain hundreds of small structural domains or areas that can be evaluated by the immune system as self to be tolerated, or non-self to be attacked. The discrimination between self and non-self is not simply a matter of the immune system finding all molecules not built by cells and destroying them. Rather, the immune system examines all molecules that it comes in contact with and evaluates, based on circumstances, whether to accept that molecular structure as self or to reject it as non-self. This evaluation process includes those structures made by cells of the body. This evaluation is not perfect. The immune surveillance at times can ignore structures that it should destroy, resulting in infections or cancer. The immune system can also err by classifying a molecule that really is self as non-self, resulting in autoimmunity. The immune system must learn to tolerate self molecules, but at times this learning is imperfect.

Immune protection involves a complex set of responses by many different cell types. Some immune cells are "stationed" in specific places in the body. Their job is to maintain surveillance in that local area. Other immune cells roam the body. They can be found in the blood. These roaming cells can also release chemical messages that allow them to leave the blood and enter the water environments of the tissues. These chemical messages literally open up the capillary barriers between the blood and the tissues, making it possible to enter the space around the tissue. Immune responses can be divided into two general categories. One is non-specific "natural" immunity. This is the first and more general line of defense as the body fights off an invader. It is not directed towards a particular kind of infecting microrganism. The other category is specific or "acquired" immunity. This is a "learned" response which is mounted against a single type of invader, such as a particular kind of bacteria or virus.

Natural immunity is an inborn intolerance: One aspect of evolution is the interaction between different forms of life. If one form of life infects another, natural selection favors the ability to resist or fight off that infection. Individuals who are capable of fighting off an infection are more likely to survive and reproduce. Humans and the organisms from which they evolved have existed for millennia with infecting extracellular bacteria, intracellular bacteria, parasites, fungi and viruses. Because of this long relationship, selection has favored the evolution of an inborn protective response to such infections. Collectively these protective responses

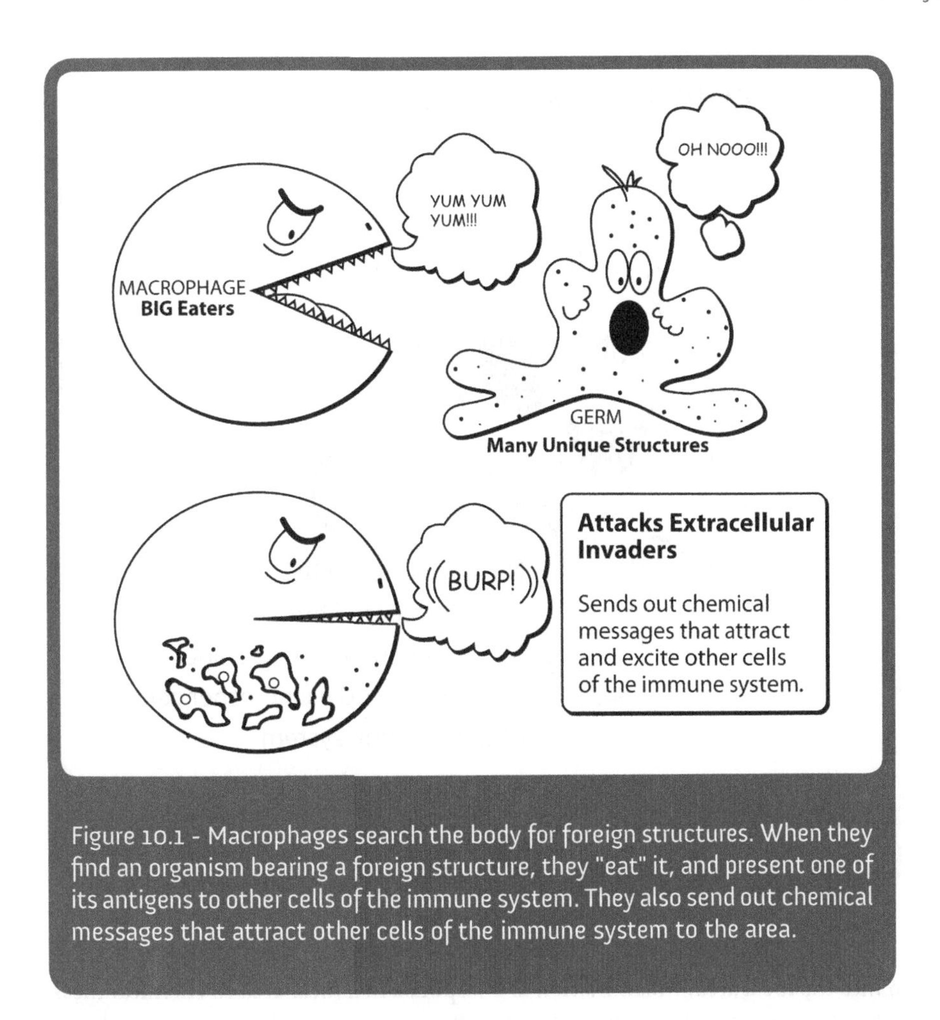

Figure 10.1 - Macrophages search the body for foreign structures. When they find an organism bearing a foreign structure, they "eat" it, and present one of its antigens to other cells of the immune system. They also send out chemical messages that attract other cells of the immune system to the area.

are known as natural immunity. Certain cells of the immune system are "born" with the capability of making a generalized response to structures associated with infective organisms that prey on humans. This response not only attacks the invading organism but also it focuses the attention of the cells involved with specific immunity on the location of the invaders.

Two basic types of cells are involved in natural immunity. The first encompass cells capable of phagocytosis. The term phagocytosis derives from the Greek word "phagein," meaning to eat. That is precisely what these cells do: they engulf and break down or digest organisms and foreign substances carrying certain marker molecules. A very important cell type with this capability is the macrophage ("big eater") (Fig. 10.1). These cells not only reside in large numbers around the barriers of the body such as the skin, lungs, and digestive system, but they also roam the body searching for organisms bearing the marker molecules that "turn them on". There are a variety of such marker molecules, most of which are associated with frequently encountered invaders. For example, one such marker is a molecule in the coat of gram negative bacteria known as lipopolysaccharide (LPS). When they encounter a bacteria bearing LPS they engulf it and digest the entire organism into smaller structural pieces. They also send out chemical signals that cause inflammation and attract other cells of the

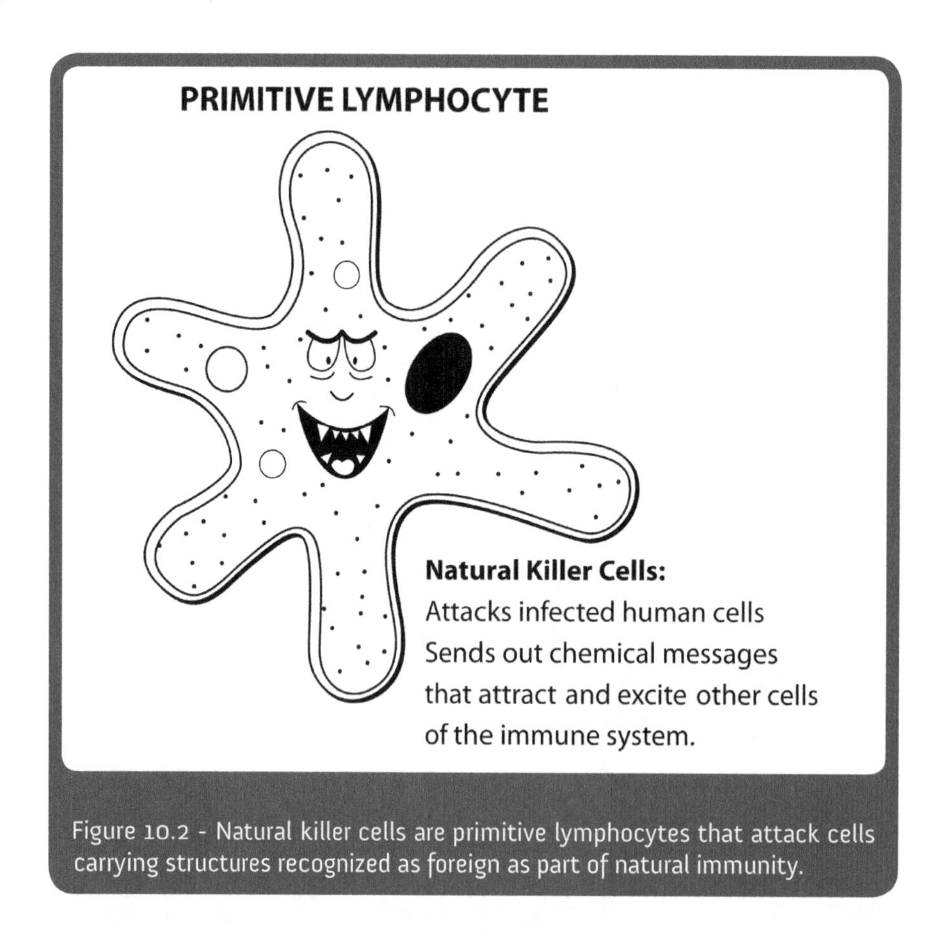

Figure 10.2 - Natural killer cells are primitive lymphocytes that attack cells carrying structures recognized as foreign as part of natural immunity.

immune system into the local area. Inflammation is a response in a local area to chemical messages that open up the capillary barrier. Opening the capillary barrier allows immune cells and water to leave the blood and enter this area. The result is swelling and "redness" in the area.

Some of the cells that enter through the opened up capillaries are also involved in natural immunity while others are involved in setting up learned specific immunity. Macrophages are very important in preventing extracellular bacterial infections such as meningococcus which causes meningitis. They also have some use in protecting against intracellular bacteria, viruses, parasites and fungi.

The second type of cell involved in natural immunity is called natural killer cells. Natural killer cells are attracted by the chemical signals released by the "turned on" macrophages (Fig. 10.2). They are also attracted by the chemical signals released by a variety of cells when they are infected by viruses. Natural killer cells are large primitive lymphocytes (white blood cells) that respond to general markers on the surface of target cells. They attack the target cell and break open its outside membrane. These cells can be effective in destroying cells of the body that have become infected with intracellular bacteria such as those that cause tuberculosis and leprosy. They can also be effective in destroying cells infected with viruses, cancer cells, and in attacking parasites and fungi.

This non-specific natural immunity, from an evolutionary point of view, is primitive and is even found in lower animals such as sharks. It evolved as we evolved with our natural invaders. It serves a very important role as the first line of defense in preventing many infections and cancers from becoming established in the body. It also serves the very important role of bringing into a localized area those cells involved in setting up specific immunity. Specific immunity is a learned response to structures that are unique to an individual type of foreign cell or virus. The cells of specific immunity learn to discriminate "self" from "non-self."

Tolerance—Discriminating "self" from "non-self": Discrimination between "self" and "non-self" is learned by cells of the immune system. The cells of the specific immune system are called lymphocytes. All mature lymphocytes produce a unique protein capable of binding to some unique molecular structure. There are a number of different types of lymphocytes, and a number of different kinds of these binding proteins. Depending on the type of lymphocyte, these binding proteins (antibodies) can reside in the outside membrane and act as a receptor for the cell or they can be secreted into the environment. All antibodies have an affinity (a tendency to specifically "stick" or bind) for some unique relatively small molecular structure called an antigen.

Antibodies are interesting proteins in that they are "hybrid" molecules. A part of the molecule is the same in every antibody (this is called the constant region). But the other part of the molecule is different in antibodies within the same human (this is called the variable region). All proteins made by every cell in the body are coded for by inherited gene (DNA) sequences. This is true of antibody proteins, but their coding is more complex than that of most proteins. The binding proteins of the immune system are constructed of a constant region which is coded for by discrete gene sequences in the DNA. The rest of the protein, called the variable region, is still coded for by DNA, but the DNA is reorganized in different ways with different antibodies. The complete code for the protein is made by "splicing" the constant region gene to a random (variable) DNA sequence. The result is a DNA sequence that codes for a totally unique protein. Only that particular lymphocyte and all of its daughter cells will contain the code for that unique protein. The part of the molecule coded for by the reorganized DNA (variable region) forms the part of the binding protein that recognizes the antigen. The extent of the ability to reorganize the DNA and produce unique binding proteins is quite extraordinary. Each individual is born with the ability to produce billions if not trillions of unique immune binding proteins. Each binding protein will recognize a different molecular structure. Such diversity allows the immune system to learn to recognize and bind to the myriad of structures associated with the vast array of organisms that cause diseases. With this diversity also comes the problem of recognizing the vast array of structures made by one's own body. Discriminating "self" from "non-self" requires that the immune system learn to tolerate one's own structures while attacking all others. The process by which the immune system learns to tolerate self involves either killing or inactivating lymphocytes that make binding proteins that bind to self structures.

There are two basic types of lymphocytes involved in specific immunity: the B-lymphocytes and the T-lymphocytes. Both are produced in the bone marrow. B-lymphocytes mature into cells able to recognize a molecular structure while still in the bone marrow. In contrast, T-lymphocytes leave the bone marrow as immature cells that are incapable of recognizing a molecular structure. They travel to the thymus, where they mature into a cell capable of recognizing such a structure. Both the B- and T-lymphocytes travel throughout the body in the circulatory and lymphatic paths. However, they are present in especially high concentrations in lymphoid tissue such as the lymph nodes, tonsils, and spleen. When a mature lymphocyte encounters a molecular structure that its antibody will bind to, one of two things will happen. On the one hand, the encounter can cause the cell to be killed or inactivated, resulting in tolerance of the molecular structure. On the other hand, the encounter can cause the cell to be activated and begin to divide resulting in many cells capable of producing the unique binding protein (antibody) that recognize that molecular structure. Which of these two things happen depends on the circumstances of the encounter.

B-lymphocyte activation: Once a B-lymphocyte begins to mature in the bone marrow, it becomes capable of making its unique binding protein called an antibody. This antibody is inserted into the outside membrane of the lymphocyte. The antibody inserted into the lymphocyte membrane acts as a "receptor." If, in the bone marrow, the lymphocyte encounters a structure that binds to its antibody it will most likely be either killed or inactivated. This leads to tolerance of the structure. The lymphocyte is killed or inactivated because generally molecular structures encountered in the bone marrow are "self" structures. B-lymphocyte encounters with antigens in the environment of the bone marrow during maturation generally result in tolerance. If it does not have an encounter with its antigen, it completes maturation, leaves the bone marrow and enters circulation carrying its antibody on its surface. If it encounters its antigen elsewhere in the body, there are again the same two possibilities: tolerance by inactivation or activation. Which of the two possibilities occurs depends on the environment of the encounter. If the encounter takes place in an environment rich in activating chemical messages produced by a natural immune response or other cells of specific immunity then the cell is activated. If the encounter occurs in a non-stimulating environment then tolerance can develop. If the B-lymphocyte is activated it begins to produce daughter cells capable of making the same antibody. Most of these daughter cells produce a form of the antibody that is released by the cell into the environment (Fig. 10.3). The release of soluble antibody into the environment is the attack phase of a B-lymphocyte immune response. It will continue as long as the daughter cells encounter the molecular structure that binds its antibody.

T-lymphocyte activation: The other class of lymphocytes mature and are activated differently than B-lymphocytes. There are several different types of T-lymphocytes. The two that are most clearly understood are helper T-cells and cytotoxic T-cells. Both types of T-cells only produce binding molecules that are inserted into their surface membrane and act as receptors. Similarly, both types of T-cells only recognize the molecular structure to which their receptor binds when it is expressed on the surface of another cell. Immature T-lymphocytes which do not make any binding protein receptors leave the bone marrow and migrate to the thymus where they begin to mature. Like the antibodies generated by B-lymphocytes, T-lymphocytes generate their receptors by random rearrangements of a region of the DNA.

Once the rearrangement has occurred, the individual T-lymphocyte begins a sequence of events that leads to maturation. In some way during this process T-cells that make receptor binding proteins that recognize self structures die, while those that do not continue maturation. T-lymphocytes that are allowed to mature, leave the thymus and take up residence in other lymphatic tissues such as the spleen and lymph nodes. It is usually

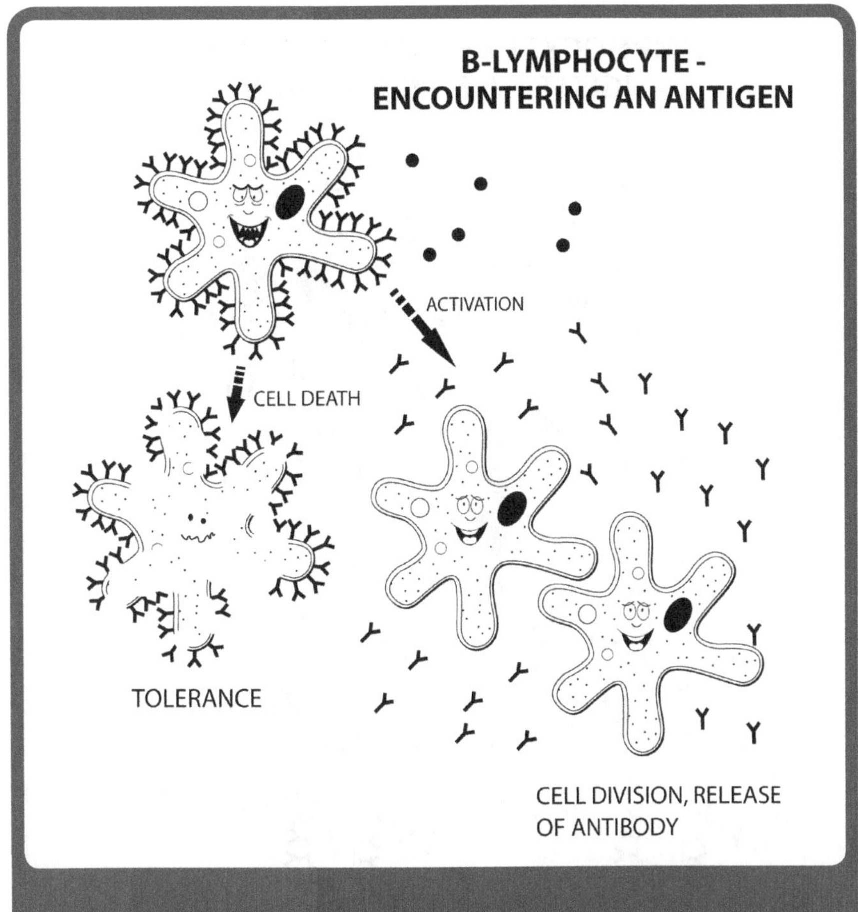

Figure 10.3 - When a mature B-lymphocyte encounters an antigen that binds to its antibody, it either dies and the antigen is tolerated as "self" or it is activated. Activation causes that B-lymphocyte to divide and release activating chemical signals. The daughter cells make and release free antibody molecules into the local environment.

in a lymph node that T-cells encounter a structure that binds to their antibody receptor (an antigen). When they encounter their antigen they are activated.

Helper T-cells are activated by other cells of the immune system such as macrophages, which present the molecular structure to them (Fig. 10.4). As discussed earlier, macrophages of the natural immune system eat foreign molecules and break them down into smaller structures. The macrophage then inserts a small piece of the foreign molecular structure into its surface membrane. When a macrophage presenting such a foreign molecular structure on its surface encounters a helper T-cell that binds to the foreign structure it releases chemical signals that activate the helper T-cell. Activation causes the helper T-cell to release more chemical signals and also to rapidly divide.

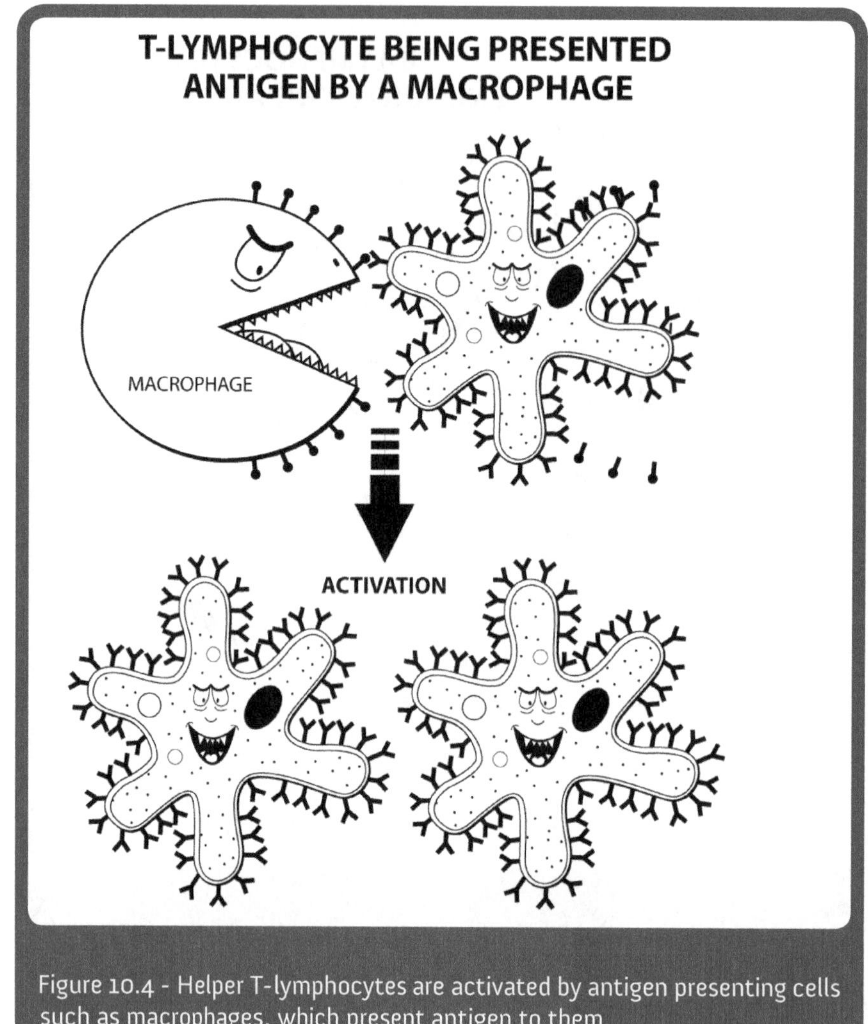

Figure 10.4 - Helper T-lymphocytes are activated by antigen presenting cells such as macrophages, which present antigen to them.

Cytotoxic T-cells are activated by a different process. Virus infected cells and tumor cells often express unusual proteins on their surface. In the case of cells infected by viruses these are usually foreign viral proteins. In the case of tumor cells, these are often either mutant proteins or proteins that should only be expressed by fetal cells (fetal proteins are often made by tumor cells that "forget" that they shouldn't be making these proteins). When a cytotoxic T-cell encounters a cell expressing a protein to which it receptor binds it is activated. Like the helper T-cell, activation of the cytotoxic T-cell causes both release of chemical messages that are activating to other cells in the area and rapid cell division.

Defense mechanisms of the immune system: The various cells of natural immunity and specific immunity along with the mechanisms that they activate provide a comprehensive host defense against the range of foreign invaders and cancer cells. Besides the cells of the immune system, there are a set of very destructive proteins called complement proteins which circulate in the blood (Fig. 10.5). Under normal conditions

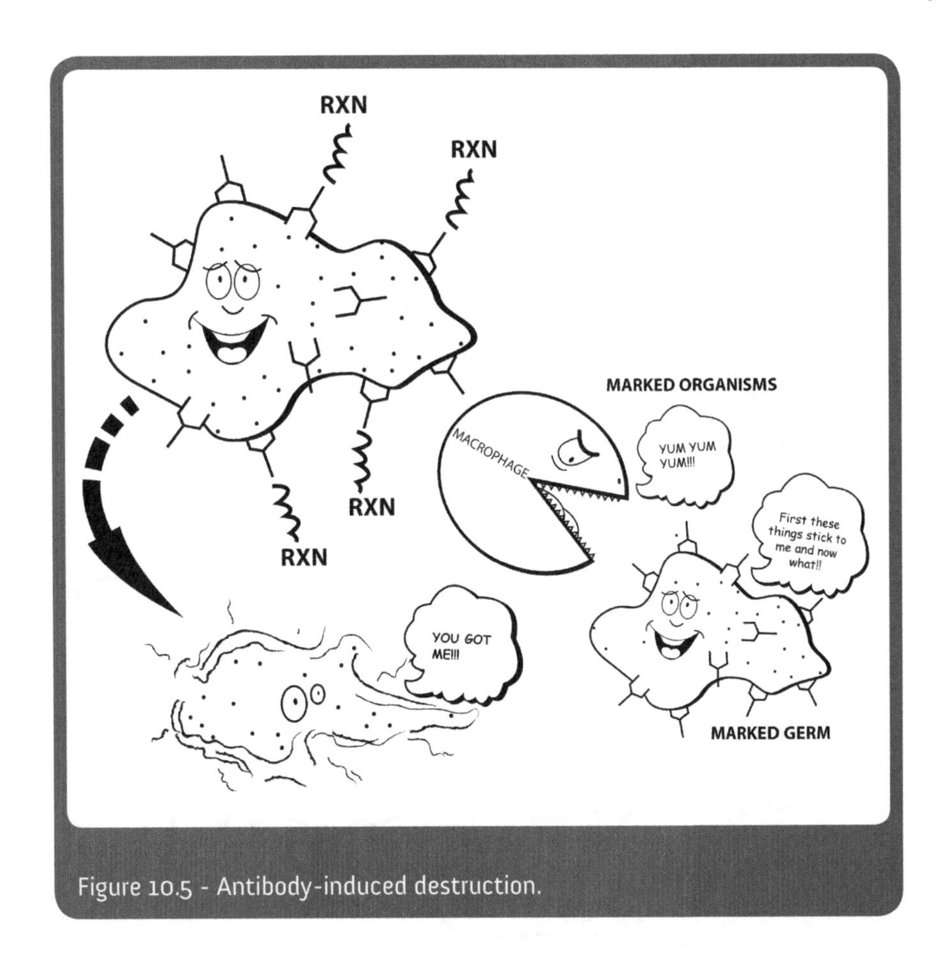

Figure 10.5 - Antibody-induced destruction.

these are inactive. However, when activated, they become very destructive digestive enzymes that break down targeted molecular structures. For example, if a soluble antibody from a B-lymphocyte binds to its antigen on the surface of a foreign cell, the complement cascade is activated and the surface of the foreign cell is digested, killing it. In general, defense against a particular invading microbe involves more than one mechanism. Extracellular bacteria excite natural immunity and the release of chemical messages that attract cells of specific immunity.

If there is already sensitivity to structures on the bacteria, helper T-cells will excite B-cells to release antibodies that will mark the bacteria for increased phagocytosis and complement attack.

Similarly cells infected with intracellular bacteria and viruses attract cytotoxic T-cells. The first step in such encounters is usually the release of chemical messages that attract the attention of other cells of the immune system. The cytotoxic T-cell destroys the infected cell. This destruction releases the foreign structures that were on the inside of the cell. These foreign structures are picked up by macrophages which initiate helper T-cell and B-cell responses which in turn activate the complement system. In general such cascading responses are self-limited by elimination of the foreign invaders (Fig. 10.6). However, when the invader is destroyed, a small number of both B- and T-cells that produced unique proteins that bind to molecular structures of the

Figure 10.6 - T-lymphocyte-induced destruction.

invader will remain. These remaining cells provide the immune system with a "memory." If the same organism returns again the encounter causes release of chemical messages that cause these memory cells to rapidly divide and resist the new invasion.

Immune responses can be destructive to the body: The mechanisms of immune defense are destructive to invaders, but they can also be quite destructive to the tissues of the host. The initiation of immune responses sets in motion cascading sequences of events. Of particular importance is the release of chemical messages that magnify and expand the response. These chemical messages both attract and activate other immune cells. They also change the characteristics of the blood vessels in the area to allow immune cells and fluid to leave circulation and enter the extracellular environment.

Bacterial meningitis and septic shock are two examples where major damage can be caused by an immune response. In bacterial meningitis, the immune response allows fluid and cells to leave circulation and enter the environment of the brain. Because the brain is encapsulated in the bony skull, pressure in the head builds. This increase in cranial pressure kills brain cells. In septic shock, a generalized immune response in the body reduces vascular tone and circulating blood volume. The result is that blood pressure can drop precipitously, preventing the heart from pumping blood to the brain and other vital organs.

Active immunization: How does immunization work? The idea behind immunization is that once the immune system has been exposed to a particular foreign structure, it is now "primed" and ready to fight that foreign structure when it sees it again. Thus, if you were able to introduce the foreign structure into a person in a benign way (that is, in a way that does not make the person sick) and still were able to expose the immune system to it, the immune system would develop memory T- and B-cells. The body would be ready to react quickly to a foreign invader carrying that structure. The result would be that the person would never get that particular disease because the invader is killed before it can reproduce enough times to actually cause the disease. Thus, the process of immunization is a protection built up in anticipation of exposure to the disease.

In active immunization, either a live or a dead microbe is introduced into the body of a healthy individual along with a structure like LPS that excites natural immunity. This causes the immune system to respond to the molecular structures associated with it. If a live microbe is used, the organism must be modified so that it does not actually cause the disease. A dead microbe, of course, is itself unable to cause the disease. The Sabin vaccine for polio is of this type. It is a dead virus vaccine. The pieces of the dead virus are introduced into the individual, and the immune system responds to the molecular structures of the virus. However, the dead microbe vaccines are not quite as good as the live vaccines. Dead microbe vaccines provide only a limited stimulus to the immune system because they do not reproduce. Live microbe vaccines are able to produce a better immune response because they produce a mild infection that reacts more extensively with the immune system. The Salk vaccine for polio is an example of a live virus vaccine. However, the "teeth" of the virus have been pulled chemically, and it does not itself cause polio. The virus has been changed into something that will still make the immune system respond, but it does not cause a severe disease in the individual. It will grow as a limited benign infection, but it doesn't cause the disease. Because the virus is still able to multiply, the immune system can respond better to it. However, it should also be kept in mind that for the same reason, a live microbe vaccine is more dangerous than a dead microbe vaccine. It is possible for the microbe to change back and become harmful to the individual again. The longer the microbe remains in the body, the higher is the probability that this might happen. That is why it is not a good idea to get immunized when one already has an infection of some type. The immune system is busy with the first infection, and the microbe introduced by the immunization process stays around in the body for a longer period of time. The second problem with these vaccines is one of sterility. Because these microbes are live, they have to be grown up outside of the body in order to make the vaccine. It is possible that mistakes happen and the vaccines may not always be pure. There have been several instances in the past where batches of vaccine have had to be recalled because of contamination by other microrganisms. In addition, a person can have an allergic reaction to something that is transferred into his or her system along with the live microbe during the vaccination. Serious allergic reactions can occur during the immunization process. Despite the problems with live microbe vaccines, they have been extremely beneficial in preventing many serious diseases. Live microbe vaccines are used for both bacteria

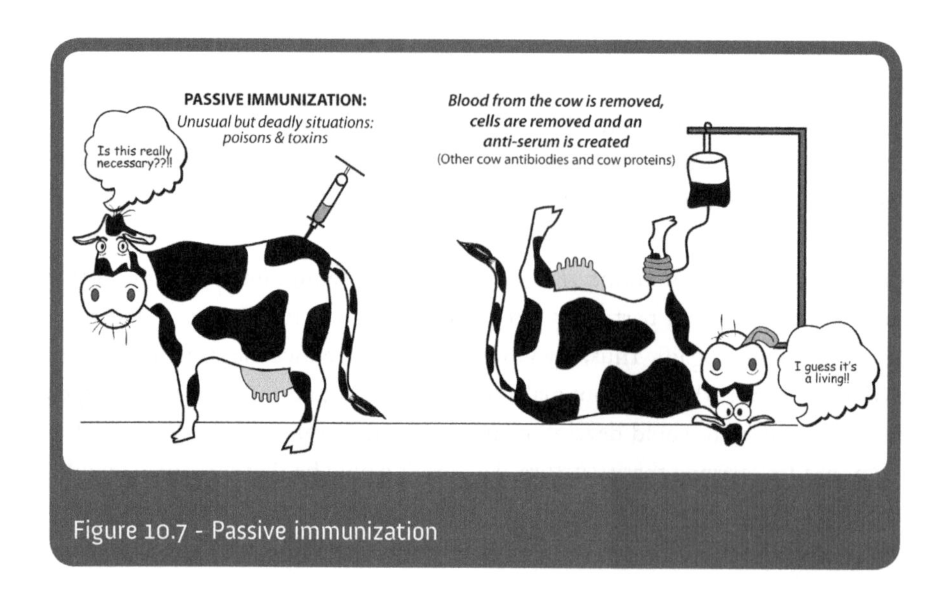

Figure 10.7 - Passive immunization

and viruses. For example, live bacteria vaccines are used for tuberculosis, salmonella, cholera and pertussis. Similarly, live viral vaccines are used for polio and measles.

Passive immunization: In passive immunization, the immune system of the individual does not play a role in the process. Passive immunization is used for situations that are rather rare, such as snake bites or botulism (food poisoning). It is not likely that many people will be bitten by a certain type of poisonous snake in their lifetime. Therefore it is unwarranted to actively immunize people against such an event unless it actually happens (that is, the risk/benefit ratio is too high). Instead, situations like a snake bite are treated by passive immunization.

In passive immunization, another species of animal (for example, a cow or sheep) is immunized with the foreign substance (for example, the venom of a rattlesnake). The toxin is given to the animal slowly over a period of time so that the animal is not killed outright by the poison. In the process, the animal's immune system builds up a high level of antibody to that toxin. The antibodies and associated serum proteins from that animal are then isolated. This is called an "antiserum" (Fig. 10.7).

When an individual is bitten by the rattlesnake, it is this antiserum that is given as an antidote. The antibodies to the toxin that were made in the other animal protect the individual from the harmful effects of the snake bite; they detoxify the toxin. The results of this type of immunization are also variable. The variability is in a large part determined by the length of time that it takes to administer the antiserum. For example with a rattlesnake bite, the antiserum must be administered very quickly, because the toxin is hemolytic: it destroys the circulatory system very rapidly. Another major problem with passive immunization is the risk of a serious allergic reaction to the antiserum. Such a massive allergic reaction is called anaphylaxis. In anaphylaxis, blood volume and vascular resistance drop precipitously.

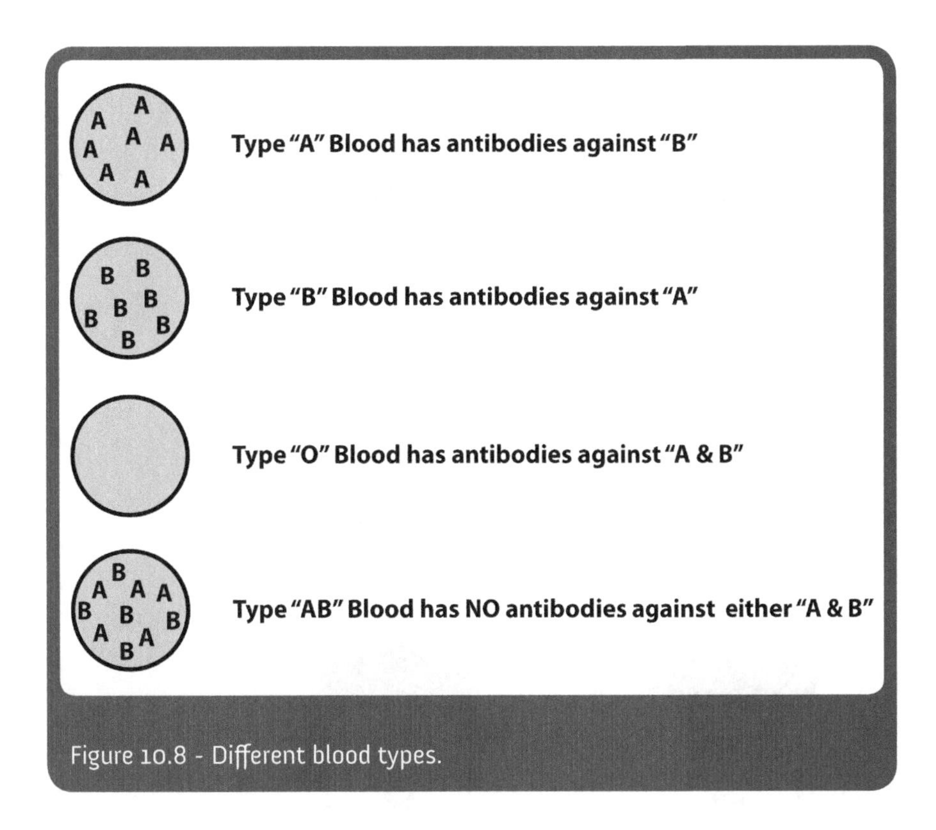

Figure 10.8 - Different blood types.

During passive immunization, foreign proteins from the serum of the animal used to produce the antiserum are also introduced into the individual. Since these proteins are foreign, the immune system develops antibody producing cells against them. This poses no problem the first time an individual is passively immunized. However, if the individual is passively immunized a second time (for the same thing or for something different) and is given an antiserum produced in the same species of animal as the first, the immune system may mount a massive attack against these proteins. Because of the severity of the immune attack, it can have dire consequences.

Hemolytic disease of the newborn: Hemolytic disease of newborn children is caused by blood group incompatibility between a mother and the fetus. The first form of transplantation was blood transfusions that began in the 19th century. At times blood transfusions were very effective while at other times they killed the recipient. The reason for the success or failure of transfusions was not understood until blood groups were discovered.

The first blood group system to be defined was the ABO system. The ABO system contains two major antigens that give rise to four basic blood types: A, B, AB, O. More reasonably the ABO system should be called the "AB zero" system because type O blood is the absence of both the A and B genes. The A and the B genes code for enzymes that add sugars to structures on the surface of red blood cells. For some unknown reason humans are born with antibodies to the type of blood antigen that they don't have (Fig. 10.8). Therefore, a person with

Figure 10.9 - The Rh factor is not a problem if the mother is Rh-positive. The Rh factor may cause a problem with the second child if the mother is Rh-negative and the father is Rh-positive.

type A blood is born with antibodies to the B antigen, people with type B blood are born with antibodies to the A antigen, and people with type O blood are born with antibodies to both the A and B antigens. Of course, people with AB blood carry neither antibody; if they did, those antibodies would attack their own red blood cells. This is an unusual situation, in that the development of antibodies does not require prior exposure to the antigen. If a person with type A blood is given type B blood as a transfusion, the antibodies will bind to the transfused red blood cells and initiate the complement cascade and macrophage activation. The result can be anaphylactic shock and death.

Incompatibility between the mother and the fetus can cause hemolytic disease in newborns. If, for example, the mother has type O blood and the fetus has type A blood then the mother has antibodies that will attack the baby's blood. However, this type of hemolytic disease is not usually severe because the A and B antigens are not fully expressed until after birth. In contrast to hemolytic disease caused by ABO incompatibility, incompatibility in a second blood group, the Rh system can be very severe and even fatal to the fetus. Rh stands for rhesus monkey. The Rh protein is a gene some of us share with these monkeys. This is a gene on chromosome one. When expressed the person is referred to as Rh positive, and when absent the person is Rh negative. As discussed previously, a person will inherit one chromosome one from each parent. If a person inherits even one chromosome carrying this gene, it will be expressed as the protein. If the mother is Rh negative and the fetus is Rh positive there is the potential of hemolytic disease in the newborn. The problem is sufficiently common that it is called "Rh disease" (Fig. 10.9).

Unlike the ABO system, people without the Rh antigen are not born with antibodies to the Rh antigen. They must be exposed to the Rh antigen to develop B-lymphocytes that make antibodies to the protein. The common time that a Rh negative woman will develop such sensitivity to the Rh protein is the first time she gives birth to a Rh positive child. It is not uncommon at birth for a small amount of the baby's blood to mix into the mother's circulation. If this happens she will be immunized with the Rh protein leading to sensitivity. This is not a problem for the first child. The problem occurs if the mother becomes pregnant with a second Rh positive child. In this case the antibodies will cross the placenta and attack the baby's blood. If the fetus survives it will be born with jaundice, anemia, and an enlarged liver and spleen. Fortunately, Rh hemolytic disease can be prevented. The solution is to prevent the Rh negative woman from developing antibodies to the Rh protein. This is accomplished by passively immunizing a Rh negative woman carrying a Rh positive child with antibodies against the Rh protein. The passive immunizations should occur at about the 28th week of pregnancy and within 72 hours after birth. The passively introduced antibodies bind to the Rh protein and prevent the mother's immune system from becoming sensitized.

Transplantation: The problems posed by blood transfusion are multiplied many-fold in organ transplantation. Red blood cells are relatively simple cells carrying relatively few antigens. Livers, kidneys or hearts, for example, are very complex structures carrying a multitude of antigens. It is an immune reaction to the "foreign" structures of these organs that will cause a person to reject a transplanted organ. Even though the organ is from a human, there is sufficient genetic diversity for an organ from one human to be viewed as foreign by the immune system of another. Tissue typing is a process of trying to match as closely as possible donor and recipient tissues for the same antigens. In addition, people who have organ transplants are usually treated with high doses of drugs that suppress the immune system.

There are two basic types of drugs used for this purpose. The first are non-steroidal anti-inflammatory drugs such as Cyclosporine A. These drugs have their major effect by interfering with the release of the chemical messages that activate the immune system. The second are synthetic glucocorticoids. These are synthetic steroids that mimic the natural steroid cortisol that is released from the adrenal gland in response to ACTH from the pituitary gland. These drugs not only interfere with the release of the chemical signals that activate the immune response, but also interfere with immune cell division. Besides suppressing the immune system, these drugs have the same effect on the body as cortisol. They shift muscle into a net degradation of proteins. They induce enzymes in the liver and kidney to convert the amino acids released from the muscle into sugar. They make muscles resistant to insulin. The result is excessively high blood sugar, a condition known as steroid diabetes. The high circulating sugar concentration strains the kidneys and can lead to kidney failure. In this circumstance the sugar is taken up by fat cells and converted to fat. The net effect is a substantial transfer of amino acid carbon from the musculature to fat. Another problem with the use of immune-suppressive drugs is that the entire immune system is generally suppressed. As a result, the person is highly susceptible to all kinds of infections.

Immunodeficiency: The proper functioning of the immune system involves a complex interaction between a variety of cells and molecules. A defect in any of these cells or molecules reduces an individual's defenses against infections and cancer. There are a variety of genetic defects that impair parts of the immune system. Genetic defects exist in T-cells, in B-cells, in the complement system, in specific types of antibodies, in chemical messages, and in the cells of natural immunity. The consequences of these genetic defects depend on the nature of the defect. In general any defect in the immune system will result in an increased susceptibility to a particular type of infection or cancer.

Besides being inherited, immunodeficiency can be either acquired or be a consequence of other disease states. For example, starvation causes the immune barriers of the intestinal tract to degenerate allowing invasion of bacteria. Drug therapies for cancer and immuno-suppression can cause immunodeficiency and an increased susceptibility to infections and cancer. In addition, there are a number of viruses that directly attack some cells of the immune system causing them to dysfunction. The result is immunodeficiency. The most notable one of these is the Human Immunodeficiency Virus, HIV, that causes human Acquired Immune Deficiency Syndrome, AIDS.

HIV is a type of virus known as a retrovirus. Unlike many viruses and all other forms of life, these viruses use RNA, not DNA as their genetic material. HIV consists of two identical strands of RNA coding for about 10 proteins. These two strands of RNA are surrounded by a protein coat and packaged in a lipid membrane. By itself HIV is like all viruses: it is incapable of making proteins by itself. HIV, like all viruses, is simply a bit of information wandering around in the world without the ability to carry out any life processes. What HIV does contain is a protein that is capable of binding to a particular structure that is found on helper

T-lymphocytes and macrophages. This allows the virus to bind to these cells, fuse with them and introduce its bit of genetic material into them.

Once the virus is inside the cell its proteins become active. The virus is broken apart, releasing the viral RNA. One of the viral proteins is an enzyme known as a reverse transcriptase. This enzyme does the opposite of RNA polymerase which makes RNA from DNA. This enzyme makes DNA from RNA. The reverse transcriptase copies the viral RNA genetic material into the corresponding DNA. The DNA then enters the cell nucleus and another viral protein integrates the viral DNA template into the DNA of the cell. In this state, the virus can remain dormant for a long time. No one knows why, but at some point in time the virus genes are activated and the normal RNA and protein synthetic machinery of the cell is used to produce mature infectious virus. These mature viruses then infect additional helper T-cells and macrophages. Many of the symptoms of AIDS are caused by depletion of helper T-cells and macrophages. Depletion of helper T-cells compromises aspects of both cell mediated immunity and humoral immunity (antibodies). Depletion of macrophages compromises natural immunity. The result is that the individual becomes susceptible to a variety of infections ranging from intracellular bacteria such as those that cause tuberculosis to other viral infections and fungal infections. The individual is also more susceptible to certain types of cancers.

Identification of individuals infected by HIV is based on the presence in their blood of antibodies to the viral proteins. However this can be complicated by the fact that the reverse transcriptase makes a lot of mistakes. As a consequence of these mistakes there are many mutations in the viral proteins. The mutant proteins often are not recognized by antibodies to the original proteins. This also reduces the effectiveness in attacking the infection by those parts of the immune system that are not compromised. Several drugs have been developed to treat HIV infections. These drugs, such as AZT and DDI are designed to interfere with the reverse transcriptase. However, neither has been shown to be particularly effective in treating the disease. One problem is that the high mutation rate also applies to the coding sequence for the reverse transcriptase. The enzyme can mutate producing a reverse transcriptase that is still active but unaffected by the drug. If this happens then those mature viruses that make the mutant enzyme are selected during the course of the infection. In essence the drug becomes a selective pressure for evolution of the infection. A similar problem arises in the development of a vaccine for HIV. Because of the high mutation rate of the genes coding for the viral proteins, antigenic sites on the proteins change. This "leopard" changes its spots.

Autoimmune diseases: The cardinal requirement of the immune system is that it be self tolerant. If for any reason, self tolerance breaks down a class of diseases known as autoimmune diseases develops. In autoimmune diseases some part or parts of the specific immune system attack structures of the body. As we have figured out more about how the immune system works we have discovered that autoimmunity is quite widespread as a cause of or component of diseases. Autoimmune diseases include rheumatoid arthritis (where the joints are attacked), multiple sclerosis (where the myelin sheaths around nerve cells are attacked), mysathenia gravis (where the receptors for a neurotransmitter on muscles are attacked) and many others. A number of different factors and mechanisms have been shown to be involved in the breakdown of self tolerance. Although not understood, there seems to be a genetic propensity to the development of at least some autoimmune diseases.

These genetic propensities do not mean that the diseases will occur; they simply mean that there is a higher probability of finding the disease associated with certain genetic backgrounds.

Tolerance is a learned process. T- and B-cell clones that rearrange their DNA to form binding sites that recognize self structures should normally be either destroyed or inactivated. In order for this to happen, the clone must encounter the structure. If a clone never encounters its antigen then it will be neither inactivated nor destroyed. One factor that clearly seems to be associated with the development of some types of autoimmunity is tissue injury. Tissue injury is usually associated with natural immunity-induced inflammation. It is also liable to expose self structures that normally would not be encountered by the cells of the immune system. Exposure of a self structure to a clone that recognizes it in the context of an inflammatory response may activate it. Such activation may cause additional damage and inflammation exposing additional self structures that usually may not be encountered by cells of the immune system. There is some evidence that this type of mechanism may be involved in the development of some types of arthritis.

Another mechanism that has been associated with development of autoimmunity is similarities between structures of microbes and self structures. Perhaps the most well known example of this type of "molecular mimicry" is associated with rheumatic fever. Structures on the bacteria that cause rheumatic fever have molecular domains very similar to structures on myocardial tissue. The immune response to the infection can lead to activation of lymphocyte clones that react to and damage heart tissue. As with tissue damage, the immune response can feed on itself. The immune response causes damage that exposes structures that normally would not be encountered by immune cells. Since the encounter of these normally hidden structures occurs in the context of an inflammatory response, the probability that a self recognizing clone will be activated is increased.

Increasingly, autoimmunity is being found associated with disease processes. These diseases can be tissue specific or quite general. They can involve either both T- and B-cell responses, or just one of the two. Their expression can take many forms. For example, systemic lupus erythematosus is associated with the presence of many different self-recognizing antibodies and causes widespread damage to the body. In contrast, myasthenia gravis is an autoimmune disease in which self-recognizing antibodies against only a single protein have been found. That protein is the muscle acetylcholine receptor that receives the signal from the motor nerve that directs a muscle to contract. In this case, the presence of the antibodies causes weakness and fatigue because of interference with muscles receiving instructions from their motor nerve. In another case, Celiac disease, the antibodies are to a dietary protein called gluten. Ingestion of this protein causes autoimmune mediated damage to the lining of the small intestine. At present the only solution to autoimmune diseases is to attempt to treat the symptoms of the particular disease and to suppress the immune system with drugs.

The immune system and cancer: The immune system also plays an important role in protecting us against an "invasion" of a somewhat different sort, cancer. Cancer is a group of diseases in which some type of normal cell becomes abnormal. There are many different types of cancer. Cancer can develop in almost any tissue of the body. Cancer cells can also migrate and set up colonies in other tissues. These colonies are called metastases. For example many tumors found in the brain have actually migrated there from tumors that arose in other tissues such as the lungs. Even when living in the brain, a lung cancer remains a lung tumor. No matter what type, cancer cells are cells of the individual that malfunction in a very special way. Cancer cells are unresponsive to the normal signals that limit cell division. Cancer cells reproduce uncontrollably.

Although the "cause" of cancers is unknown, there are some factors that are closely linked to the development of cancer. The first are certain viruses. Although a direct link has only been found in one case in humans, in animals many examples of cancer-causing viruses have been identified. The second is tissue damage. Chronic tissue damage is closely related to the development of many cancers. Two examples are chronic sunburns linked to skin cancer and chronic damage due to smoking linked to lung cancer. Tissue damage as a cause of cancer has several good rationales. As discussed previously, tissue damage often removes the normal constraints on cell reproduction in order to allow for repair. If such damage is chronic then the local population of cells is continuously in a situation of reduced restraint of cell division. This could very likely select for cells that stop expending effort in making proteins associated with the function of the particular cell type and devote all of their effort to cell division. From a point of view of evolution, tissue damage imparts a selective advantage to cells that direct their effort to reproduction. The second rationale for damage-causing cancer is that it causes inflammation, bringing increased attention of the immune system. It is very clear that a tumor expresses unusual structures on its cell surface. These include both fetal antigens and mutant or otherwise unusual antigens relatively unique to the tumor. Although inflammation should promote immune responses to aberrant cells, chronic inflammation should increase the probability that a single aberrant cell capable of eluding immune surveillance will arise. In all probability tumors arise from a single cell.

Drugs to manipulate the immune system: We have already discussed the use of relatively high dose immune-suppressive drugs in transplantation and autoimmunity. However, synthetic glucocorticoids are also available in much lower dose preparations. Corticosteroid containing creams are available over-the-counter. Their indicated use is for allergic skin rashes. The potential problem with their use is that it is not always clear whether or not a rash is due to an allergic reaction or if it is due to an infection. These preparations suppress the immune system. If the rash is due to an infection then their use inhibits the ability of the immune system to deal with the infection.

A second common drug which is used to inhibit aspects in immune function is antihistamines. Histamines are chemical messages released by immune cells to increase capillary permeability. Histamines greatly reduce the capillary barriers in a localized area which allows cells of the immune system to leave circulation (inflammation). Inflammation reflects an immune response to a foreign structure. It also allows a significant amount of fluid to leave circulation in a local area, causing swelling or in the case of the lungs, congestion. Antihistamines are drugs that block the receptor for histamines. They are used very commonly for lung congestion. A problem with the use of this type of drug is that histamines are used other places in the body as chemical messages. For example, histamines are used as neurotransmitters in some parts of the brain. For this reason, some antihistamines cause drowsiness. However, as the study of histamine receptors has progressed, it is becoming more and more possible to design antihistamines that selectively block one type of histamine receptor and not others.

Perhaps the most exciting area of immune-related drugs derives from a combination of the areas of immunology and molecular biology. The immune system uses a large number of chemical messages called cytokines. Molecular biology has provided the tools to produce many of these cytokines commercially in large quantities. Techniques in molecular biology allow one to insert the gene coding for the cytokine protein into bacteria. Then the bacteria are "turned on" to express the protein. Such cloned cytokines have the potential to be used as drugs to treat a variety of diseases including both cancers and autoimmunity.

The immune system holds the solution for many problems: The area of immunology is progressing at an incredibly rapid pace. Advances in immunology already have and will continue in the future to contribute advances in two other medically related fields. The first is diagnostics. Antibodies are very specific, high affinity reagents. As we have seen, many different diseases are associated with the expression of unusual molecules. Antibodies are becoming increasingly useful in diagnosing diseases and other conditions. For example, home pregnancy tests are a very common diagnostic use of antibodies. As discussed previously, very early in pregnancy, HCG appears in a woman's blood and urine. Home pregnancy tests are based on the use of antibodies to indicate the presence of HCG in a woman's urine. Similarly the presence of several types of cancer can be diagnosed using antibodies. The second area is drug targeting. To the degree that the entire dose of a drug goes to the intended target (the place in the body where the drug needs to act) and to nowhere else, side effects are reduced. If for example one could produce an antibody to an antigen that is only expressed by a tumor cell, one could attach to it one of the powerful poisons used in chemotherapy. The antibody would then selectively deliver the poison to the tumor, sparing the individual from destructive side effects.

chapter eleven

infections and cancer

The immune system is not always dependable: Multicellular organisms such as humans are complex societies of cells. They exist only because a high level of organization provides control of all individuals in the society. Survival of the organism is the predominant priority of the body. The actions and existence of each individual cell in the society must subserve this predominant priority. Mechanisms exist for distributing, on a priority basis, the necessities of life to individual cells. All cells are not treated equally, and often the needs of some cells are sacrificed for the benefit of the society. A body that is functioning appropriately is one in which each cell is a slave to the organized priorities of the system. Individuality is not tolerated.

The function of the immune system is to cleanse the body of individuals that express independent priorities. The immune system is designed to destroy individuals that operate independently of the organized priorities of the system. Infections represent such individuals in the society. The cells of an infection do not belong to the society and do not pay attention to the communications and directions which express the priorities of the organism. Similarly, cancer cells function as individuals. Although the cancer cell is derived from a cell which was once a normal part of the society of cells, it has stopped "paying attention." Much like a psychotic individual, it has its own unique reality and set of priorities.

When such individuals, regardless of whether they are microorganisms or cancer, escape destruction by the immune system the result is disastrous to the organism. It is disastrous because the most common characteristic of such individuals is an unrestrained rate of reproduction. As we have discussed, the rate of reproduction of normal cells in the society is controlled to meet the needs of replacement and growth. Infections and cancer, which start as a few individuals in a localized area, reproduce at uncontrolled rates. They can spread throughout the body using resources, producing waste, and disrupting the functions of many different types of cells.

Why does the immune system fail? The immune system sometimes fails to protect an individual against one or more aberrant individuals, whether they are "homegrown" cancers or microbe invaders. This may occur for a variety of reasons. Some of the reasons are peculiar to a particular type of invader and some are due to the general failure of the immune system. There are a variety of genetic defects that cause deficiencies in the immune system. These can cause generalized deficiencies, or they can cause a quite specific deficiency such as the lack of a particular chemical message or type of antibody. The consequences of these genetic defects can range from susceptibility to certain types of infections or cancers to total dysfunction.

The immune system functions through proteins. If a person inherits a defect for any of these proteins specific to the immune system, the result is some type of immunodeficiency. Genetic reasons for failure also can be quite subtle. Evolution is a process of selection. We have evolved with our diseases and our diseases have evolved with us. Although extensive, the ability of the immune system to generate unique molecules to fend against unique invaders is limited. We are all different; we are different in the complex collection of our genetic makeup. Until very recent times, humans were distributed among isolated groups of people. Within these isolated groups were particular diseases. Diseases, like many aspects of existence, are a selective pressure. Those within an isolated group of humans that had the genetic makeup to resist a particular infection survived to reproduce while those that didn't perished. Since the various groups of humans did not mix, their diseases did not mix. Particular diseases were a selective pressure on only a limited population of humans. As these small groups of humans began to mix, each disease worked its selective pressure on the larger group. The result was death of many individuals and resistance in the part of the population that survived. Cogent illustrations of the power of disease as a selective pressure on a population are provided by 15th and 16th century European explorers visiting isolated island populations of humans. They introduced the array of microbes from a population of many millions into a small genetically isolated population of humans. The result was disease in a "naive" population and death of large percentages of people. The introduction of measles as a disease into the people living in Hawaii is a good example of an old disease in a new venue. From a point of view of resistance to disease, genetic mixing favors survival of the human race. Mixing expands our protection and dilutes our vulnerabilities. For millennia humans, their animals, their plants, and their diseases have been mixing and selecting.

Deficiencies in the immune system can also be acquired. There are a number of viruses like HIV that attack cells of the immune system and cause deficiencies. Similar to genetic reasons, the immune system can be deficient in its protection. The deficiency can be limited or quite generalized. But diseases that attack the immune system are not the only diseases that cause deficiencies in the immune system. Almost any infection or cancer suppresses the immune system to some extent. The suppression can be part of a general inability to provide resources to the immune system. For example, any disease process that depletes nutrient availability suppresses the immune system. Starvation, diabetes, anorexia, and nutrient depleting cancers suppress the immune system. Suppression of the immune system can also be quite specific. Any immune response causes the immune system to release chemical messages that act on the brain to generally suppress the immune system. Cells involved in the immune response release chemical messages that act on the brain and hypothalamus to increase the output of ACTH which in turn increases the output of glucocorticoids. The increased output of glucocorticoids suppresses the immune system. This feedback loop to the brain is designed to keep the destructive power of the immune system in check. It also makes the individual more vulnerable to a new infection.

The last reason for the immune system becoming deficient is drugs. As we have seen there are a variety of circumstances such as autoimmunity and transplantation in which drugs are used to suppress the immune system. A common consequence of drugs used to treat infections and cancer is immuno-suppression. Immune cells must divide and reproduce quickly in order to respond to the rapidity of the reproduction of an invader. An immune response is a cascade of reproducing immune cells. Reproduction magnifies the response just as it magnifies the infection. It is very difficult to interfere with the reproduction of microbes and cancer without similarly impeding reproduction in immune cells. Selectivity is difficult because all forms of life work on the same principles and codes. Drugs to treat invasions of the body by outlaw cells that don't follow directions must exploit the subtle differences between the cells of the society and the invaders.

All of life has the same requirements. In a minimalist sense, life is reproduction. Copy yourself and you are "life." The range of agents capable of infecting humans is quite remarkable. We are in fact still discovering previously unrecognized modes of infection. Perhaps the newest, most curious forms of infection to be identified are infective proteins called prions. Prions are proteins that are resistant to digestive enzymes. When eaten they can survive the digestive system and enter intact into the body. Although the nature of these infections is not clearly understood, a prion infection has been linked to the brain degenerative diseases referred to as "mad cow disease" and human Creutzfeldt-Jakob disease in man. The transmission of prion disease is illustrated by another disease, Kuru, that became epidemic in parts of New Guinea in a tribe that had cannibalistic funeral rites. People ate the bodies of the dead. They also developed a degenerative brain disease. It is now believed that the more recent spread of prion disease may be related to the practice of mixing in the feed of farm animals such as cows the remains of other animals.

Much better understood are viruses, bacteria, parasites, fungi and cancers. When such infections escape the immune system, the solution is to use a class of drugs known as antibiotics. Antibiotics are "anti-life" drugs. These drugs are designed to exploit the subtle differences between humans and the infective organism. There are two areas of differences that these drugs exploit. The first area focuses on small molecules. There are some differences in small molecule building blocks used by human cells and by microbes. The differences can be absolute in that the microbe requires the small molecule and the human cell does not. However, more commonly the difference is relative in that the microbe is more dependant on the molecule than is the human cell. In this case, the drug is simply more toxic to the microbe than to the human cell. The second target of antibiotics is the large molecules made from small molecules. These are the molecules involved in making proteins, making DNA, making RNA, and making structure. To the degree that these large molecules are unique to the infection they can be exploited as targets for drugs. As with the small molecule targets, often the differences are not absolute, resulting in relative toxicity.

There are many types of infections: Infections in humans can be caused by prions, viruses, bacteria, fungi and a variety of parasites. Each of these classes contains a large number and variety of different organisms. Most of the organisms in a particular class are not involved in human infections. For example, a mushroom is a fungus. Among those in each class that do cause human infections, the range in characteristics of the different types is very large. There are many different types of infective viruses, many different types of infective bacteria, many different types of infective fungi and many different human parasites. Each different type in a class will cause a different disease. For example, one type of bacterium will cause pneumonia, another will cause a type of gum disease, and another will cause a sore throat. Similarly, measles, flu, AIDS, and polio are all caused by different viruses. Because of this range of differences within each class, no drug is effective against all infections of a type.

A good offense is the best defense: There are many ways that humans can be infected. Although exceptions clearly exist, viruses tend to be rather species specific. These exceptions such as bird flu derive from close contact between humans and some type of animal. However, for the most part, viruses tend to move from human to human or from human waste to human. In contrast, bacteria and fungi have much less species specificity and therefore move both across species and into humans from environments contaminated by other species. Parasitic infections are quite different. Parasitic infections are caused by a range of organisms that live part of their life cycle in humans and part of it in some other organism. The infection of humans with parasites is quite unique to the particular parasite.

The simplest way of dealing with an infection is to prevent it. If the organism cannot invade the body, dealing with an infection becomes a moot concern. Although this may seem like a facile statement, in truth it has been little more than 100 years since the medical profession has accepted the concept that infections are caused by invading organisms. The general use of surfactants (soap), antiseptics, and disinfectants in order to prevent diseases dates only to the later part of the last century. Prior to that time, the medical community often served as the means of transmitting disease from one individual to another. However, as is often the case, what works is often in use before it is scientifically recognized. The use of soaps in order to clean dates to antiquity. Soaps prevent infections in two ways. First, soaps dissolve oils such as those of the skin, therefore removing potentially infective organisms contained in these oils. Second, soaps have a structure that is very similar to the phospholipids that forms membranes (Chapter 2). Because of this similarity in structure, they can destroy the surface membrane barriers of microbes. Similarly, the use of preservatives dates to ancient history. Dead plants and animals rot. Such rotting, which is called putrefaction, is caused by microorganisms. Long before the recognition of the existence of microorganisms, it was understood that certain agents such as spices, wines, vinegars and oils could prevent putrefaction. In modern times, such agents are in wide use for the prevention of disease. Their applications range from the purification of water to the cleansing of wounds to the sterilization of surgical instruments.

Antiseptics are agents applied to living tissue in order to prevent the growth of microorganisms. Many types of chemicals have been used as antiseptics. These include reactive atoms such as iodine or chlorine; alcohols, such as ethanol; heavy metals, such as silver nitrate; acids, such as acetic acid and boric acid; oxidizing agents, such as hydrogen peroxide and benzoyl peroxide; and phenol derivatives, such as hexachlorophene. In general, these agents have their effect because of their ability to either react with biological material or to destroy biological structures such as membranes. Because of the general nature of their mechanism of action, their value is based on their ability to kill microorganisms relative to the tissue damage they cause. Control of this ratio is usually a function of the concentration of the solution used. A second consideration is the question of penetration: how deep does the agent penetrate the layers of the skin? Penetration is usually a function of the relative solubility of the agent in both oil and water. Finally, because of the toxicity of many of these agents, it is important to know the degree to which the agent is absorbed into general circulation. For example, the phenol derivative hexachlorophene in a soap solution gained popular use on newborn infants. It was later discovered that the agent was absorbed into circulation and caused damage to the nervous system.

Disinfectants are agents or procedures that are applied to nonliving material in order to kill microorganisms. Because disinfectants are applied to objects rather than to people, tissue damage is not a concern. Often, disinfectant solutions are simply much higher concentrations of the same agents used as antiseptics. The general principle of action is the same: reactivity with biological material. In addition, two physical methods are often used to disinfect objects. These methods involve heat and light. High temperature in a moist environment can kill all living organisms. This procedure is called "sterilization." The heat must be applied in the presence of water because some microorganisms can dehydrate, become dormant and withstand even very high temperatures. However, when moist heat is applied the organism is prevented from dehydrating. Similarly, the use of radiation energy in the form of light, especially short wave ultraviolet light, is quite destructive to living organisms.

Viral infections: As discussed previously, viruses are a bit of genetic information (either DNA or RNA) that has "learned" to make its way in the world by itself. Their structure varies somewhat but in general they consist of a small number of relatively short strands of DNA or RNA which are the genetic material of the virus. Packaged with the genetic information are a limited number of proteins. These proteins are involved with targeting the cell to be infected and initiating the infection. The genetic information and the few proteins are generally packaged in a type of membrane structure called the viral coat. Often, the membrane package is taken from the host cell membrane. What makes the virus unique as a class is that they are not cells. They do not carry out life processes.

The list of diseases caused by viruses is quite extensive. DNA viruses cause diseases ranging from smallpox and herpes to warts. RNA viruses cause diseases ranging from flu, measles, and rabies to AIDS, hepatitis and at least some types of cancer. Viruses, as they work their wiles with a cell, cause the cell to become dysfunctional. When a motor nerve cell is infected with a polio virus, it doesn't work. Therefore, the muscle that is controlled by that nerve doesn't work. As a consequence, the bones that should be moved by that muscle don't move. This virus, which attacks nerve cells, causes paralysis. When a lymphocyte is infected by HIV, it doesn't work. The protection that it should provide against infection is not there. As a result, the human develops other infections. Viral infections can spread from cell to cell by keeping the cell alive and releasing infective viruses into the host organism, or the virus can kill the cell and release a large number of infective viruses into the system all at one time.

The difficulty with treating viral infections is that outside the host cell they are doing no life process that can be attacked. They can only be attacked by drugs when they are active in the host cell. To date, the most effective way of dealing with viral infections is a good offense. That offense is immunization (Chapter 10). The use of immunization to prevent viral diseases such as polio and measles has been extremely effective. Immunization produces immune cells that recognize viral structures. When a previously immunized individual is exposed to the virus, the immune system destroys the virus and prevents the infection. However, some types of viruses change their surface quite rapidly and as a result effective vaccines for some diseases produced by viruses have remained elusive. For example, a "flu shot" is only good for one season. Flu viruses mutate rapidly, making the vaccine ineffective. This is an even bigger problem with the HIV virus, and to date no effective vaccine for AIDS prevention has been developed.

Development of drugs to treat (rather than prevent) viral diseases has had only limited effectiveness. Viruses must copy their genetic material in order to reproduce. They must synthesize either DNA from deoxyribonucleotides or RNA from ribonucleotides in order to copy their genetic information and be an infection. Often the virus uses its own enzymes in this process. Their enzymes are different than the host enzymes in some but not all ways. There are a variety of anti-viral drugs that are "phony" nucleotides. When they are mistakenly picked up by the viral enzymes that make viral DNA or RNA, they interfere with its action. Acyclovir for herpes and AZT and dDI for HIV are such drugs. The selectivity of these drugs is only partial. They bind more tightly to the viral enzymes than to host cell enzymes that use the same substrate. The consequence of some binding to the host's enzymes is toxicity. A common consequence is bone marrow suppression resulting in anemia and immunosuppression. Liver and kidney damage are also common. In addition many of these drugs are carcinogenic: they can cause cancer. Another problem with these drugs is the development of resistance. Often such drugs are used for extended periods of time, and because of toxicity to host cells these drugs cannot be used in very high doses. The consequence is that the drug acts as a selective pressure on the virus population. Small mutations which cause changes in the structure of the viral enzyme can reduce the ability of the drug to bind to and inhibit the enzyme in most of the viruses but not all. Those viruses that are least affected by the drug then begin to dominate the infection. Viral infections can evolve as they progress, making initially effective drugs ineffective in the long run.

Other than these "phony nucleotide" drugs, there are a few experimental drugs designed to interfere with specific proteins associated with specific viruses. These are focused treatments for particular viruses. Finally, since genetic technology has made most of the chemical messages of the immune system available in quantities that make them practical as drugs, a variety of approaches are being developed to use those chemical messages to stimulate the immune system to destroy viral infections.

Bacterial infections: Unlike viruses, bacteria are complete life forms. Bacteria are cells. They have the ability to generate energy, make their own molecules and reproduce themselves. Reproduction by bacteria is asexual:

it occurs by simple cell division. The primary genetic material of bacteria is a circular, double stranded DNA chromosome. In addition to this chromosome, bacteria contain small circular pieces of DNA called plasmids. Plasmids are not necessary to the life of the bacteria. Rather, plasmids contain genes that code proteins that can be expressed under particular situations. Depending on circumstances, the expression of plasmid genes can be either advantageous or disadvantageous to the bacteria. In many respects, plasmids are like viruses in that they are independent pieces of information that are dependant on their host for reproduction. However, unlike viruses, they do not move independently from one cell to another. Plasmids are transmitted between bacteria both during reproduction and as a method of exchanging genetic information without reproduction.

Like viruses, bacteria come in many forms and cause a large variety of diseases. Most bacteria invade and live in the extracellular water outside cells of the body. Their effect on the body is closely related to how they enter the body and how they spread through the body. In this case the damage is caused by the bacteria taking up space, using nutrients, and producing waste. They demand resources in a local area. In addition, some bacteria produce toxins that poison the host. These toxins can do a variety of things such as cause fever, paralyze muscles or destroy the vasculature. Finally, the immune response to bacterial infections can itself cause substantial damage to the host. For example, much of the damage caused by bacterial meningitis is caused by the immune response. There are also bacteria like tuberculosis and leprosy that infect and live inside cells. These intracellular bacteria cause rather intransigent chronic infections. They are difficult to treat because, like a virus, they "hide" inside the cell. Drugs to treat these infections must meet the challenge of entering a host cell to attack the bacteria.

There are two basic ways to treat a bacterial infection. One way is to prevent bacteria from reproducing. This approach buys time for the immune system. It slows down bacterial cell division and thus slows the spread of infection. The second way is to kill the bacteria. In the last 50 years, many drugs have been developed which are effective in treating bacterial infections. Because there are many types of bacteria and many locations in the body where bacterial infections can occur, many different antibacterial drugs are necessary. Each one of these drugs is effective against certain types of bacteria. In addition, each drug has its own distribution in the body. A drug must be able to enter the body in the location of the infection in order to be effective in treating that infection. In general, the effectiveness of a drug in treating a particular infection is based on the effect of the drug on the infection relative to its effect on the cells of the host. As implied by this "relative damage" principle, such drugs are often associated with side effects.

There are some distinct differences between bacterial cells and human cells. These differences are in both the small and large molecules involved in life processes. Unlike a human cell, bacterial cells have no internal membrane-enclosed compartments such as nuclei or mitochondria. Like human cells, bacteria do have an outside membrane. However, unlike human cells (but like plant cells) they also have a cell wall outside this membrane. Some bacteria even have a second membrane outside the cell wall.

Bacterial antibiotics exploit these differences between bacterial and human cells. The first antibiotic was a type of small molecule called a sulfonamide. This drug demonstrated that it was possible to therapeutically exploit the differences between microbes and humans. Interestingly, the action of the drug was initially misunderstood. Although this type of drug no longer enjoys the premier place it once held as an antibiotic, it is still an effective class of antibacterial drug. The history of sulfonamides illustrates the types of problems associated with antibiotic therapy. Sulfonamides were developed in Germany in the early 1930's. The actual compound that was patented was a blue dye containing a sulfonamide group. The drug, called Prontosil, was quite effective in treating certain types of bacterial infections. However, it was later learned that the "active" drug was not Prontosil. The "active drug" was a sulfonamide metabolite of Prontosil. This illustrates a very important aspect of drug therapy. Not only does the drug act on the body, but also the body acts on the drug. Such actions of the body on a drug can produce less active, more active or even toxic molecules. The liver is generally the place in the body where drugs are converted to other structures. Problems arise because of the actions of the liver on drugs for several reasons. First, there is significant genetic variation in the liver enzymes involved in the process. The livers of different people are different, and as a result people respond differently to the same dose of the same drug. Secondly, many disease processes can modify the action of the liver on drugs. As a result there are variations in response between different people depending on their health. Finally, the consumption of a variety of foods and other drugs can affect the way that the liver acts on a drug. The result is that the individual responds differently to the drug at different times. All of these factors influence predictability. Predictability, knowing how the drug will affect the body, is the single most important factor associated with the use of drugs.

Sulfonamides are bacteriostatic. This means that they do not kill the bacteria, they simply slow down or prevent reproduction. The result is that this drug is ineffective or far less effective if the individual's immune system is impaired in some way. This is not true of all antibiotics. Sulfonamides are only bacteriostatic because they block the synthesis of a small molecule that is important to DNA and RNA synthesis in all cells. The small molecule is vitamin B12 or folic acid. Folic acid is necessary for making nucleotides which are the building blocks of both DNA and RNA. The difference between human cells and bacterial cells is that human cells obtain this vitamin from their diet while the bacteria must make it. If a cell cannot synthesize DNA, its genetic material, then it cannot reproduce. Sulfonamides "look like" another small molecule necessary for the synthesis of folic acid. This molecule is para-aminobenzoic acid (pABA). Like the anti-viral drugs discussed earlier, sulfonamides compete with pABA for an enzyme important to folic acid synthesis. As in the case with any new drug, the use of sulfonamides taught a very important lesson. Specifically, it showed that antibiotics serve as a pressure for genetic selection on the population of bacteria that they affect. There exist plasmids that contain a gene for the enzyme that sulfonamides bind to. The plasmid enzyme binds pABA with high affinity, but only binds sulfonamides with low affinity. A bacterium having this plasmid is therefore resistant to sulfonamides. This plasmid can be transmitted from one bacterial cell to another during the course of an infection. The result is that in a relatively short period of time, the individual has a sulfonamide resistant infection.

World War II provides an illustration of the importance of understanding how antibiotic use selects for antibiotic resistant bacteria. Sulfonamides were the major antibiotic available for most of the war. Many soldiers were wounded and developed infections. A potentially logical approach to the problem of infection was tried for a period of time. Healthy soldiers were treated with sulfonamides to protect them from infection "just in case they were wounded." The result was that when they were wounded, they rapidly developed infections which were resistant to sulfonamides. Although society was provided with this lesson more than 50 years ago, it still has not been learned. Today we use the antibiotics that we have excessively and indiscriminately, and the result is that many of them are becoming ineffective. Perhaps the most blatant example of such use of antibiotics is in animal feed. In the case of antibiotics and infectious microorganisms, the old saying that "familiarity breeds contempt" is true. This is well illustrated by infections contracted in the hospital environment ("nosocomal infections"). The most difficult infections to treat are those which one develops in a hospital. Antibiotics are used extensively in hospitals. As a result, bacterial infections obtained in a hospital are often resistant to a variety of antibiotics and are difficult to treat. The microorganisms that exist in hospitals have evolved in an environment of antibiotics. For this reason, hospitals are dangerous places.

Generally, sulfonamides are taken orally and are absorbed primarily in the small intestine. After passage through the liver, they enter the general body circulation. Much of the sulfonamide in circulation travels bound to serum proteins. Binding to components of the blood is a very important and quite common characteristic of drugs in general. It is important because it is only "free" drug, that which is not bound to something, which can leave circulation to enter the water around cells or be removed by the kidney. In general, there develops a balance between "free drug" and "bound drug." As free drug leaves circulation, it will be replaced by bound drug in order to maintain the balance between free and bound drugs. This will continue until all bound drug is gone from circulation. As a result, the propensity of a drug to bind to components of the blood will influence both how concentrated a drug can become in the tissues and how long the drug remains in the body.

Sulfonamides distribute quite uniformly throughout the body. This is not true of all drugs. It is not uncommon for a drug to be excluded from certain tissues. The barriers between circulation and the water environments around different groups of cells can be quite selective with respect to the drugs that they allow to enter. However, the exception to the uniform distribution of sulfonamides is the kidney. Sulfonamides are concentrated by the kidney. This too is a common occurrence with drugs in general. The kidney often concentrates drugs as part of the process of removing them from the body. In the case of some sulfonamides, this created problems. When they are concentrated, sulfonamides form insoluble crystalline precipitates in the kidney. Because the kidney has a tendency to concentrate drugs, potential kidney damage by one mechanism or another is always a concern with drugs.

A variety of other side effects are also associated with the use of sulfonamides. These include potential liver toxicity and bone marrow suppression. In addition, hypersensitivity reactions can occur with sulfonamides. Like an allergy, they occur because the body has been exposed to the drug, becomes sensitized, and a reaction occurs when the drug is taken again. This type of hypersensitivity is also not uncommon with drugs in general. In many respects sulfonamides, with their assets and liabilities, are quite typical of drugs in general and antibiotics in particular. We now have more than 60 different antibiotics available. However, most of them were not produced by the effort and intelligence of humans. Most antibiotics are compounds "developed" by the microbes themselves.

Biological warfare amongst the microorganisms: Humans compete with each other and with other animals for territory and for resources. For this purpose we have developed weapons and defenses. Such is also the case with microbes. They compete for the available territory and resources, and have developed weapons and defenses for that purpose. Although it has been known for thousands of years that certain concoctions made of fungi, plants and other forms of life have curative effects, it has only been since the middle of this century that humans have scientifically begun to exploit microorganisms as a source of antibiotics. The first and perhaps the best known antibiotic obtained in this way was penicillin. Penicillin is a small organic molecule produced by a fungus. It was discovered in the late 1930's, and by the early 1940's it was in use as a human drug. The story of penicillin is an excellent illustration of the impact that just one biomedical discovery can have on the general health of society. Bacterial pneumonia was a major contributor to infant mortality. The advent of this single drug changed that situation.

Penicillin works by interfering with the synthesis of a large molecule that is necessary for the assembly of most bacterial cell walls. As discussed previously, bacteria have an outside barrier that consists of both a membrane and a cell wall. Without the wall, the bacterium is unprotected and dies. Like all antibiotics, penicillin is not effective against all bacteria. Some are insensitive because their external structure inhibits the penetration of penicillin. Other bacteria are insensitive because they have an enzyme, called penicillinase, which inactivates penicillin. The genes for these penicillinases are often found in plasmids. Because these penicillinase genes can be transmitted between bacteria, the extensive use of penicillin has lead to extensive resistance to penicillin and derivative drugs.

Penicillin has its own distribution characteristics in the body. The distribution of penicillin in the body is not as general as sulfonamides. For example, penicillin is excluded from the brain under normal circumstances. This exclusion is quite fortuitous because if penicillin is allowed to enter the environment of the brain it will cause seizures. Like all drugs, the use of penicillin has its risks as well as its benefits. However, in general, penicillin has fewer risks than most drugs. Two general types of problems can arise with the use of penicillin. The first is allergic-type hypersensitive reactions. Probably because of its relatively complex structure and its binding characteristics to blood components, penicillin has a tendency to cause allergic reactions. Such allergic reactions can be quite severe and lead to anaphylaxis and death. The second type of reaction involves destruction of "good" bacteria. For example, human digestion involves bacteria that live in the intestine. Penicillin has a tendency to kill such bacteria and cause significant disturbance of human digestive processes.

After more than 50 years of work, penicillin is no longer one drug. The basic core of the molecule has been synthetically added onto in order to produce a group of penicillin drugs. Some are "designed" for different methods of administration such as oral or injection into muscle. Some are modified for different rates of absorption or distribution in the body. Some are modified to change how long they stay in the body. Some are even modified so they are able to "evade" penicillinase. Even after more than 50 years, the group of drugs which are derived from the penicillin core remain some of the most effective antibiotics available.

Other antibiotics: For more than 50 years, there has been a concerted effort to discover other natural antibiotics. This effort has been quite effective. At present more than 60 agents and their derivatives are available. Each such drug has its own specificity as to what microorganisms it is effective against. Each drug has its own particular absorption and distribution characteristics. Some stay in the body longer than others. All have some degree of risk. Certain risks are quite common and are observed with many antibiotics. Liver problems can occur because the liver is responsible for detoxification. Similarly, kidney problems can also occur because all antibiotics are removed from the body by the kidney. Effects on the rapidly reproducing cells of the body such as those of the blood and skin are also quite common. These groups of cells are most often affected because, like the cells of the infection, they are quite active and are reproducing rapidly. Such cells become "targets" due to commonality of function. Destruction of "good" bacteria is common for the same reason.

Many antibiotics also have their own unique toxicities. For example, the group of drugs which includes streptomycin can cause hearing damage. Use of these drugs is destructive to the hairlike projections in the inner ear that are responsible for generating the nerve signals to the brain. In effect, these drugs produce damage to the hearing process which is quite similar to that caused by excessive noise. Regardless of the risks, antibiotics have in the past and will continue in the future to save millions of lives due to their impact on bacterial infections. The major problem impacting the effectiveness of antibacterial drugs is the rapidity with which bacteria are developing resistance to these drugs. The problem is exacerbated by their extensive use in immune compromised individuals. In immune compromised individuals the antibiotic can become the only protection that individual has. The consequence is extensive chronic use of antibiotics which accelerates the evolution of resistant microbes.

Fungal infections: The human point of view is kill the invaders whomever they may be. The microbial point of view is competition for the spoils of war. They fight over us. This competition is nowhere closer than between bacteria and fungi. After all, penicillin was made by a fungus to allow it to "kill" the competition. Humans have some mutually beneficial relationships with other forms of life. In the process of fighting over us they often protect us. Fungi that are potentially infective to humans are generally held in check by bacteria that live "on us" rather than "in us." Bacteria live on our barriers to the outside world and protect us from fungi. Like the beneficial bacteria that live in the large intestine, bacteria live on our barriers: on the membranes of the respiratory system, the digestive system, the mouth, and the vagina. These bacteria do not normally invade us: rather they protect our borders. They generally protect us from fungi. Bacteria and fungi fight for the bottom of the organic food chain. Fungi live on organic molecules that were once a part of a living organism. Fungi digest the world and suck up the products of digestion. Fungi release certain proteins similar to the enzymes we have in our stomach that break up big molecules into small building blocks. They break down the world around them into simple sugars, simple amino acids, simple fats and a vast array of simple molecules. They then suck up these simple molecules to use as "food." Along comes a bacteria that makes its way in the world by using simple molecules. There are bacteria that live on our barriers and generally protect us because they "eat" the simple molecules produced by fungi faster than the fungi can suck them up and use them. Bacteria protect us by living on the fruits of the labor of fungi.

Even with bacterial protection, humans are susceptible to a variety of fungal infections. Fungal infections are in fact increasing in frequency as serious diseases in humans. There are a variety of reasons for this increase. One reason for this increase is the use of antibiotics. Antibiotics can kill the bacteria that protect our barriers. A second reason for the increase is immunodeficiency. Individuals with suppressed immune systems due to AIDS, transplantation, and chemotherapy are far more susceptible to fungal infections than are normal people. Fungal infections fall into two basic groups: those that attack the surface barriers of the body and those that invade inside us. Common barrier fungal infections range from ringworm and athlete's foot to vaginal yeast infection. Increasingly common in immune compromised individuals are internal fungal infections of organs such as the brain, heart, and lungs.

There are a variety of antifungal drugs available. Many of these drugs interfere with the synthesis of a component of the fungal membrane. As discussed earlier, membranes are constructed of fat-like molecules. In humans a major component of membranes is cholesterol. However, in fungi the place of cholesterol is taken by a different steroid, ergosterol. Most antifungal drugs interfere with enzymes involved in the synthesis of ergosterol. Generally, side effects of these drugs are associated with kidney and, to a lesser extent, liver toxicity.

Parasitic infections: There are a variety of organisms that live part of their life cycle as parasites in humans. These organisms cause unique and varied diseases. Perhaps the most common parasitic infection in the world is malaria. Malaria is caused by an organism known as a sporozoan. These organisms have characteristics that are somewhere between a protozoan and a fungus. They live part of their life cycle in a type of female mosquito and part of their life cycle in humans. In humans they live as intracellular parasites in red blood cells. Malaria is an excellent example of a disease acting as a selective pressure. Sickle cell anemia exists because of malaria. Sickle cell anemia is caused by a single mutation in the gene for hemoglobin. Red blood cells containing this defective hemoglobin will collapse and assume a sickle shape in veins where the oxygen level is low. As discussed previously, every individual inherits two copies of each gene. Individuals born with two defective copies of the hemoglobin gene are quite ill and usually they do not live long. Individuals born with only one copy of the defective gene only develop problems in stress situations. The reason that the defective gene is so common is that individuals with one copy of the gene have a selective protective advantage with respect to malaria. Having one copy of the defective hemoglobin gene protects individuals against the parasite that lives inside red blood cells. In the areas of Africa where malaria is very common these people were more likely than people with two good copies of the hemoglobin gene to survive malaria and reproduce.

Besides malaria, there are other protozoan-like microbes that are human parasites. Of increasing interest is the organism that causes pneumocystis pneumonia. Before AIDS, this type of pneumonia, which is caused by a protozoan/fungi-like organism, was virtually unknown. Now this type of pneumonia is often the first indication that an individual is HIV infected.

There are also a variety of worms that invade and infect humans. Some of these, like tapeworms and intestinal roundworms invade and live in the intestinal tract. Others live and reproduce in the blood, the lymphatics, connective tissue, and a variety of other tissues. In general, the treatment of parasitic infections is quite complex and tailored to the particular infection.

Cancer, an "infection" from within: Cancer is a condition in which one is "infected" by cells of his or her own body. In effect, cancer is an infection in every sense of the word. The cancer cells reproduce at a relatively rapid rate and compete with the normal body cells for space and resources. What causes cancer? What causes cells of one's own body to become "renegades" and ignore the laws of society? Do chemicals in food cause cancer? Does smoking cause cancer? Does chemical contamination of the environment cause cancer? Does radiation cause cancer? Do viruses cause cancer? Is cancer inherited? All of the above are simple objective questions. However, the answers provided by science are rarely if ever absolute truths. The answers which come from science are always partial truths. That is, they are truths within limits. Unfortunately, the limits are usually rather complex, and much of the effort of science is involved with figuring out and defining those limits. The answer to all of the above questions is yes, within complex limits. Chemicals, smoking, radiation, viruses and many other things cause damage to various groups of cells in the body. When damage occurs to a population of cells in the body, repair begins. In most cases, repair requires cellular reproduction. It is necessary to make more cells in order to repair the damage. The more damage, or the more often the damage, the more often cells must reproduce to repair the damage. Chronic sunburn requires chronic repair of the skin. Chronic smoking requires chronic repair of the lungs. Chronic exposure to toxic chemicals causes chronic damage to the liver by those chemicals which are absorbed into circulation and chronic damage to the digestive system by those that are not absorbed. Viruses infect, damage, and change or kill host cells. However, no process is perfect. Such is the case with cellular repair and reproduction. Mistakes are sometimes made as cells repair damage and reproduce. The more often cells reproduce, the more mistakes are likely to be made. The more mistakes that are made, the more likely a mistake will be made which leads to a cancerous cell. Damage to cells of the body from whatever source increases the chances of cancer. In a sense, damage is a selective pressure on a population of cells for reproduction—the thing that cancer cells do best. Is cancer inherited? Of course it is, in that the tendency to develop cancer is inherited to the degree that all potential characteristics and processes of the body are inherited from a person's parents. A simplistic example of the inheritability of developing cancer is the color of one's skin. Some people have light skin and are quite sensitive to ultraviolet radiation from the sun. Other people have dark skin and are less sensitive. Skin color is an inherited characteristic. The color of a person's skin can influence the tendency or likelihood of a person developing skin cancer. Skin color is nothing more than a biochemical process involving enzymes. All of the myriad of processes of the body are inherited as a mix of inherited levels and forms of enzymes. Like sensitivity to the sun, individuals inherit the strengths and weaknesses of their body from their parents. Because of the human genome project, the genetics of cancer is a rapidly developing area of knowledge. Specific genes have recently been shown to be related to the development of specific types of cancer (for example, certain types of breast cancer). As knowledge in this area increases, there is hope that gene therapies may be developed to prevent specific types of cancer.

Treating cancer: In some way, the immune system has failed when cancer develops. A tremendous effort has been expended in the last 20 to 30 years to "cure cancer." This effort has lead to a better understanding of the processes involved in developing cancer cells, better detection methods, and a more effective use of drugs and procedures available for treating cancer. However, cancer chemotherapeutic drugs are still poisons. Because cancer cells belong to the body, it is extremely difficult to introduce a drug that will kill or damage cancer cells without killing or damaging normal cells of the body.

The most singular characteristic of cancer cells is that they grow and reproduce uncontrollably. Drugs to treat cancer are designed to attack the processes associated with cell growth and reproduction. In the most simplistic sense, the approach is to attempt to kill the cancer before you kill the person. Most of the damage to normal cells of the body by cancer drugs is sustained by the rapidly growing cell populations. The lining of the digestive system, the skin, and cells involved in blood formation all sustain substantial damage during chemotherapy. In most cases, the result of chemotherapy remains experimentally unpredictable. At present, the knowledge is simply not available to do much more than increase the efficiency of the currently available tools and procedures. However, great promise is associated with two developing approaches. The first approach involves the specific targeting of poisons. Tumor cells express certain unusual characteristics, such as particular proteins on their cell surfaces not present on normal cells. To the degree that these proteins can be used as "targets" to deliver drugs specifically to the cancer cell, side effects are eliminated. The second approach involves the immune system. More and more knowledge is accumulating concerning the messages that control the immune system. Understanding these messages is the key to directing the immune system to do what it was supposed to do in the first place: kill cancer cells.

It is always a problem of selectivity: Infections, whether they are invasions from outside the body by microbes or from within by cancer, are destructive because they disorganize the body. The body can only function when all cells carry out their essential roles in the organized operation of the body. Treating infections with drugs always reduces to introducing into the body a substance that kills living cells. The degree of damage to the body by such substances is always a function of selectivity. Damage to the normal cells is minimized when a drug selectively disrupts processes necessary to the life of the infecting cells. Damage to the body is also minimized when the drug is selectively delivered to the infecting cells.

chapter twelve

stress

Life is a balance of forces: A living organism is a collection of balances. For example, nutrient molecules are brought into the body, stored, and released at just the proper rate in order to maintain their correct amounts in circulation. Heat is generated by the muscles and lost through the skin in order to maintain proper body temperature. Oxygen is brought in and carbon dioxide lost through the lungs in order to maintain both the ability of cells to generate organizational energy and the proper pH of the blood. Skin cells are born at the rate that they die in order to maintain a surface cover for the body. Ion and water loss from the body are coordinated in order to maintain proper concentrations. An imbalance in any aspect of the system causes disease and ultimately death (Fig. 12.1). We take most of our balances for granted. Maintaining all of these balances is done automatically: we don't have to think about them.

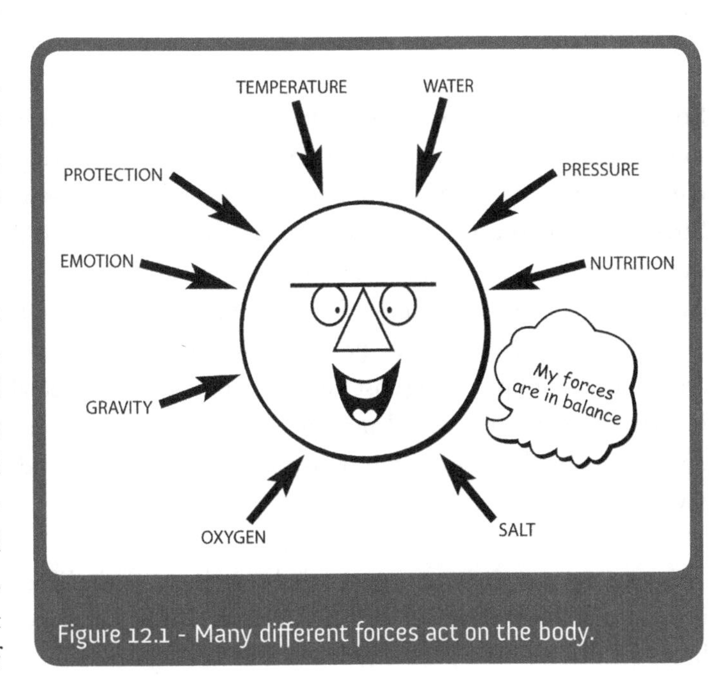

Figure 12.1 - Many different forces act on the body.

If you run, your muscles work. If you drink a glass of water, blood pressure and ion concentrations are maintained. If it is cold outside, your body temperature remains 98.6° F. The amount of every atom and molecule in circulation is a balance. The addition of components to blood is balanced with removal of components from blood. Because of gravity, blood pressure must adjust to the change in the orientation of

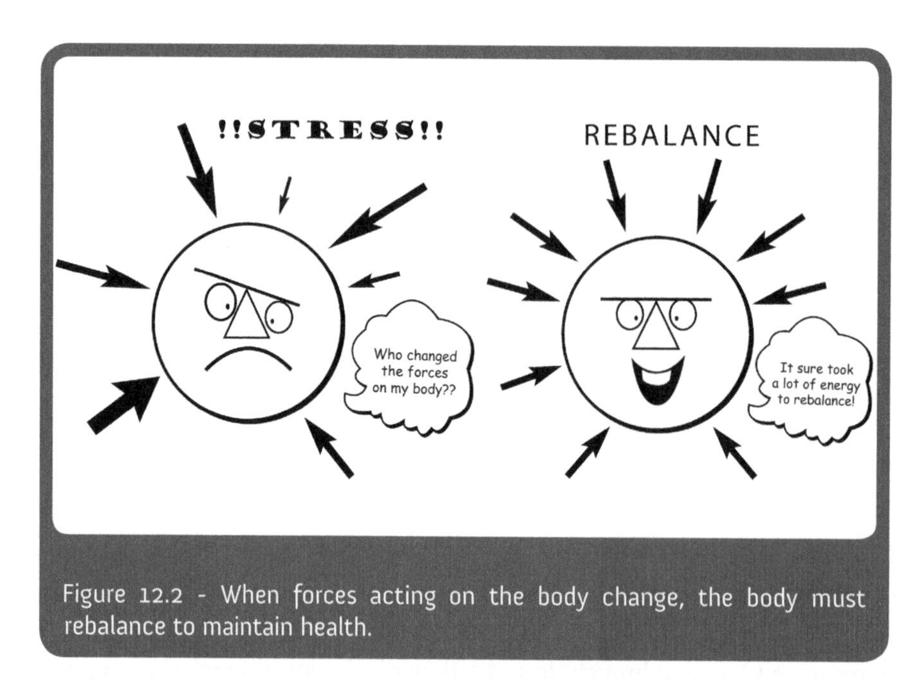

Figure 12.2 - When forces acting on the body change, the body must rebalance to maintain health.

the body every time one moves. In addition to the many physical balances necessary for maintaining the quality of the internal environment, drives and motivations are also balanced. Hunger, thirst, fear, anger, and aggression can either be valuable to survival or destructive depending on circumstances. At all levels, life requires balance.

Stress—a disruption of balance: Stress is change. Any change requires that the system responds in order to maintain balance (Fig. 12.2). Changes can be small requiring small adjustments, or changes can be large requiring large adjustments. Stress responses are those adjustments necessary in order to maintain balance. Stress responses occur at all levels of organization. If the skin is damaged in a local area then the stem cells in that area are allowed to reproduce and differentiate faster in order to affect repair. However, response to the damage occurs not just in the local area. The entire society of cells in one way or another supports the repair of local damage. Nutrients are provided, signals directing repair are sent from the brain, and the immune system invades to protect against infection. If the damage is small then the involvement of the entire system is small. Alternatively, severe damage in a local area can involve a major response involving every system in the body. For example, wasting of the musculature in order to provide resources for repair is a common aspect of severe trauma.

This wasting can even be severe enough that it threatens the life of the individual. The society will dysfunction if the demand for change exceeds the capacity of the system to respond. Stress responses have limits. The limits are determined by how much and how fast adjustments are required to maintain balance (Fig. 12.3).

However, limits are not absolute. Limits are a function of the individual at the time. Running a mile requires a small stress response in a marathon runner. Walking a mile may exceed the response limits in a 400 pound "couch potato."

Response to stress is a balance between demands and the ability of the system to meet the demands. In every sense, stress responses are a lesson in evolution. The ability to respond to change has evolved with the demand for change. Organisms that were required to change beyond their capacity perished. The capacity of humans to respond to change is the product of trial and error, involving the success or failure of all the organisms that went before us. Stress responses are constantly going on all over the body all of the time. Information describing the collection of demands comes into the brain constantly. Some demands are more important than others. Collectively they set the level of "stress" for the society of cells which makes up the body.

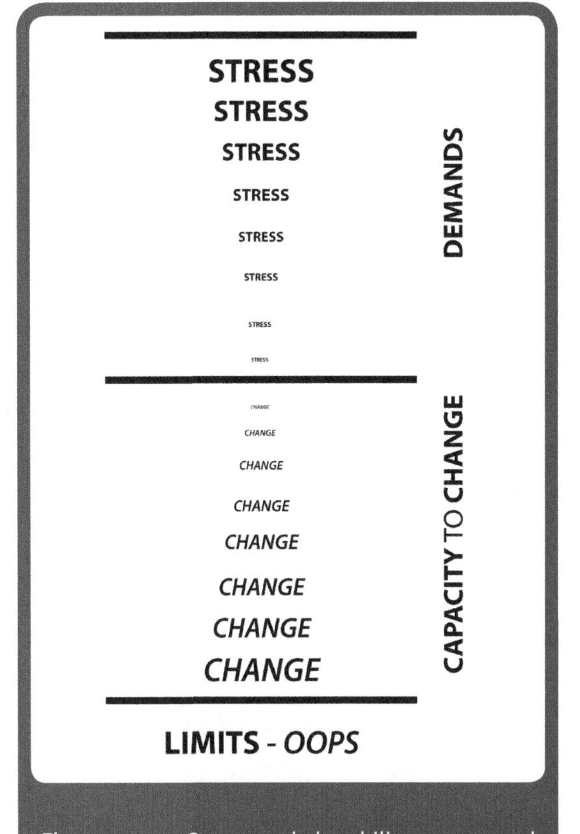

Figure 12.3 - Stress and the ability to respond to stress are not "all or none"—there are degrees of stress which require different degrees of response by the body.

Think about the body that went on a walk on a nice warm sunny day. Other than the walk, demands were minimal. The walk required that more resources be allocated to the muscles, and perhaps a bit more to the heart and the lungs. As you walk on, you become hungry—it's about noon and you had no breakfast. Your liver is becoming depleted of sugar. The drive of hunger from your limbic system is being allowed to be expressed. The hypothalamus begins to direct the pituitary to direct the adrenal gland to increase the output of glucocorticoids. There is an increased breakdown of muscle, and the liver and kidney increase the production of sugar from amino acid carbon released by the musculature. As you start across a road a bus carrying students runs a STOP sign and careens toward you. The sight of the bus coming towards you causes strong signals of fear to descend from your limbic system.

Your attention is focused as the signals descend on the reticular system. The sympathetic nervous system is activated. Your blood pressure increases. Blood flow to the heart, the lungs, the skeletal muscles, and the brain increases. Your liver dumps sugar into circulation. Fat is dumped into circulation. Your heart rate increases. Your airways open up and you gulp in air. You jump back as the bus flashes past. You stand there on the curb seething with fear and anger. Calming down is now your problem, since you survived.

A stress response can be like a ripple in a pond or like a hurricane. Whatever the origins and magnitude of the demand for change, the brain must balance the totality of all demands. It uses the resources available to meet the collection of demands on the system. Stress is a problem for the brain. The demand for change can have many origins. You can walk outside. You can get hit by a car. You can have a cold. You can have a baby. You can eat dinner. The brain must use the resources available to it in order to meet the totality of demands for balance by the system. When you think of your brain, remember that it spends most of its time running the body. Life is a constant emergency of balance. If you don't breathe for a few minutes, or if your heart doesn't work for a few minutes, or if your kidneys do not work, then you dysfunction. When this happens your brain is no longer taken care of and you die.

There are many different situations that require adjustments in order to maintain balance, but all are similar in the responses they require. In all cases the response requires organizational energy. Injury requires mobilization of energy for repair. Disease requires energy for the immune response. Pregnancy requires a major stress response by the woman's body. Growth is very demanding, even to the extent that if sufficient nutrients are not available then the output of growth hormone by the brain will be reduced and the process will be retarded. A significant aspect of aging is a reduced capacity to respond to change. Stress responses are a constant part of life. The feature common to all stress responses is that they require energy. Nutrient molecules containing energy in their bonds must be mobilized in order to provide the energy needed for the stress responses. As discussed previously, energy is stored in the body in three forms: sugar, fat, and amino acids. All three classes of nutrient molecules play a role in stress responses.

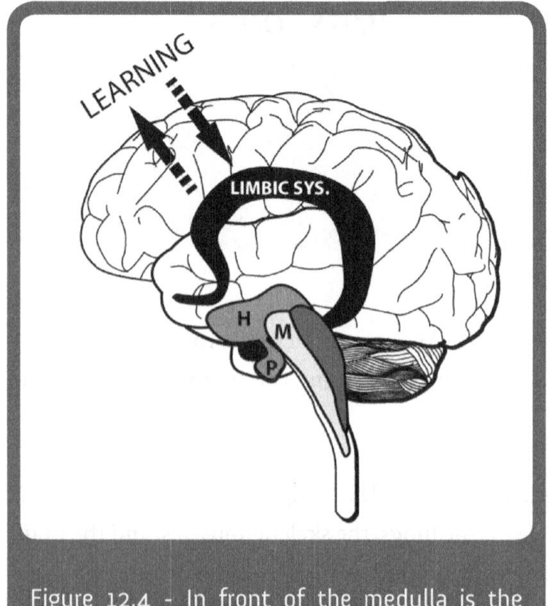

Figure 12.4 - In front of the medulla is the hypothalamus, which controls the hormones of the pituitary gland. Wrapped around the hypothalamus and below the neocortex is the limbic system, the originator of drives and emotions.

There are two basic components to a stress response. The first component consists of short-term adjustments. Short-term adjustments involve the preparation for action and the action itself. These are immediate, virtually instantaneous responses. The second component involves long-term adaptive changes. These two components are very closely related to each other. From a functional point of view, long-term adaptive changes are designed to maximize the capacity of the individual to make future short-term adjustment responses. The relationship between the two appears to be based on the frequency and intensity of the signals for the short-term adjustment responses. The more frequent and intense the demands for short-term adjustment response then the greater are the longer-term adaptive changes in the individual. Chronic long-term adaptive changes in the body are designed to assure the capacity for the continuing short-term adjustment responses.

Short-term stress responses are mediated by the sympathetic branch of the autonomic nervous system located in the brainstem. As you recall from Chapter

8, the brainstem is involved in the fundamental operation of the body. This axis includes the medulla, which contains the control for both the sympathetic and parasympathetic nervous system. It also contains some basic reflex centers for coughing, vomiting, and breathing control. In front of the medulla is the hypothalamus, which controls the hormones of the pituitary gland. Finally, in front of the hypothalamus and below the frontal cortex is the limbic system, the originator of drives and emotions (Fig. 12.4).

In response to incoming demand signals, the control of the body is switched from the parasympathetic branch of the autonomic nervous system to the sympathetic branch. Such switching in control systems can be minor and transient, or they can be intense and long lasting. For example, if one simply stands up from a sitting position, there will be a transient shift to sympathetic control in order to accommodate for the change in orientation of the body with respect to gravity. If you stand up, gravity tries to pull the blood down from your head.

In contrast, a real or perceived danger can cause an intense, sustained shift to sympathetic control. "The bus is about to run me over and I'd better get the heck out of the way." Signals for switching to sympathetic control can take many forms. In the example of standing up, the switch is initiated by a brief drop in blood flow to the brain. In the case of real or perceived danger, the switching is initiated by the limbic system. In both cases the quality of the response is the same. Signals are sent out through the sympathetic nerves. These signals go

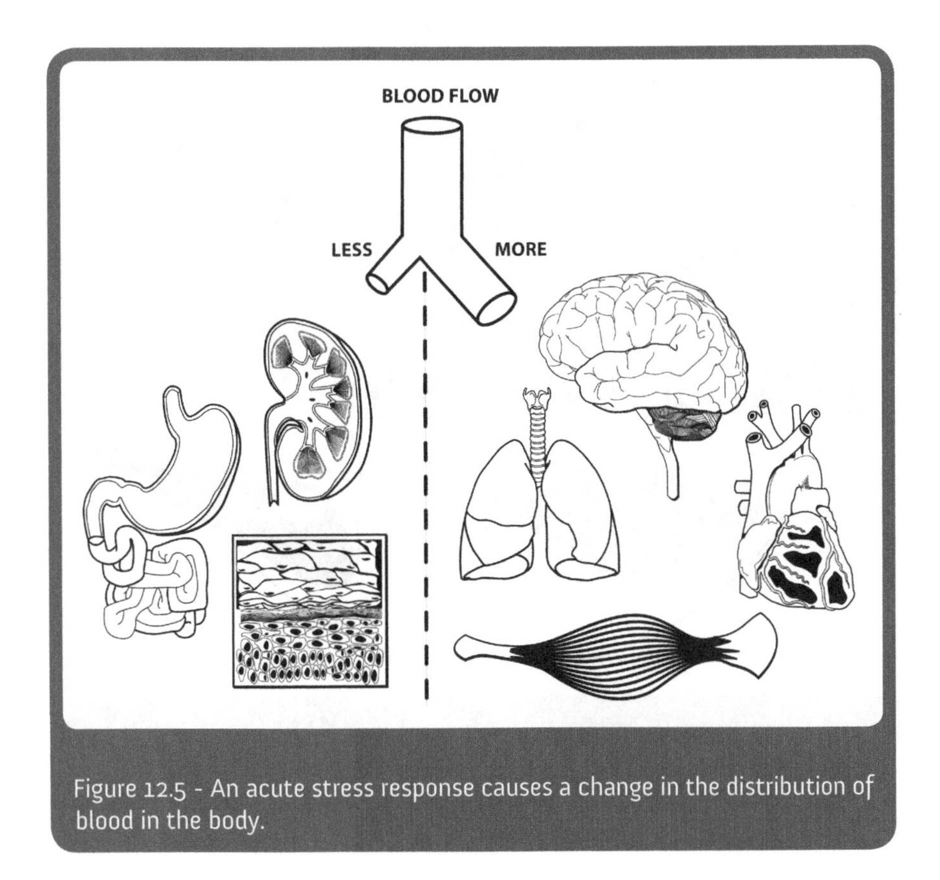

Figure 12.5 - An acute stress response causes a change in the distribution of blood in the body.

to all organ systems but especially to both smooth muscles and the adrenal gland. These signals to the organ systems adjust their function to the response. For example, the sympathetic signal to the heart increases the strength and rate of the heart's contraction: as a result, more blood is pumped. The signals to smooth muscles around blood vessels to the digestive system and kidney will contract, reducing blood flow to these systems. The signals to smooth muscles around blood vessels to the heart, lungs, skeletal muscles and brain will relax, increasing blood flow to these tissues (Fig. 12.5).

Similarly, smooth muscles around the airways will relax, allowing more air to be taken into the lungs. In addition, the signal to the adrenal gland causes it to release adrenalin (epinephrine) into circulation. Adrenalin is a chemical message. This chemical message travels through the blood and reinforces the responses of all organ systems to direct sympathetic activation by nerves. In addition adrenalin acts on the liver, muscle, and fat to cause mobilization of energy containing molecules (Fig. 12.6). Adrenalin directs the liver to break down glycogen into glucose and release the glucose into circulation. Similarly, adrenalin directs fat cells to release fat into circulation. Adrenalin also directs muscle to break down glycogen into glucose. However, in contrast to

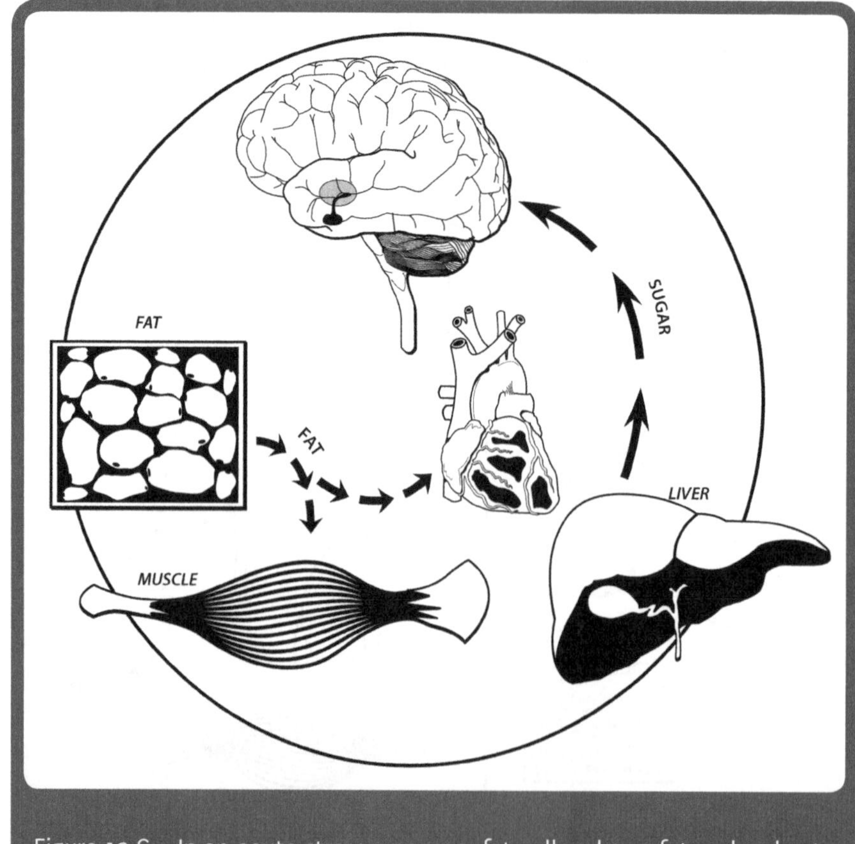

Figure 12.6 - In an acute stress response, fat cells release fat molecules to provide energy for the heart and skeletal muscles, while the liver releases sugar to provide energy for the brain.

the liver, muscle does not release the resulting sugar into circulation. Muscle keeps the sugar and uses it for its own energy requirements for the stress response.

Adrenalin also relaxes some muscles and increases the efficiency of others. In particular, the muscles which maintain posture are made more relaxed and the muscles involved with general movement are made more efficient. This allows for more agile movement of the body. Finally the adrenalin, along with internal nerve signals in the brain, influences both the reticular formation and the hypothalamus of the brain. The effect of sympathetic activation on the reticular formation focuses attention. Your mind does not "wander" when you are in danger. The effect of sympathetic activation on the hypothalamus is to set up the long-term adaptive stress response.

These quick responses deplete resources necessary to making such responses. The capacity to make these responses must be replaced and enhanced if the system is to survive. Stress responses generally cause an attempt by the system to increase its ability to make such responses. That is why working the body strengthens the body. After all, response to the same change most likely will be required again. As the old saying goes, "the squeaky wheel gets the grease."

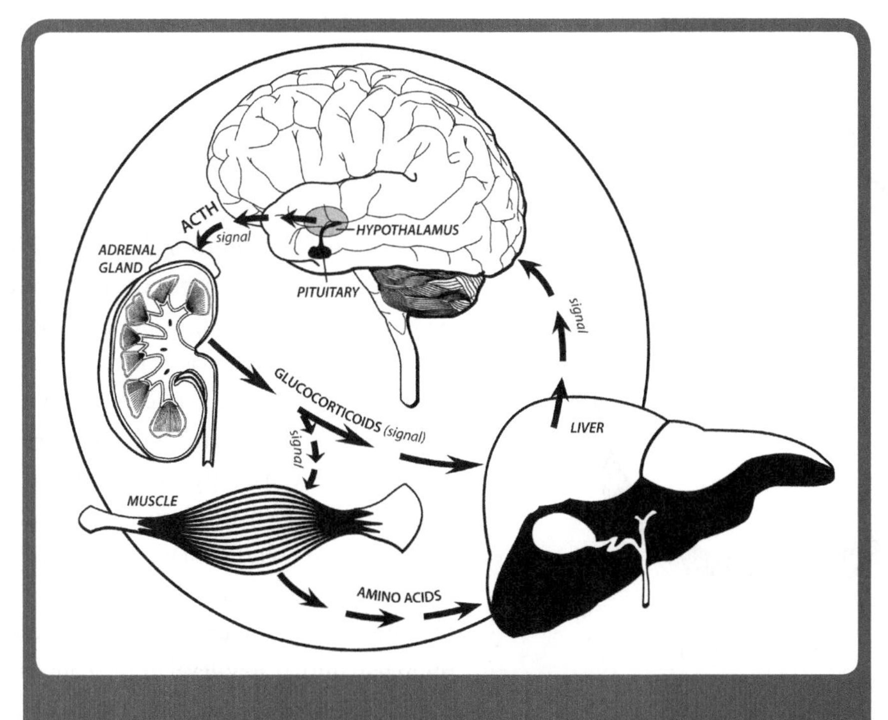

Figure 12.7 - Short-term stress responses initiate a longer-term adaptive response. The hypothalamus directs the pituitary to release ACTH. ACTH directs the adrenal glands to release glucocorticoids. Clucocorticoids are the long-term stress hormone.

Systems that are stressed often are given extra resources in anticipation that they will be stressed again. This is not a prescient act on the part of the brain, the distributor of all largesse in the body. Part of the stress response usually includes an increase in the sensitivity of responding tissues to signals that allow the cells greater access to nutrients.

Adaptive responses to change are longer-term: The longer-term stress response involves changes in the release of hormones from the pituitary. This release is directed by the hypothalamus of the brain. Of particular importance is the release of ACTH. ACTH from the pituitary travels through circulation and acts on the adrenal gland. ACTH directs the adrenal to increase the output of cortisol, the natural glucocorticoid of humans. Glucocorticoids both limit demands and provide resources on a longer-term basis (Fig. 12.7). Almost every tissue in the body responds to glucocorticoids. Glucocorticoids limit demands on the system of tissues not essential to the stress response by reducing their sensitivity to signals like insulin which provide resources. Glucocorticoids provide resources by their effect on skeletal muscle, the liver, and the kidney. In skeletal muscle, cortisol shifts the balance between protein synthesis and degradation. The shift is such that more proteins are being broken down than are being made. The result is free amino acid carbon which is released into circulation. At the same time, cortisol is directing the liver and kidney to make more of the specific enzymes that are necessary for using amino acid carbon to make sugar. Cortisol also suppresses the output of insulin so that most cells do not take up the sugar made and released by the liver and kidney. Finally, as you will recall, cortisol (like all glucocorticoids) suppresses the immune system. This limits immune responses and their demand on the system. Even immune protection must be a balanced priority.

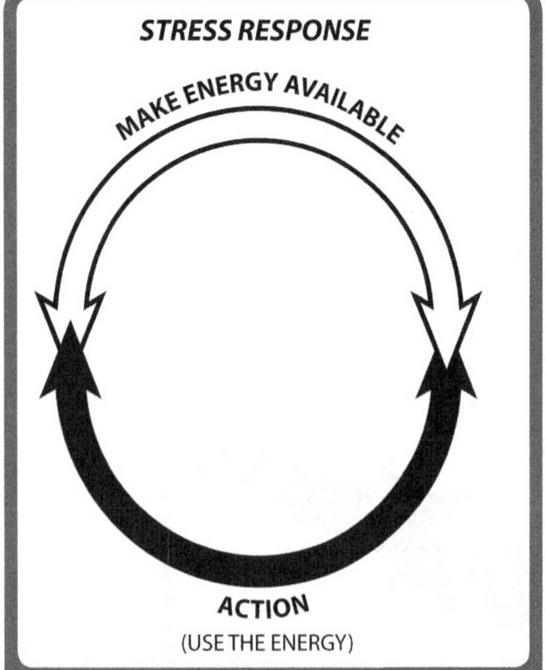

STRESS RESPONSE

MAKE ENERGY AVAILABLE

ACTION

(USE THE ENERGY)

Figure 12.8 - All stress responses do not affect the body the same way.

Stress is not "all or nothing," nor is it always bad: The word stress invokes images of major responses of the body that are damaging to long-term health. In fact, this is not always the case. First, stress is generally a healthy, highly graded phenomenon. More or less of both the short-term and long-term adjustments to stress are constantly occurring in the body. The hormones involved are always present at some level in the blood and even fluctuate with the sleep/wake cycle of the individual. What is important is how much is present. Simply standing up and walking around the room will initiate low-level immediate stress responses. Such a walk requires energy and awareness. Similarly, a slight annoyance will initiate low-level immediate stress responses.

However, there is a difference between the two examples. In the case of the walk around the room, the response is complete. The response provides energy and attention to the task; the process is completed because the energy is used in performing the physical task of moving the body around against the force of gravity. In the case of the annoyance, the response is incomplete. The response mobilizes the energy, attention is focused on the situation, but in general the energy is not used to perform an energy requiring physical task. The problem of the use or lack of use of the mobilized energy containing molecules is a major determinant of the value of the stress. If the energy containing molecules are used, then stress can be quite good for the body (Fig. 12.8).

If they are not used, then stress can be quite detrimental The stress response elevates the level of sugar, fat and amino acids in circulation and prohibits many tissues from using them. The elevated levels of sugar and amino acids strains the kidney, while the elevated fat can clog arteries.

Exercise—a "good" stress: Exercise is an example of a "good" stress. Exercise alerts the brain, mobilizes energy, and uses the mobilized energy. In addition, chronic exercise increases the efficiency of the system such that less demand is made on the system as a whole in order to make the same magnitude of response each time. Often, just the anticipation of exercise activates the short-term stress response. The brain activates in anticipation of the need for the response. This activation mobilizes energy resources, directs the flow of energy resources and oxygen to tissues vital to a physical response, raises the blood pressure and reduces body support of nonessential functions. The liver breaks down glycogen and releases the resulting sugar into circulation. Blood sugar rises. Muscles begin to break down glycogen so that the resulting sugar can be used by them for mechanical work. At the same time the sympathetic signals decrease resistance to blood flow to the muscles so that more oxygen is made available and the heat generated by the movement can be removed. In anticipation of more sustained muscle work, adrenalin also causes the fat cells to liberate fat molecules for use by skeletal and heart muscle. Sympathetic nerve signals and adrenal chemical messages open up the airways to allow for more oxygen to be taken in and for more carbon dioxide to be removed. Similarly, the brain is provided with more blood by reducing the resistance to blood flow up to the brain. Blood pressure is increased to accelerate flow to tissues vital to the response. In order to aid in the increased blood flow to the brain and the muscles, blood flow is reduced to the digestive system and the kidney. In addition, the tone of the muscles around the digestive system is reduced. This reduces the movement of food in the digestive system. In essence, the digestive system is put "on hold" while the activation response is in progress. The same type of activation is also occurring in the brain itself. Chemical messages, like those from the adrenal, also serve as neurotransmitters in a variety of communication paths in the brain. In the brain, these messages are released in paths to the reticular activating system which increases the level of activity. Internal brain messages, along with those in the blood circulating to the brain, set up the longer-term stress response through the hypothalamus and the pituitary. ACTH is released from the pituitary and acts on the adrenal gland to cause it to release cortisol. In addition, another chemical messenger, endorphins, are released along with the ACTH. Endorphins are referred to as the "natural opiates" of the body. These chemical messages reduce both the flow of pain messages to the brain and reduce the perception of pain by the brain. The incoming short-term stress signals to the hypothalamus also modify other pituitary hormonal signals. For example, the output of the two reproductive hormones, FSH and LH, is reduced. This reduction can be mild and unnoticeable in response to

Figure 12.9 - Many situations that cause stress responses do not allow for action.

normal low-level stress. Alternatively the reduction can be significant and limit reproduction such as can occur in female athletes. Similarly, thyroid hormone output is reduced. This in turn reduces general body energy usage. Body water is conserved by increasing the output of antidiuretic hormone. Various changes in other pituitary hormones such as growth hormone, prolactin and oxytocin also occur. The net result of all of these changes is the adaptation of the body to the level of stress commensurate with the level of exercise.

However, the body also develops a tolerance to the stress of exercise. As the individual's body is subjected to the same amount of exercise over and over, the system becomes more efficient. As the body becomes more efficient the same amount of exercise requires less central support and induces lower-level short-term stress signals. The heart, the lungs, the vasculature and the muscles become more efficient in accomplishing the same amount of exercise, so the demand on the whole body is less.

Exercise increases the efficiency of the short-term stress response system, and therefore fewer longer-term adaptive responses are elicited each time. Quite often, exercise is viewed as being beneficial to the heart or the muscle or some other part of the body. In fact, exercise provides a workout of the entire physiological stress response system of the body. Of particular importance are the adaptive adjustments in the hormonal response mechanism. Continued demand for adrenalin alters the synthesis pathways of these hormonal signals in the adrenal gland. Less adrenalin is released when less is required. The range of hormonal output is broadened.

Similarly the response sensitivity of target tissues is altered such that the net result is that the range of responses is greater. They respond less because they need to respond less. This allows for a greater quantitative variation in the response to stress. The value of such an increase in range is that it allows for lesser responses to small changes and greater responses to greater changes. The critical aspect of exercise is that it represents a complete response. The body prepares for activity and follows up with activity. Such is not always the case with learned stress responses (Fig. 12.9).

Incomplete stress responses can cause disease: Given changes in our society, exercise is probably the modern substitute for the sustained physical work of the past. However, this same stress response system also is designed to provide the individual with protection from danger. Initiating such responses in humans is generally a learned process involving the relationship between the neocortex and the limbic system. We learn to apply emotions to decisions. In general, novel situations are stressful to some degree, depending on the individual's experience with broaching novel situations. In addition humans learn to associate danger, anger, and threat with various situations.

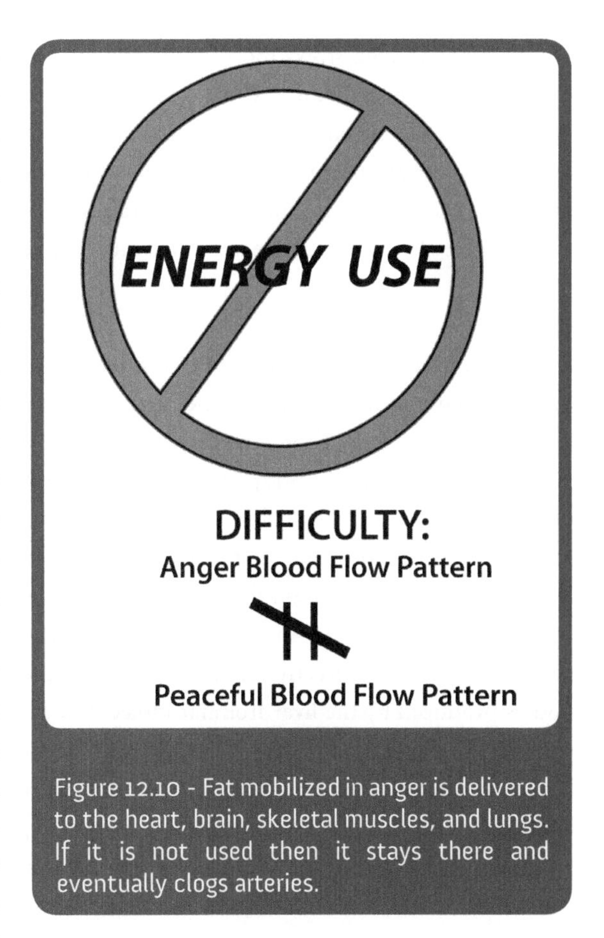

Figure 12.10 - Fat mobilized in anger is delivered to the heart, brain, skeletal muscles, and lungs. If it is not used then it stays there and eventually clogs arteries.

Returning to the example of "the tiger in the woods": the second time that the individual walks through the woods a learned stress response is activated, even when the tiger isn't there. The reticular core of the brain is activated, the sympathetic paths from the medulla are activated, the general body is activated (including the mobilization of energy), and changes occur in blood flow, heart function, lung function and muscle function. Even the longer-term hypothalamic responses are set in motion. The body is ready to respond to the danger with "fight or flight" if the tiger should appear. If the tiger appears, the stress activation is followed through with muscle activity. Activation in anticipation of danger is highly useful for survival of the individual. In addition, stress activation in that situation is reinforced if the individual escaped the tiger. However, if the tiger does not appear, the body has the problem of either using all the molecules mobilized or putting them back where they belong. Of particular importance is the fact that high levels of fat have been liberated, and because of changing patterns of blood flow this fat has been delivered to the brain, the heart, and the skeletal muscles (Fig. 12.10).

Humans have evolved a very efficient set of procedures to deliver fat to tissues. Unfortunately, humans have NOT evolved an efficient mechanism for returning unused fat to where it belongs: in fat cells. In general this fat must be returned by the same mechanism involved when the body stores fat absorbed by the digestive system. Digestion and absorption occur during "peaceful" times, when the body is not stressed. During calm periods, blood flow facilitates storing fat where it belongs. In contrast, stress focuses blood flow much differently. Somehow the fat must find its way back into the "peaceful channels" when it was mobilized for stress but not used. Often the fat does not make it back. A consequence of not returning this fat is often cardiovascular disease.

Variations in chronic stress responses do occur: The stress response system evolved as a mechanism for increasing survival in face of threats to balance. The set of responses to stress is activated by both learned situations and environmental elements such as starvation and cold. There are variations in the character of the response depending on the nature of the stress.

For example, in cold stress most of the responses are quite similar to those that occur with exercise. Fat is liberated to the muscle and the muscle uses the fat as a source of energy to do the mechanical work of contraction. In this case the purpose of the work is to provide heat for maintaining body temperature. However, with cold stress thyroid hormone output is also increased. The increased thyroid hormone output raises general metabolic rate which facilitates the generation of heat. If one considers that the central regulation of body temperature also resides in the hypothalamus, then such specific adjustments in the stress response seem quite reasonable.

Starvation is also a stress and as such also induces a stress response. As in the case of cold stress, the response has both general elements and elements that are specific to the nature of the stress. In starvation the problem is nutrient availability. Therefore, thyroid hormone output is reduced. This reduces the rate at which nutrients are used by all cells of the body. During the initial phase of starvation, the system attempts to protect its supply of the full range of energy substrates: sugars, amino acids, and fat. Fat molecules are mobilized for use by the muscle, and glucocorticoids mobilize amino acids for use by the liver for the purpose of making sugar. Sugar is spared for use by the brain through a glucocorticoid induced decrease in insulin output and a glucocorticoid induced insulin resistance in many tissues including in the muscle. However, if starvation continues, gluconeogenesis in the liver is greatly reduced and the muscle is spared. The body, including the brain, lives on ketone bodies produced by the liver from fat molecules. The body will remain in ketosis, sparing muscle, until all fat is gone. When the body begins to run out of fat, muscle degradation once again increases, liver again begins making sugar from amino acids and the brain receives its final support from sugar just before death. When about one-half of the musculature of the body is depleted, the person dies. There is a tremendous variation in the ability of humans to withstand starvation. A person with much fat can live with starvation for a month or two, while a thin person may die within a week or so.

In the examples of both cold and starvation, certain aspects of the stress response are tailored to meet the nature of the stress. Such is also the case with the stress of danger and threat. The stress response to danger and threat focuses on the energetic response of the musculature ("fight or flight"). The stress of danger and threat is a learned response in humans. As society has become more complex and mobile, individuals are exposed to more novel situations. The uncertainty of novel situations is by itself a stress. Success or failure in a novel situation determines the degree to which particular situations or novel situations in general assume and maintain the "charge" of threat and fear. At the same time, the number of these situations that require a physical response is declining. In addition, the number of general life activities that require physical activity is also declining. The result is more and varied learned stressful problem situations within the context of far less physical activity. Learned problem solving by the brain is dependent on the stress activation of the brain. The activation and the focusing of the brain on the problem generate the solution to problems. A student preparing for an exam learns faster with a bit of stress activation and focusing of attention. Similarly, a student taking an exam does better with a little stress activation and focusing of attention. However, what happens if the student continually fails? The answer to this question is that more stress activation is applied to the situation. Activation is increased and problem solving focuses on more and more aspects of the situation. Such a cycle of increasing activation and searching for solution continues until one of three things occurs. The individual may

Figure 12.11 - We learn to apply stress to particular situations. We can also learn not to apply stress to those situations.

be removed from the situation. Alternatively, the cycle can continue to the point of frustration and the central activating stress response mechanism shuts down. This is sometimes referred to as a "nervous breakdown." The brain withdraws. The third possibility is that the cycle is broken by success. The individual then learns that solutions are provided by the stress activation mechanism. However, the side effects of stress activation are the mobilization of energy substrates and putting the digestive system "on hold." Thus stress related diseases of modern life are set up even in the successful situation.

Stress can be dealt with in three ways: Three types of solutions have been developed for dealing with stress. The first is education, the second is exercise, and the third is drugs. Just as the education provided by experience teaches an individual to apply stress activation to different situations, education can teach a person to remove or at least minimize the stress activation associated with particular situations. Desensitization is the process by which a person learns to remove or reduce the "emotional charge" of threat and fear from situations (Fig 12.11). In a simple sense, a person can learn to anticipate or imagine success in a situation. This approach, in effect, taps the natural mechanisms by which the brain removes situations from the "threat" category. Returning to the tiger in the woods: if the individual returns many times to the place in the woods where the tiger was first seen and the tiger is never there, then the "threat charge" of the situation decays. However, the same process of decay can be accomplished by returning to the same place in the imagination and never finding the tiger. In essence, what can be learned can be unlearned. Desensitization has been used quite successfully in treating phobias such as fear of heights.

Another but related strategy for dealing with stress activation is biofeedback (Fig 12.11). In this case, the emphasis is on teaching the individual to control the physical elements of the response. If a person is provided with the information concerning certain functional aspects of the body, then it is often possible to learn to control the physiological function. For example, if an individual is connected to a monitor for blood pressure, EEG (brain electrical activity), or heart rate and allowed to watch the monitor, then the individual can learn to control these reflections of body function. Practically, the individual "plays" with images and learns to control the readings of the monitor. In learning to control the monitor, the individual learns to control the body. The result is that a person can develop a learned set of images and thoughts that elicit a response of relaxation. Just as an individual can learn to apply stress activation to thoughts and images, an individual can also learn to apply relaxation to thoughts and images. Biofeedback has been used with some success in controlling high blood pressure and certain types of headaches.

The second approach to dealing with stress is exercise. This solution is simple: use up the mobilized fat molecules so that they don't need to be put back into the fat cells. Exercise also increases the capacity of the stress response system such that lower-level stresses need only require lower-level responses. Finally, exercise has the added attraction of providing the person with natural opiates, endorphins. Exercise gives the person a shot of the natural drug for dealing with stress and pain. The last approach to dealing with stress is drugs. Drugs are used to block the signals, block the effects, or to just make the individual unaware of the situation. This is the subject of the next few chapters.

chapter thirteen

pain, a fundamental stress

Pain is a signal to the brain: Pain is a signal sent from somewhere in the body to the brain by way of nerves (Fig. 13.1). Pain is an informational signal. It enters the brain and is interpreted by the brain.

Figure 13.1 - The brain "knows" you are feeling pain.

Pain is localizable: in most cases, the brain can identify the origin of the signal. The individual "knows where it hurts." However, sometimes pain is referred. For example, often pain signals from the heart are felt in the left shoulder. The reason that the brain "mistakes" pain in the heart for pain in the left shoulder is that the nerves from both the left shoulder and the heart enter the spinal cord at the same place.

There are a variety of situations, conditions, and stimuli that can cause pain. Internal pain of the viscera (such as the structures of the digestive system) is often caused by stretch. Surface pain is often caused by noxious stimuli such as a cut or a burn. Damage releases chemical messages that stimulate, or "turn on," pain receptors. Finally, the pain associated with muscles, including the heart, is often caused by reduced blood flow. In any case, the purpose of the pain is to inform the brain of damage or impending damage. Pain is a beneficial protective signal designed to warn the brain of a problem

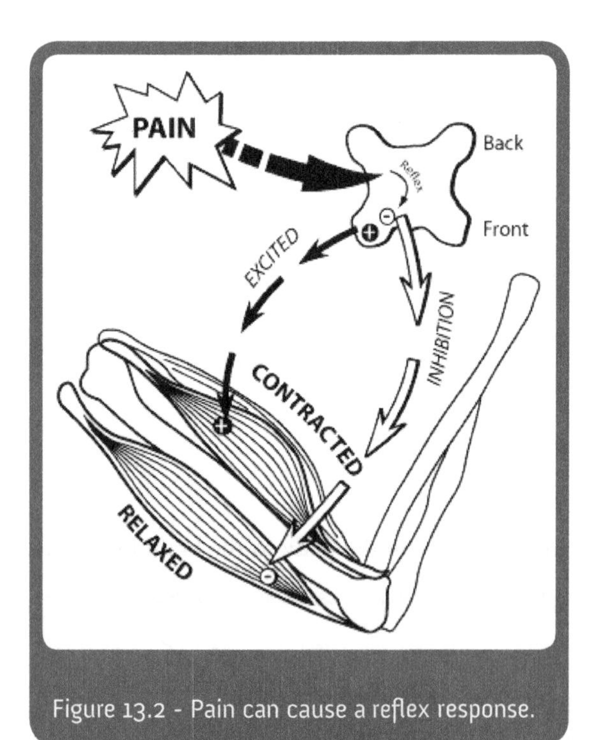

Figure 13.2 - Pain can cause a reflex response.

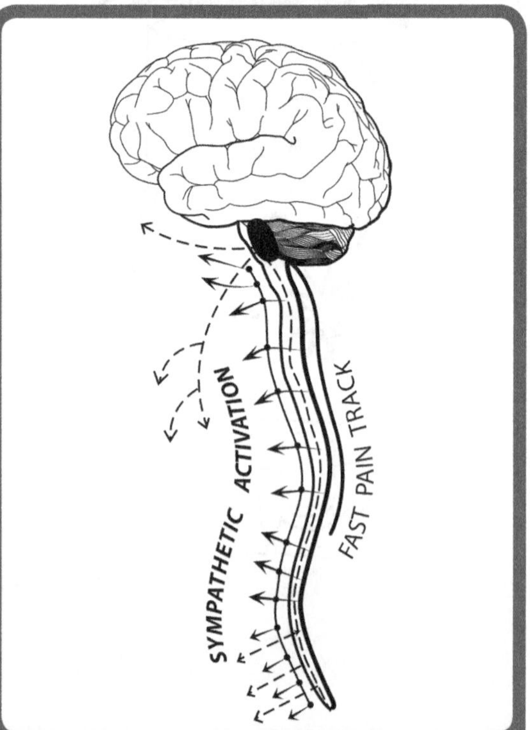

Figure 13.3 - Pain quickly causes an acute stress response.

in some area of the body so that action can be taken. Pain is a fundamental alerting stress which, by dint of the discomfort, demands attention. Pain is also a major medical and social problem, and is commonly treated with drugs.

Pain signals are directed to three places for action: Pain signals originate in structures called pain receptors. The pain receptor is a naked nerve ending (dendrite) with its cell body located in the spinal cord. Painful stimuli cause the release of chemical messages which act on pain receptors and initiate signals that are sent to the spinal cord. At the level of the spinal cord, pain signals are sent in three directions. First, they are sent to nerve cells which are present at the same level in the spinal cord. These nerve cells have the ability to send out signals to structures in the area of the incoming pain signal. These local connections allow for reflex responses to pain signals (Fig. 13.2).

An individual removes his or her hand from a hot stove without thinking about it. A reflex response to pain occurs rapidly at a local level, without any input by the brain. Second, pain signals are directed by a "fast" path to the brain. In the "fast" path, signals move at 20 to 30 meters per second. This allows for more complicated rapid reflex stress responses mediated by the sympathetic nervous system (Fig. 13.3). Finally, pain signals are directed by a "slow" path to the brain. In the "slow" path, signals move at one to two meters per second. These signals are directed to the reticular activating system, the cerebral cortex, and the limbic system (Fig. 13.4). Thus, within less than a second, the brain will receive two types of signals associated with the original pain signal. (Remember, reflex responses are processed in the spinal cord—they never go to the brain).

How the brain handles pain: The first fast track connections are to the medulla of the brainstem. This provides the pain information to the part of the brain that controls the autonomic nervous system. Based on this information, immediate stress signals are sent out through the sympathetic nervous system. As a result the heart speeds up, the airways of the lungs open wider, and blood flow in the body changes. Incoming pain provides the body with an immediate "stress shot." Like most other stress signals, the magnitude of the "shot" is a function of the magnitude of the signals. Next, slow track pain signals are directed to the reticular core of the brain. Activity in the reticular activating system is increased, and the increased activity is projected up to the cerebral cortex. In addition, these pain signals are projected to a representational area of the cerebral cortex. This allows for localization of the pain (Fig. 13.5).

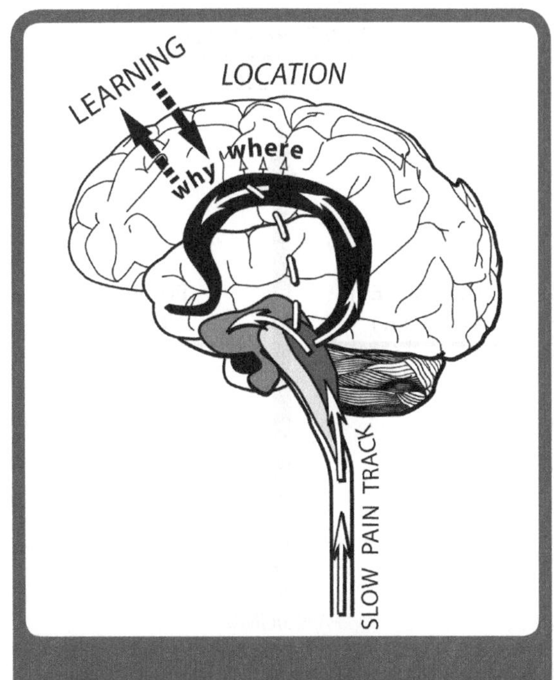

Figure 13.4 - The slow pain signals activate the limbic system and inform the cerebral cortex.

Often in the case of internal pain, the location of the pain is perceived as being on more superficial parts of the body. The perceived surface location is generally a part of the body that sends signals into the spinal cord at the same place where signals from the actual internal location of the pain enter the spinal cord. For example,

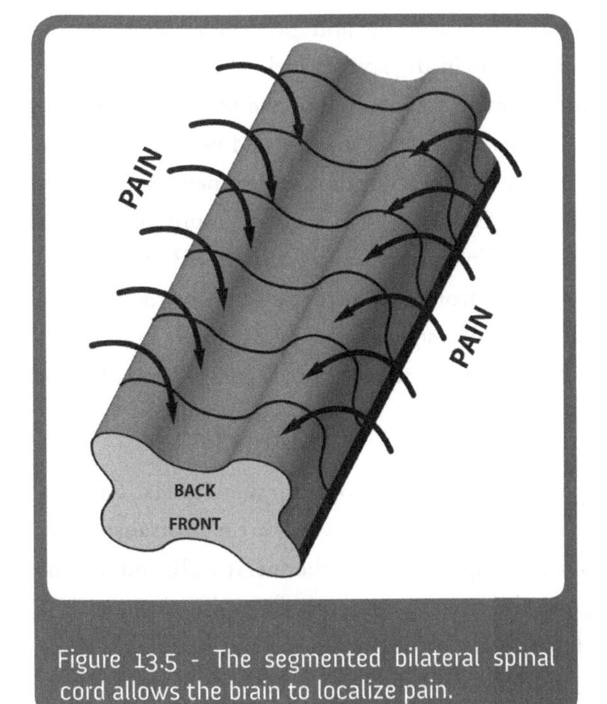

Figure 13.5 - The segmented bilateral spinal cord allows the brain to localize pain.

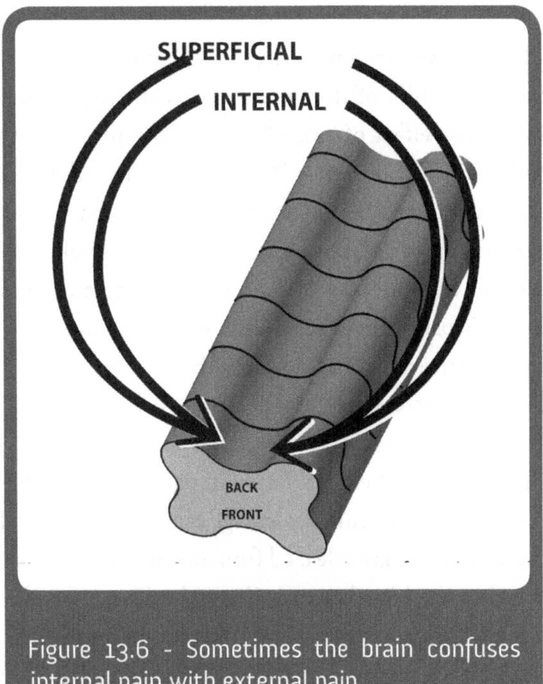

Figure 13.6 - Sometimes the brain confuses internal pain with external pain.

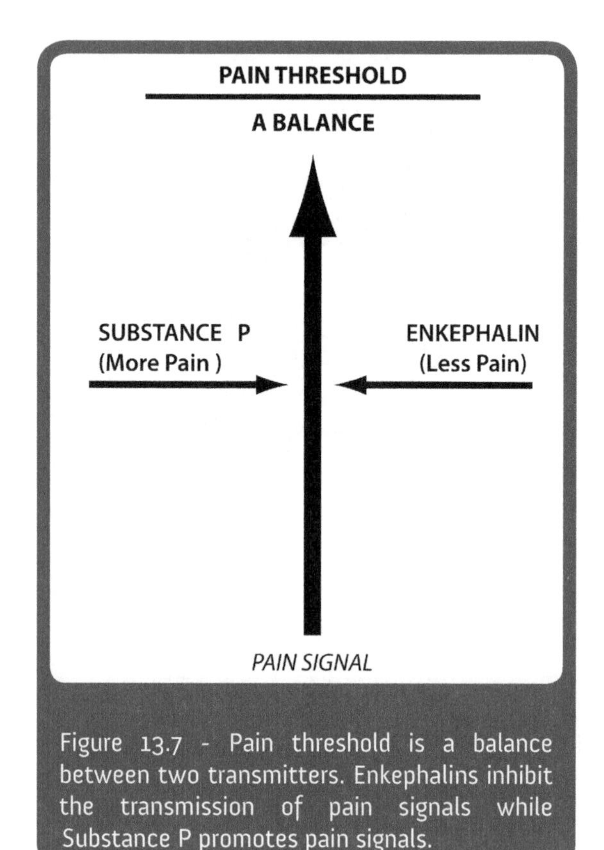

Figure 13.7 - Pain threshold is a balance between two transmitters. Enkephalins inhibit the transmission of pain signals while Substance P promotes pain signals.

like the left shoulder and the heart, kidney or ovarian pain is often interpreted as lower back pain (Fig. 13.6). The signals are also directed to the association areas of the brain. This allows for integration with other incoming signals from the rest of the body. Finally, pain signals are directed to the limbic system. The conversation between the limbic system and the frontal cortex, where decisions are made, allows for learning. For example, one quickly learns not to put his or her hand on a hot stove. Pain is a good teacher. Thus, very rapidly, many parts of the brain have been notified of the pain and an immediate stress response has been initiated.

"Natural" painkillers act on the brain: Pain is an incoming sensory signal that demands attention. However, like most signal processes in the nervous system, pain signals are a balance of excitation and inhibition. There exist nerves in pain paths that contain neurotransmitters called enkephalins. Enkephalins inhibit both the transmission of pain signals to the brain and inhibit the flow of pain signals within the brain. Nerves containing enkephalins operate at all levels of the paths transmitting and processing pain signals. These nerves are found around the paths transmitting pain signals in the spinal cord. They are also found in the medulla of the brainstem, in the hypothalamus, in the limbic system, and in the frontal cortex. The inhibitory effect of enkephalins on the movement of pain signals is balanced by a second neurotransmitter, Substance P. Substance P is produced by other nerves in the pain paths. Substance P is a neurotransmitter that facilitates the transmission of pain signals (Fig. 13.7). When Substance P is "squirted" onto nerve paths carrying or processing pain signals, the signals move more easily. Enkephalins and Substance P provide balance to the intensity of pain signals. There are many pain signals from all over the body being sent to the brain. Most of them are minor and don't make it to the brain. Even if a signal does make it to the brain, most signals are extinguished before they can cause a response. Only the most intense signals gain the attention of the brain's stress response system.

In addition to the modulation of pain signals in the nerve paths, a second pain control system exists. This system involves a chemical message called endorphins. However, unlike enkephalins which are neurotransmitters (and therefore are released from one nerve cell and travel across the synapse to another nerve cell), endorphins travel in the bloodstream. General release of endorphins into the circulation occurs from the pituitary. It is interesting that endorphins, enkephalins and ACTH are all part of the same gene. As you recall, ACTH is the pituitary hormone that causes the adrenal gland to release glucocorticoids.

The particular parts of the gene which are made into proteins at any one time in any cell vary with the cell and the circumstances. In general, enkephalins are made by nerve cells in nerve paths. Alternatively, both ACTH and endorphins are made and released together by pituitary cells. Thus, these compounds which reduce pain appear to be both related to the stress hormone system and are released in conjunction with stress responses. However, even with a natural mechanism for dealing with pain, the control of pain is still a significant medical problem.

Non-drug methods for dealing with pain: There are a variety of strategies used to either reduce or block pain. Although most involve drugs, there are several of note which do not. One approach is to generate competing signals. For example, the "bite the bullet" strategy and the "maze on the wall in the dentist's office" represent a distractive approach. Distraction is, in fact, a real mechanism. The means by which the mechanism works probably involves the flow of incoming signals at the level of the reticular core of the brain. Strong incoming sensory signals compete for attention and focus at the level of the reticular core. Therefore if another strong incoming signal is provided there is competition between the signals for attention. The consequence of this competition is that the threshold of both signals is increased. Threshold is a very important factor in considering the perception of pain. Although pain can be "felt" and reacted to at lower brain levels, much of the more annoying aspects of pain involve focus, perception, and interpretation of the signals by the cerebral cortex.

A second non-drug approach to pain is acupuncture. Although the mechanism of acupuncture is still somewhat poorly understood, some evidence suggests that, at least in part, the insertion of pins and stimulation in certain strategic areas may cause the release of the natural pain blocking messages, endorphins and enkephalins. Beyond such natural approaches to the control of pain, there are a large variety of drugs used to block or reduce pain.

Local anesthetics: One of the most selective approaches to blocking pain involves the use of drugs known as local anesthetics. Local anesthetics are drugs that can block both the generation of the pain signal by the pain receptors and block the transmission of the pain signal by nerves. Such compounds generally have a structure which allows them to enter the environment of the membrane of the nerve cell. When they get inside the membrane, they interfere with the processes by which the ion-based electrical potentials across the membrane are controlled. By interfering with electrical potentials, they either prevent the generation of signals by the pain receptors or they interfere with the transmission of the pain signal by the nerve to the spinal cord (Fig. 13.8). Because of their mechanism of action, local anesthetics are only effective as long as they are present. Generally, their effect is terminated by circulation's removal of the drug from the local site where it was applied.

Local anesthetics are used in three general ways. First, they can be applied to the pain receptor directly. For example, a cream containing a local anesthetic can be applied to a surface area on the skin. In this way, they block the generation of pain signals in the area where they were applied. The drug penetrates from the surface into the body and affects the generation of the pain signal.

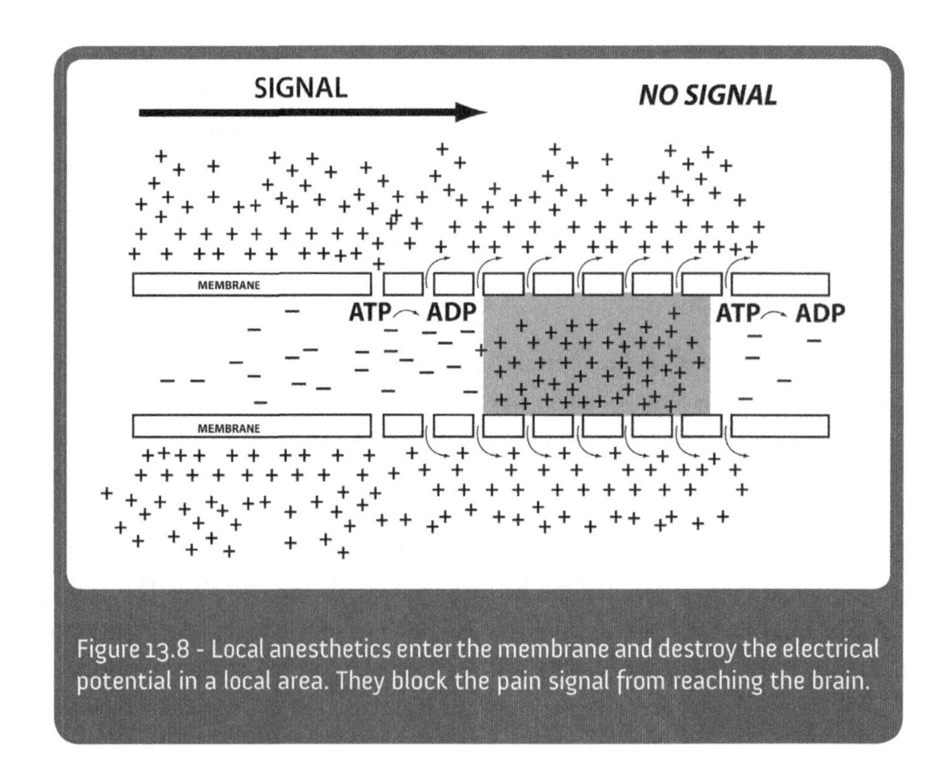

Figure 13.8 - Local anesthetics enter the membrane and destroy the electrical potential in a local area. They block the pain signal from reaching the brain.

Action is generally terminated by the drug being removed from the local area by circulating blood. The second way in which these drugs are used is to inject them into the area near a particular nerve. Dentists often introduce local anesthetics in this manner. Knowing the distribution of nerves in the mouth, they inject the local anesthetic near the nerve carrying the pain signal from a particular set of teeth to the brain. In this way, they block the transmission of the pain signal to the brain. The effect of the drug lasts until circulation removes the drug from the local area. It is not uncommon to add adrenalin along with the local anesthetic used in some areas to constrict the blood vessels in the local area and thereby slow down the removal of the drug from around the nerve. Reducing the blood supply to the area prolongs the action of the drug. The third way local anesthetics are used is to cause a "spinal block." Spinal blocks are often used during childbirth. For a spinal block, a needle is inserted into the area around the spinal cord and a local anesthetic is introduced. The effect is to block sensory signals to the brain from all areas below the block. As with the other two approaches the drug effect is terminated by removal of the drug from the cerebral spinal fluid. If the individual moves around much, one side effect that can occur with a spinal block is headaches. Movement pumps the local anesthetic from the localized area in the cerebral spinal fluid up to the brain.

Although not common, there are potential side effects in the use of local anesthetics. The side effects derive from the fact that the drugs are absorbed into general circulation. The potential for side effects is therefore based on the balance of how fast the drug is absorbed into circulation, relative to how fast it is destroyed by enzymes in the liver and to a lesser extent by enzymes in the blood. The side effects occur because of the ability of the drugs to block both the generation and transmission of ion based electrical signaling activity. The consequences of excessive absorption of local anesthetics are depression of skeletal muscle function, smooth muscle function, and heart muscle function. The effect on the central nervous system is initially hyperexcitability followed by general depression of excitability. This apparent paradox is fairly common. The explanation of the mechanism involves the balance in the nervous system between inhibitory signals and excitatory signals. Balance of electrical activity in the central nervous system involves the convergence of excitatory signals with inhibitory signals. If either the excitatory side of the balance is increased too much or the inhibitory side of the balance is reduced too much the end result is too much excitation, even to the point of seizures. On the other hand, if the inhibitory side of the balance is increased too much or the stimulatory side is reduced too much the result is depression of electrical activity, even to the point of coma. Local anesthetics initially inhibit the inhibitory side of the balance more than the excitory balance. The result is increased excitability.

Drugs in this class, which include compounds such as novocain, lidocaine, and tetracaine, vary in penetrance, potency, and metabolism. These properties will determine how fast and how far the drug moves from the local site of application, how much drug must be used to cause the effect, and how long the drug effect lasts. In general, local anesthetics are useful for the type of transient pain associated with certain surgical procedures.

Non-narcotic analgesics—a very common drug: Over-the-counter non-narcotic analgesics represent one of the most common types of drugs used in this country at the present time. Analgesics by definition block or reduce pain. This class of drugs includes aspirin and aspirin-like drugs. These include acetaminophen and more recently a new class of compounds that initially required a prescription. This newer class of compounds includes ibuprofen and naproxen. The common denominator among these three types of drugs is that they all are available in non-prescription preparations for the treatment of low-level pain. In addition, all three types of drugs deal with pain without altering the consciousness of the individual. The three types are different in chemical structure and in side effects. Therefore, it is necessary to consider each separately.

Aspirin-like compounds have a history that goes back to home remedies derived from willow bark. For a long time, aspirin-like compounds were the primary non-prescription analgesic available. Aspirin is quite effective in relieving pain of headaches, muscle aches, and joint pain. However, aspirin is not effective in relieving the pain associated with the internal organs such as the digestive system. Although the mechanism by which aspirin reduces such pain is not entirely clear, much evidence suggest that the effect is associated with the ability of aspirin to block an enzyme necessary to the synthesis of an oily class of chemical messengers known as prostaglandins. Prostaglandins appear to be involved with the actual production and magnification of pain signals. By blocking an enzyme necessary to prostaglandin synthesis, aspirin and related compounds interfere with the pain signal.

Although aspirin is relatively selective in blocking a particular enzyme involved in prostaglandin synthesis, it is not selective in its function as only an analgesic. The lack of selectivity stems from the fact that the same enzyme and closely related enzymes are used in other functions around the body. One such function involves the regulation of body temperature. As discussed previously, body temperature is controlled by balancing heat generation with heat loss. There are a variety of compounds produced by bacterial and viral infections that can influence this balance. These compounds can either affect the balance directly or indirectly, by causing macrophages to release a chemical message that influences the balance. The cells that control temperature balance are in the hypothalamus of the brain. Fever occurs when the infection or macrophage-derived compounds act on these cells in the hypothalamus and disrupt the thermostat. In some way the disruption of body temperature control involves the synthesis of prostaglandins. Aspirin will block such rises in body temperature, presumably by blocking the synthesis of prostaglandins. However, aspirin does not have an effect on body temperature in the absence of fever. The reduction of fever is often a very useful side effect of aspirin.

Another process that involves prostaglandins is inflammation caused by the immune system. As discussed previously, the immune system can release chemical messages that alter the permeability of capillary beds. This allows water and cells of the immune system to leave circulation and enter the local area. The increase in fluid and immune cells in the local area is called inflammation. Inflammation can be quite painful. One of the primary reasons that inflammation is painful is because of the increased water collection associated with the immune response. The "stretch" associated with this build-up of fluid causes pain. By inhibiting the synthesis of prostaglandins, aspirin and related compounds reduce inflammation. This effect of these compounds makes them useful for treating inflammatory diseases because they both reduce the pain and the inflammation that causes the pain.

The aggregation of blood cells known as clotting also involves prostaglandin synthesis. Therefore, aspirin interferes with blood clotting. This effect is observed as an increase in clotting time. For this reason, aspirin is not recommended in circumstances such as appendicitis where excessive bleeding may occur. However, in recent years, the effect of aspirin on clotting has been put to a beneficial use. Studies have shown that the inhibition of clotting can be used in some cardiovascular disease conditions where clotting may be a problem. For example, chronic low doses of aspirin reduce the probability of second heart attacks.

Aspirin does have a variety of negative side effects. Aspirin in higher therapeutic doses has a significant effect on respiration. The cause of this effect is twofold. First, aspirin tends to uncouple the production of ATP for energy from the production of carbon dioxide as a waste product in cells. The result is that more carbon dioxide is produced relative to the amount of ATP produced. The increased carbon dioxide results in an increased production of hydrogen ions. To blow off this extra carbon dioxide and reduce the hydrogen ion concentration of the blood, the rate and depth of respiration are increased by signals from the respiratory control area in the medulla. Second, aspirin directly stimulates the respiratory areas in the medulla. The net result of both of these effects is that the respiratory system works harder to maintain both hydrogen ion balance and sufficient oxygen for the production of ATP.

Aspirin can also have corrosive effects on the gastrointestinal system. The cause of this side effect is twofold. First aspirin stimulates emetic (vomiting) areas in the medulla. The result is disturbance of the digestive system, even to the point of nausea and vomiting. In addition, aspirin when used chronically can be erosive to the lining of the digestive system. Like every process in the body, acid secretion represents a balance. Histamines are the chemical message that causes acid secretion. Prostaglandins inhibit acid secretion. By inhibiting prostaglandin synthesis, excess stomach acid is produced.

A final side effect of aspirin pertains to the development of a rather serious disease situation called Reye's syndrome. Reye's syndrome is a rather severe disorder affecting both the nervous system and the liver. Evidence suggests that giving aspirin to children with either chicken pox or flu may promote the development of Reye's syndrome.

The second widely used over-the-counter, non-narcotic analgesic is acetaminophen. Acetaminophen is actually the active metabolic product produced by the liver from the older drug phenacetin. Like aspirin, acetaminophen blocks an enzyme necessary to the synthesis of prostaglandins. Because of this, acetaminophen relieves low-level pain and reduces fever. However, unlike aspirin, acetaminophen has very little effect on inflammation. For this reason, acetaminophen is not as effective as aspirin for treating inflammatory diseases such as arthritis. Acetaminophen is less effective than aspirin in inhibiting prostaglandin synthesis. This probably explains why it does not have the same anti-inflammatory effects. It also does not have the same side effects as aspirin. However, this drug has its own set of side effects. Of particular note are both potential liver and potential kidney toxicity. Such permanent damage appears to result from the production of toxic metabolites of the drug by the liver. The development of these toxic metabolites makes drug interactions an important consideration when using acetaminophen. If the pathways that metabolize acetaminophen are occupied with other compounds such as alcohol, then there is an increased build-up of the toxic metabolites which damage both the liver and kidney. Because of potential kidney and liver toxicity, chronic use of acetaminophen is not recommended. There is also a warning not to use alcohol with acetaminophen. Because of concerns about kidney and liver toxicity, the older drug, phenacetin, is no longer sold.

More recently, a third major class of drugs has been added to the over-the-counter, non-narcotic analgesic market. This class of drugs includes both ibuprofen and naproxen. Both drugs had been prescription drugs for quite some time and were more recently allowed into the over-the-counter market by simply reducing the amount of drug in each tablet. Their mechanism is similar to the other two classes of drugs: they interfere with the synthesis of prostaglandins. These drugs (like aspirin) reduce pain, reduce fever, and reduce inflammation. The mechanism of action appears to be similar to aspirin in that it interferes with prostaglandin synthesis in a similar manner. Therefore, it is not surprising that these drugs have many of the same side effects as aspirin. Ibuprofen appears to be less likely than aspirin to cause gastrointestinal problems, while naproxen appears to be more likely than aspirin to cause such problems. In addition, they have some of their own unique side effects. Of particular note is the relatively infrequent side effect of toxic amblyopia. Because of the possibility of irreversible damage to vision, any visual disturbance is an indication that use of this drug should be stopped immediately.

If all three classes of drugs interfere with the same enzyme, why do they have some different effects? The answer is that they interact differently with the enzyme because they have different structures. Because they have different structures, they cross-react with different molecules, they enter some different environments, and they yield different products when metabolized by the liver. At least in the case of acetaminophen, the metabolites produced by the liver have toxic effects of their own.

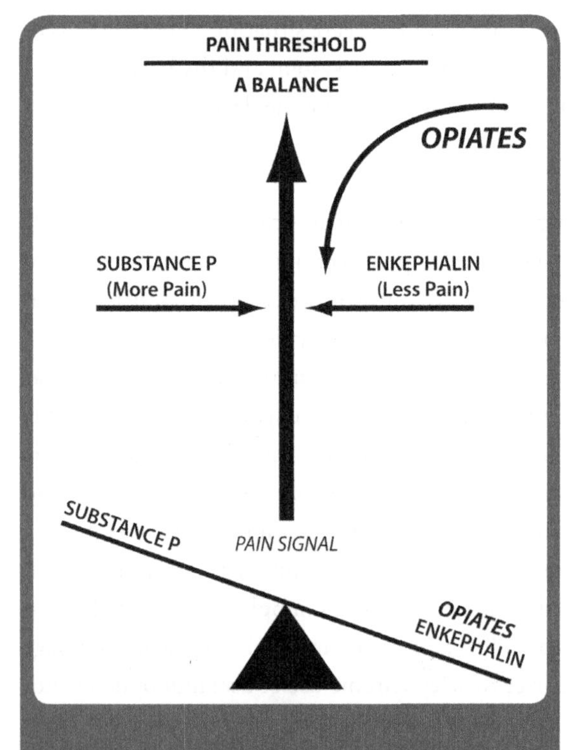

Figure 13.9 - Opiates raise the pain threshold because they add to the effects of endorphins and enkephalins on the nervous system. They tip the balance against transmission of signals in pain pathways.

Narcotic analgesics—the opiates: Regardless of which non-narcotic analgesic is being considered, all are only effective against low-level pain. However, the oldest known analgesic, opium and its derivatives, are very effective in relieving even rather intense pain. In addition, understanding this group of related drugs has taught us a great deal about pain itself. Morphine and codeine are natural analgesics obtained from the unripe seed pods of a particular type of poppy plant. In addition, there are several semisynthetic opiates such as heroin and dihydromorphine, and several totally synthetic opiates such as methadone and meperidine which have gained wide use for one purpose or another.

The analgesic effect of the opiates appears to be related to their ability to mimic the actions of the natural pain relieving chemical messages, endorphins and enkephalins (Fig. 13.9).

As discussed above, endorphins and enkephalins are a group of peptides (small proteins) which block the transmission and processing of pain signals. Although these two classes of peptides and their receptors have different distributions in the body, they do have similar structures and similar effects. The analgesic effects of the opiates appear to be related to the fact that they bind to the receptors for the endorphins and enkephalins and mimic their action. Such opiate receptors are located in many places throughout the body, with high concentrations in many areas of the nervous system. Knowing where the receptors are located coincides well with many of the actions of the drugs. Opiate receptors are found associated with the paths of the spinal cord that conduct pain signals to the brain; they are found in the brainstem, in the area of the medulla which is important for controlling autonomic function; they are found in the hypothalamus which controls many general body functions including the release of pituitary hormones. .they are found in the limbic cortex and associated substructures which are involved in basic emotions; and they are found in the frontal (decisions) and temporal (memory) parts of the cerebral cortex. These parts of the cerebral cortex appear to add intellectual control to the primitive drives which descend out of the limbic cortex. Finally, substantial amounts of opiate receptors are found in an area of the brainstem reticular core known as the locus ceruleus. The locus ceruleus is very involved in projecting activating signals mediated by the chemical messenger noradrenalin to many parts of the brain. Opiate receptors are also located outside of the central nervous system. These are places such as the digestive system where pain signals originate. Opiates control pain at many stages.

They interfere with the generation of pain signals. They interfere with the transmission of pain signals to the brain by the spinal cord. They interfere with processing of pain signals in the medulla such that the rapid sympathetic stress response to pain is not induced. They alter the longer-term stress responses to pain mediated by the hypothalamus and pituitary. Finally, the perception of pain mediated by the higher cortical areas and the limbic system appears to be modified. The result of this modified perception is that the normal fear and anxiety associated with pain is reduced or alleviated. At normal therapeutic doses, opiates do not cause unconsciousness: the individual is awake, knows the pain is there, but the pain is no longer bothersome.

There are also a significant number of side effects of the opiates. First, the sensitivity of the hypothalamus to other incoming signals concerned with the regulation of hunger, temperature, thirst and the release of pituitary hormones is altered. The result is that regulated responses to incoming signals are reduced. For example, temperature regulation is impaired because the hypothalamus is less sensitive to the relevant incoming signals. Hunger and sex drives are also reduced for the same reason. Similarly, the regulated output of hormones from the pituitary such as ACTH, FSH, and LH are also impaired. Because of the effect on the hypothalamus and pituitary, normal control of most major body functions is impaired.

Effects on functions controlled by the medulla are similarly impaired. For example, an overdose of opiates can kill a person because of respiratory failure. Respiratory depression appears to be due to an insensitivity of the medulla to the signals that control breathing. As discussed previously, an individual breathes because the brain tells the person to breathe. Instructions on how much and how fast to breathe are based on information sent to the brain concerning the hydrogen ion concentration of the blood. If the medullary centers directing respiration become insensitive to the signals, the medulla stops telling the person to breathe. The person suffocates because cells of the medulla do not tell him to breathe. The various effects of opiates are generally proportional to the amount of drug taken and how often the person has taken the drug.

Opiates also influence many other aspects of body function such as the control of pupil dilation, tone of the smooth muscles of the digestive system, and the secretions associated with the process of digestion. Another rather interesting effect of the opiates is that they cause vomiting in many individuals. The vomiting reaction occurs rather quickly and appears to be due to stimulation of the brainstem area associated with the vomiting reflex. This effect is just like aspirin except that it is more intense. The effect is interesting because it is one of the few stimulated responses caused by opiates. However, in this case it is important to keep in mind that stimulation often occurs because a natural inhibitory influence has been depressed. In contrast, opiates inhibit the cough reflex. This inhibition of the cough reflex is the basis for adding the opiate codeine to many cough syrups.

From the point of view of temporarily controlling high-level pain, opiates have been an extraordinarily useful drug. However, their long-term use for pain has always been problematical because of tolerance and dependence (Fig. 13.10). With many drugs, the more often you take the drug and the more of the drug you take, the less effective the drug becomes. The body reacts to the drug and makes it less effective. This phenomenon is called tolerance. The simplest way to understand tolerance is to view the drug as a force acting on the body. The force

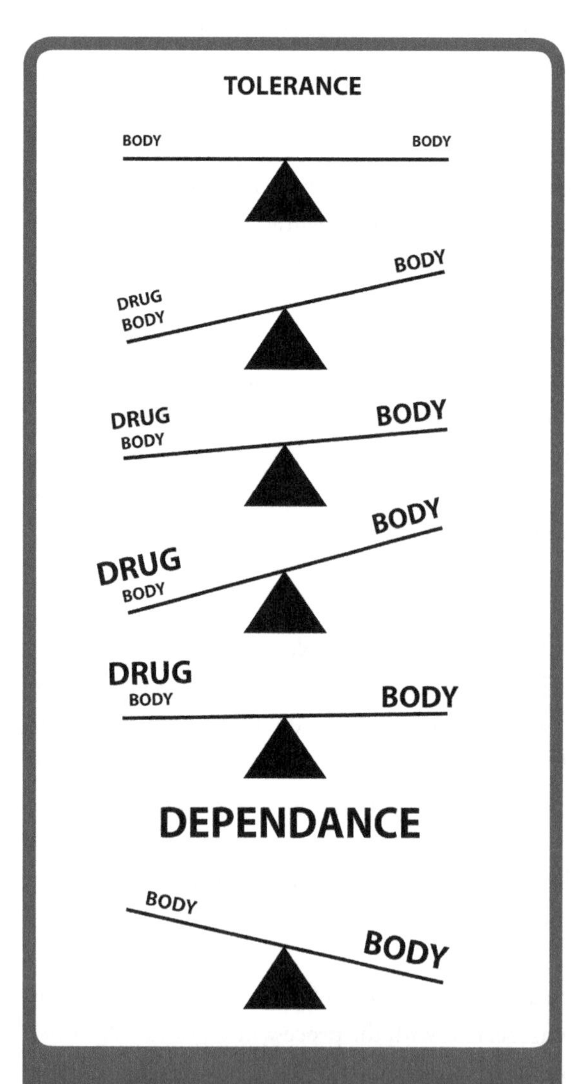

Figure 13.10 - In pain pathways, opiates tip the balance towards inhibition. The body responds by decreasing natural inhibition and increasing natural excitation. The result is that you must take more drug the next time to have the same effect: TOLERANCE. This will continue as long as you take the drug. You must continue to take the drug because the natural excitatory side of the balance has greatly increased and the natural inhibitory side of the balance has dwindled. Stopping opiates suddenly is very painful: DEPENDANCE.

of the drug is applied and the drug has an effect. The drug pushes functions of the body in some direction: it causes "more of this or less of that." However, what also often happens is that the body, through a variety of adjustments, begins to develop a counter-balancing force to the drug. The counter-balancing force is opposite in effect to the drug. The counter-balancing can take several forms. Cells can decrease sensitivity to the drug. Other cells that produce signals that have the opposite effect can increase activity.

The liver may increase its ability to destroy the drug. The result is that when the same amount of the drug is taken again, it has less effect. Force applied by the drug is balanced with force developed in the body. The body does not like its balances pushed around, so it pushes back. The consequence is that to have the same amount of effect, an individual must take more of the drug. More of the drug induces more counter-balancing force by the body, so more drug must be taken the next time to have the same amount of effect. The same amount of drug that may have no effect on an "experienced" person can kill an individual who has never taken the drug. With opiates, tolerance develops quite rapidly.

The other side of the tolerance coin is dependence. When the drug is abruptly stopped, the counter-balancing force developed by the body is still present. The result is that withdrawal often causes effects which are opposite to the effects of the original drug that was taken.

The consequence is often like one person jumping off of a seesaw. The balance is gone and the system shoots in the opposite direction. For example, many of the effects of opiates described above involve depressing one brain function or another. Upon abrupt withdrawal, hypersensitivity and excitability are observed throughout the nervous system. With all drugs, withdrawal symptoms are the force for causing dependence. With the strong, fast-acting opiates such as heroin, this force is particularly effective in causing dependence. The reason for this is the relatively rapid time frame of onset. Withdrawal symptoms begin almost as soon as the drug is being cleared from the body. Therefore, as the person is "coming down from the drug" the discomfort begins. The consequence is that the person "seeks" more of the drug immediately in order to relieve the pain of withdrawal symptoms. Opiates are specific blockers of pain. The best way to describe withdrawal from opiates is that "everything hurts." There is no check on the flow of pain to the brain.

Why do we need drugs for pain at all? Why should we use drugs like opiates to control pain when we know what the natural chemical messengers are that block pain? Why not use natural "drugs" for this purpose, and perhaps they will not have the untoward side effects and cause dependence? The answer is that the natural "drug" is really not any better than the opiates. Like morphine or heroin, they make the brain less sensitive to pain. Also like morphine and heroin, they make the brain less sensitive to many other important signals. Pain is an important signal that demands the attention of the brain. A response is required. Similarly, signals to the respiratory centers are important signals that require attention. So are signals to the hypothalamus concerning temperature regulation and many other aspects of the condition of the body. Drives are important. Hunger, thirst, sex and anxiety are important.

It is probably quite misleading to consider pain as a single entity. Actually there appears to be several classes of opiate receptors involved in the control of the flow of many types of "urgent" signals to the brain from many parts of the body. Opiate receptors are distributed throughout the body. Enkephalins can be "squirted" on many nerves throughout the nervous system, and endorphins are released into circulation coincidentally with the hypothalamic stress response. Regardless of whether the opiate is natural or artificial, the effect is a reduced sensitivity to urgent signals. In some respects, one might view endorphins and enkephalins as a neural mechanism for allowing the brain to select the most urgent messages. Blocking of pain messages is simply part of balancing the set of urgent signals. The brain must balance all priorities. The fact that tolerance develops rapidly to the endorphins and enkephalins suggests that the physiological balance involves a natural limit on the degree to which the brain is allowed to ignore urgent messages. Balance by definition must involve limits. In contrast, dependence is an artifact of drugs. Introducing the message from outside the body allows one to override the natural limits of balance. Within the natural limits, awareness is not disturbed and dependence does not develop. Beyond the natural limits, awareness becomes disturbed and dependence develops.

What of the situation where the natural limits are insufficient to deal with a chronic level of pain that serves no productive purpose? Pain only has value when it is useful. For example, cancer patients are often subjected to chronic pain that serves no purpose. The pain may have prompted the individual to seek treatment, but thereafter the purpose of the pain has been accomplished. In such a case the natural limits of tolerance reduce the effectiveness of the opiates and often the solution is general depression of the activity of the nervous system. In the next chapter we will consider additional drugs that broadly alter the level of activity in the nervous system.

chapter fourteen

manipulating the brain

The brain—It is all a question of balance: The society of cells which makes up the human body can exist because of balance, because of coordination, because of control, because "someone" is in charge. The brain is that "someone." You don't die when you go to sleep. The brain is still paying attention even when you are not. The brain constantly pays attention to everything. You are dead when the brain stops paying attention. You are dead or on your way to being dead when anything significant in the body malfunctions. Because the brain provides central control of all body functions, malfunctions of the brain have an extraordinary impact on the entire body.

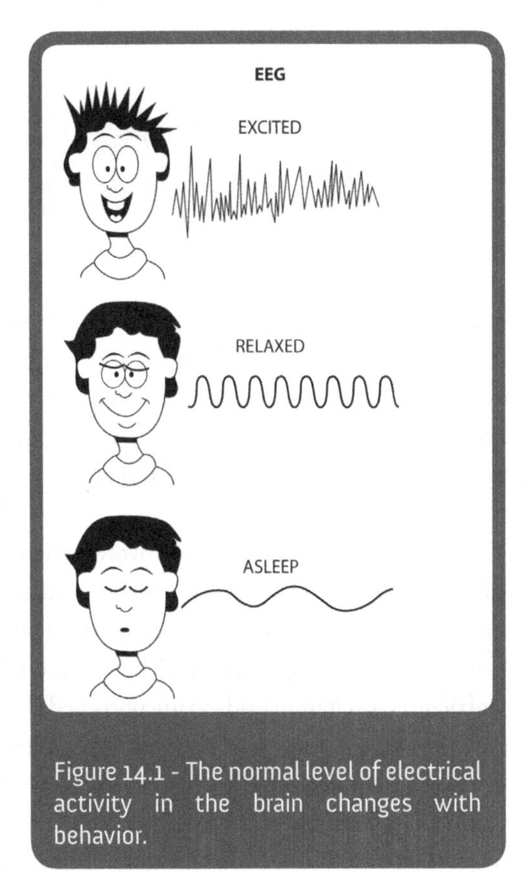

Figure 14.1 - The normal level of electrical activity in the brain changes with behavior.

The central nervous system is composed of the brain and spinal cord. The fundamental objective of the nervous system is to maintain balance in body functions. There is never a lack of activity in the brain. Electrical signals are constantly moving from one level of structural organization to another through long nerve paths, and within a level of structural organization through short nerve paths. The brain is never quiet. If one simply looks at gross electrical activity in the brain, it is quite clear that the level of activity varies considerably from one time to the next. If you measure gross electrical activity in an individual who is passively sitting and daydreaming, you see low-level highly repetitive patterns of activity. If the

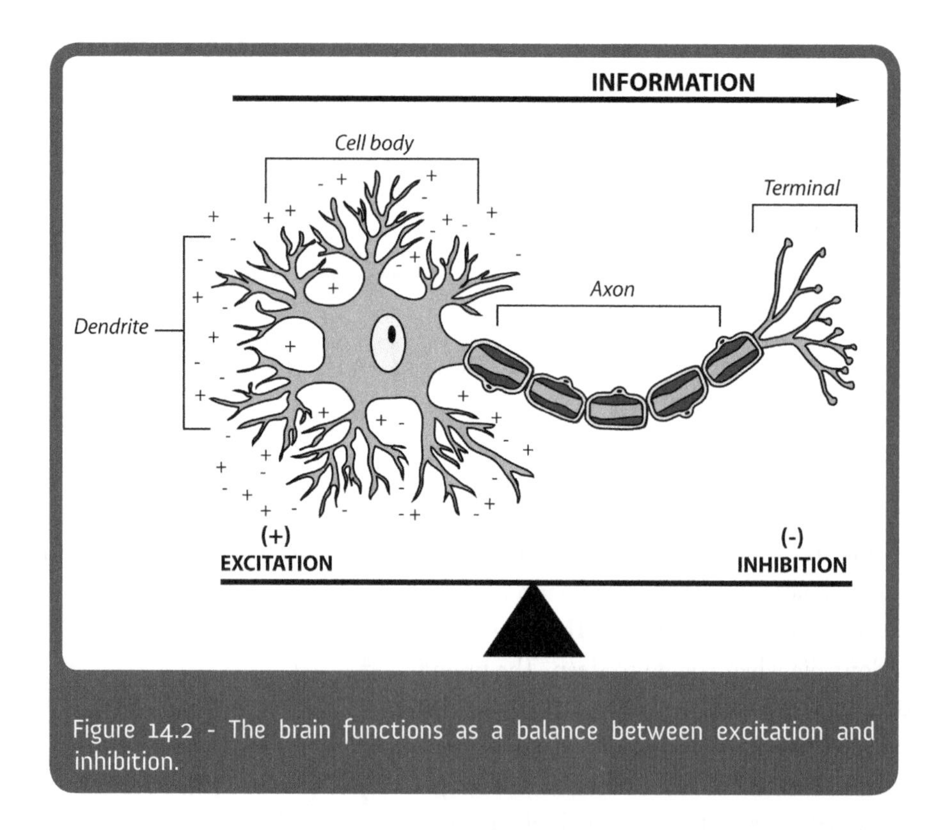

Figure 14.2 - The brain functions as a balance between excitation and inhibition.

individual simply opens his or her eyes and pays moderate attention to something, you see higher-level non-repetitive spiky patterns. There is a wide range of normal activity in the brain (Fig. 14.1). In a normal brain, depression of activity and stimulation of activity are like two people of similar weight on a seesaw, moving up and down between two different extremes, but never hitting either extreme. However, those two extremes do exist. One extreme is depression and coma, and the other extreme is hyperexcitability and seizure.

The mechanism of brain activity involves chemical messages released from one cell that traverse a minute gap (the synapse) and bind to a receptor on another cell. This passes along a signal. As discussed in Chapter 8, that signal can direct the next cell to either pass along a signal (excitation) or not to pass along a signal (inhibition). Control of every single aspect of brain function is a balance between excitation and inhibition (Fig. 14.2). Unbalanced excitation leads to uncontrolled electrical activity resulting in seizure. Unbalanced inhibition leads to depressed electrical activity resulting in depression and coma. The extremes must be avoided. If excitation increases too much the result is seizure. However, if inhibition decreases too much, the result is also seizure. Conversely, if inhibition increases too much the result is coma. Likewise, if excitation decreases too much the result is also coma. Such is the nature of the balance.

There are many types of nerve cells receiving and releasing many different types of chemical signals. Just as there are many types of nerve cells, there are many types of neurotransmitters. Some of these chemical transmitters excite the next cell all of the time. Some inhibit the next cell all of the time.

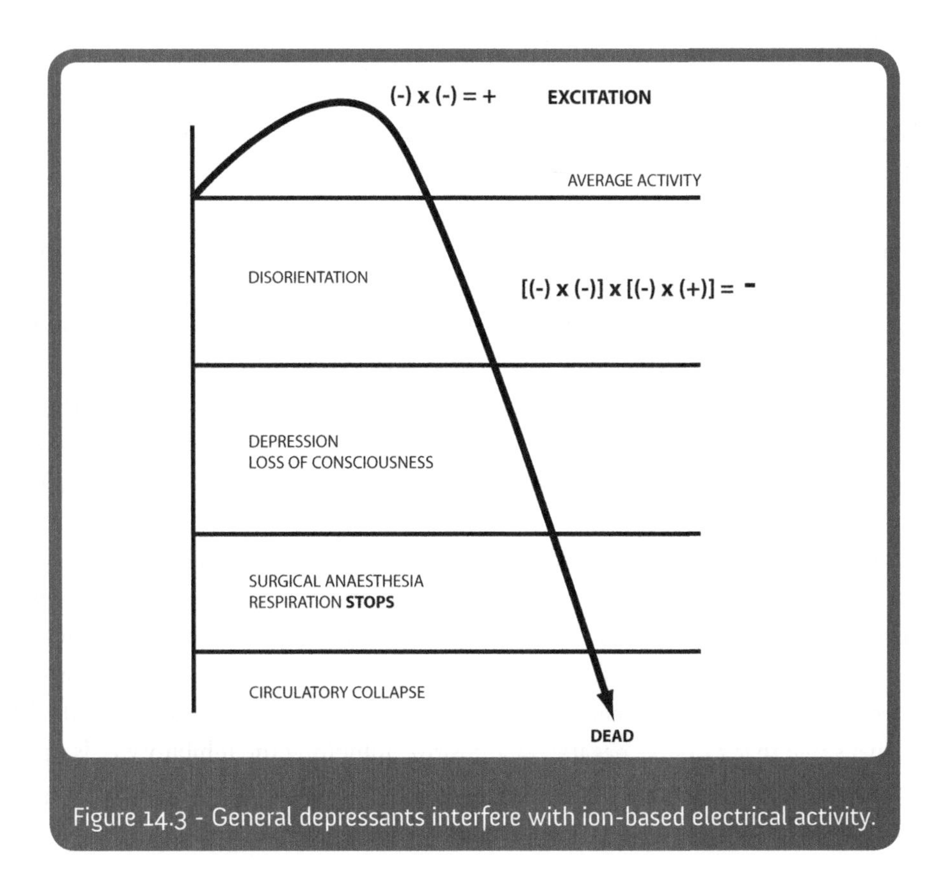

Figure 14.3 - General depressants interfere with ion-based electrical activity.

Some either excite or inhibit depending on the character of the next cell. It is quite probable that we have not as yet even identified all of the various chemical transmitters in the brain.

It is the balance between excitation and inhibition in the multitude of information paths that controls the level of activity in the brain. It is the nature of drugs acting on the brain that they modify the chemical transmission of signals. Drugs acting on the brain change the production, the destruction, the storage, or the release of neurotransmitters. In addition, drugs acting on the brain may also change the sensitivity of different populations of cells to particular neurotransmitter signals. Drugs influence the balances of the brain.

General depressants—alcohol, barbiturates, and gaseous anesthetics: One of the most common modifications of the brain neurotransmitter signaling systems involve general depressants. The most commonly used general depressants are ethyl alcohol, barbiturates, and inhalant anesthetics. Although some variations exist between and even within these three groups, their fundamental common characteristic is that they depress electrical excitability. In a real sense, all three can be considered as general anesthetics because they depress the basic activity on which the brain works: ion-dependent electrical activity. It should also be pointed out that skeletal, smooth, and cardiac muscles also function on fundamentally the same type of electrical activity and will also be depressed by these drugs to some degree. The difference is not that the brain is more sensitive and therefore it dysfunctions at lower doses. The difference is that a little dysfunction of the brain results in a big dysfunction of the body. By analogy, a little dysfunction of the thermostat in your house can have the same result as total dysfunction of the furnace. Regardless of the intent of the use of these drugs, their action is to cause general brain dysfunction.

These drugs appear to work by a complex interaction of several mechanisms. They enter and modify the surface membrane of the cell. The surface membrane is all important in establishing and controlling the ion-based electrical activity of excitable cells. When these drugs enter that membrane, they generally depress signal/response activity. They also appear to increase activity in GABA pathways. As you recall from Chapter 8, GABA is a major inhibitory neurotransmitter. In addition, they appear to inhibit glutamate pathways. Glutamate is a major excitatory neurotransmitter. When such drugs are used as anesthetics, the administrator of the drug, the anesthesiologist, carefully monitors the progression of brain dysfunction (Fig. 14.3).

Information about the progressive stages of brain dysfunction due to the action of general depressant drugs is actually quite informative concerning how the brain "works." If one slowly begins to administer a general depressant, the first response observed is an increase in general activity and excitement. The fact that a depressant drug can be excitable illustrates the fundamental concept of balance. The excitable stage is caused by the drug first depressing the activity of cells which carry out the natural inhibitory side of the balance. Like the English teacher's rule that a double negative is a positive, inhibiting the inhibitory cells tips the balance of brain activity towards stimulated excitability. During this stage, one observes a general sympathetic stress response including increased blood pressure and heart rate.

In general, drinkers try to prolong this stage and anesthesiologists attempt to take a patient through this stage as rapidly as possible.

Following this excitable period, the drug begins to depress cells involved in both sides of the balance. Consciousness is lost and vital functions begin to slow. Respiration, heart rate and blood pressure all decline. In addition, muscles begin to relax, and both kidney and liver functions decline. If administration of the drug continues, then at some point respiration will stop. The brain no longer tells the person to breathe. If this is a self-administering drinker, the person is dead. Unfortunately, many people do not realize that one can literally drink himself or herself to death in a single binge. If the situation is being controlled by an anesthesiologist, the person is on artificial respiration and is ready for surgery. If more of the drug were administered at this point, the next major transition would be circulatory collapse and death, regardless of artificial respiration. The heart and smooth muscles around blood vessels are now depressed to the point of total dysfunction. In general, surgical anesthesia is the "window" between respiratory failure and cardiovascular collapse.

For this reason, general anesthesia is always considered one of the more dangerous aspects of any surgical procedure. When considering side effects of the three types of general depressants, each must be evaluated alone because of differences in their mode of use.

General anesthetics are used primarily in the controlled environment of a surgical procedure. Most often, gaseous drug compounds are used because they can be administered rapidly to reach the necessary high concentrations, and they can also be removed from the body very rapidly. In anesthesia, rapid onset and rapid recovery is the optimum situation. Other than damage that may have occurred by accident during the acute surgical phase of anesthesia, long-term damage is restricted to a small percentage of individuals with subsequent toxic liver and kidney damage.

Barbiturates have a long history of being the dominant drug in a group also known as sedatives and hypnotics. Drugs in this group cause a nonselective, or general, depression of nervous system function. The difference between a sedative and a hypnotic is that a hypnotic is stronger. The original compound in this class of drugs was barbituric acid. However, over the years, more than a thousand chemical derivatives have been made and investigated. At present, about a dozen different barbiturates are available in this country. The major differences between the various barbiturates are how long it takes for the drug to have an effect and how long the effect lasts. For example pentobarbital (Luminal) is a relatively long-acting barbiturate, secobarbital (Seconal) is an intermediate- to short-acting barbiturate, and thiopental (Pentothal) is short acting.

Perhaps the two major medical uses of these drugs are as "sleeping pills" and in the treatment of epilepsy. In the first case, the drug is used to produce unconsciousness for people who have difficulty sleeping. In the second case, the drug is used in conjunction with other drugs to suppress seizures. However, in neither case is the drug optimal for the purpose. In the first case, barbiturates do not produce "natural" and more importantly productive sleep. Sleep is a highly complex subject. There are many phases of sleep. One very important phase of sleep is the phase called "Rapid Eye Movement" (REM) sleep. This phase derives its name from the fact that during this type of sleep, the eyes flutter back and forth. This is also the phase when people dream. If a person is deprived of REM sleep he or she acts as if he or she has been deprived of sleep. Barbiturates inhibit REM sleep. Therefore, using barbiturates as "sleeping pills" is to a certain extent counter productive.

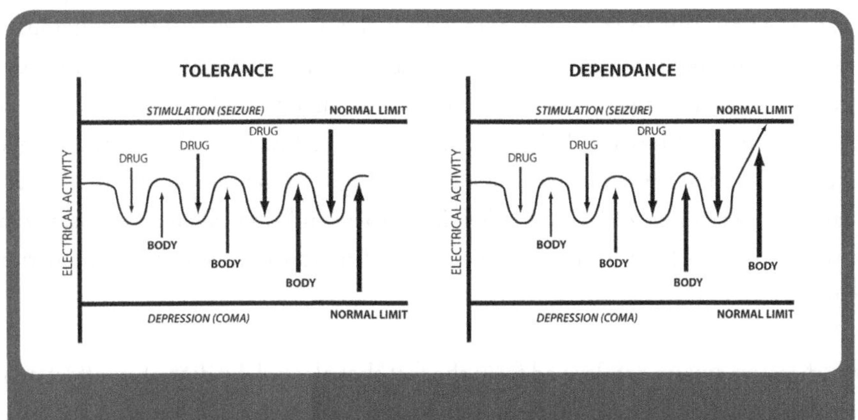

Figure 14.4 - Tolerance develops because the body adapts to the drug by developing opposing forces. Dependence occurs because the presence of the drug is necessary to maintain balance.

In epilepsy, barbiturates have been used for years and are still commonly used to depress excitability of the brain. The depressed excitability reduces the probability of the development of the uncontrolled electrical activity which causes the seizure. The initial side effect is the depression of higher brain functions such as thinking. The major side effects of chronic barbiturate use involve the brain and the liver. The brain develops tolerance and dependence occurs (Fig. 14.4). The result is that more of the drug is needed to maintain an effect. The consequence is hyperexcitability if the drug is abruptly withdrawn, even to the point of seizures.

The effect on the liver actually makes the tolerance problem worse. The liver responds to the chronic use of barbiturates by greatly increasing the capacity of the enzyme system that metabolizes barbiturates. The result is that the drug is more rapidly destroyed by the liver the longer it is taken. A consequence is that other compounds such as steroids and fat soluble vitamins are also subjected to this more rapid metabolism. This provides the potential for vitamin deficiencies and hormonal imbalances.

The last of the three general depressant drugs is ethyl alcohol, better known as just alcohol. Alcohol, although a general depressant, is widely used for recreational purposes. Like anesthetics and barbiturates, alcohol is a depressant and has the probability of developing tolerance and dependence if used on a chronic basis. Abrupt withdrawal from chronic alcohol use causes hyperexcitability in the nervous system, even to the point of seizures. Like the other depressant drugs, the degree of hyperexcitability produced upon abrupt withdrawal depends on how much has been taken for how long.

Chronic alcohol use also has a variety of other side effects. Of particular note are effects on the liver, the cardiovascular system, and the digestive system. The effects on the liver and cardiovascular system in many respects derive from the mode of administration. In general, individuals attempt to self administer in dosages that maintain them in the hyperexcitable as opposed to the anesthetized phase of the response curve to depressants. The effect is a sympathetic stress response.

One aspect of this response is the release of adrenalin and noradrenalin from the adrenal gland. These chemical messengers cause the fat cells to release fat into circulation at a time when the individual is usually not active. In addition, ethyl alcohol is a "pure" easy-to-use substrate for producing energy and sugar. Because of the stress response fat is released into circulation, but it is not highly used by the body. The result is that it collects in the vasculature of the heart and of the brain, promoting heart attack and stroke. In addition, some of this fat is sent to the liver. However, the liver has all of this high potency alcohol to use. The result is that the liver stores the fat. Chronic storage of fat in the liver leads to a condition called cirrhosis. Cirrhosis is caused by a large build-up of fat in the liver. Eventually this fat build-up causes severe liver dysfunction. If your liver dysfunctions, you die.

The effects on the digestive system are derived from the fact that alcohol, by dint of its effect on the brainstem, causes excessive release of acid in the stomach and is generally erosive to the lining of the small intestine. In chronic alcoholism, one or some combination of the above effects generally proves lethal.

Drugs which selectively depress some brain activities: As knowledge of the workings of the brain has increased, more selective drugs designed to manipulate one or more specific neurotransmitter systems have been developed. Among the most effective of those designed to selectively depress the nervous system is the group of drugs known as the benzodiazepines. The "premier" drugs in the benzodiazepine class are Librium and Valium.

In contrast to the general depressants discussed above, benzodiazepines do not cause surgical anesthesia. However, they can depress brain activity sufficiently to cause death. The mechanism of action of these drugs is related to the natural inhibitory neurotransmitter GABA. GABA is one of the two primary nervous system inhibitory transmitters which have been identified. When nerves that release GABA signal, the effect is to inhibit signaling activity on the receiving nerve cells. Benzodiazepines make the receiving nerve cells more sensitive to the inhibitory GABA action. Because of this, these drugs only have an effect when GABA is released. However, since GABA discharges represent a major part of the inhibitory component of the excitation/inhibition balance of the nervous system these drugs are quite potent. Their effect is to significantly "tip the balance" towards inhibition.

GABA-releasing nerves are found throughout the brain and spinal cord. Most GABA-releasing nerves are small nerves exerting inhibitory influences within an organizational level. In the spinal cord, GABA-releasing nerves are found in relatively high concentrations associated with the sensory paths carrying information to the brain. In the brain, GABA-releasing nerves are widely distributed, with high concentrations in the reticular formation, in motor control paths, in the hypothalamus, the limbic system, the cerebral cortex, and the retina of the eye. The effect of benzodiazepines is to increase GABA inhibition in all of these paths where the releasing nerves are located. In some respects, this makes the effect of these drugs more difficult to predict because the effect of the drug is dependent upon the level of activity in the various paths being influenced.

Like all drugs, the effect is a function of the amount of drug taken. With these drugs this is particularly true.

The benzodiazepines are used for a wide variety of purposes. These drugs are used as "sleeping pills." Their preference over barbiturates stems from their fewer side effects, especially those associated with depression of medullary centers which control basic body functions such as respiration. They are also used as anti-anxiety drugs (Fig. 14.5). This effect derives from their actions on the limbic system. Certain benzodiazepines are used in the control of seizure activity associated with some types of epilepsies. Finally, these drugs are mildly effective in controlling the uncontrolled muscle activity associated with Parkinson's disease in early stages.

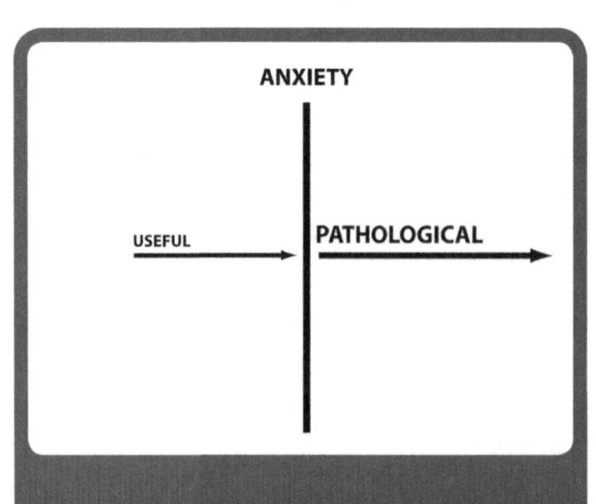

Figure 14.5 - Anxiety can be motivational and useful. However, at some point anxiety can become debilitating.

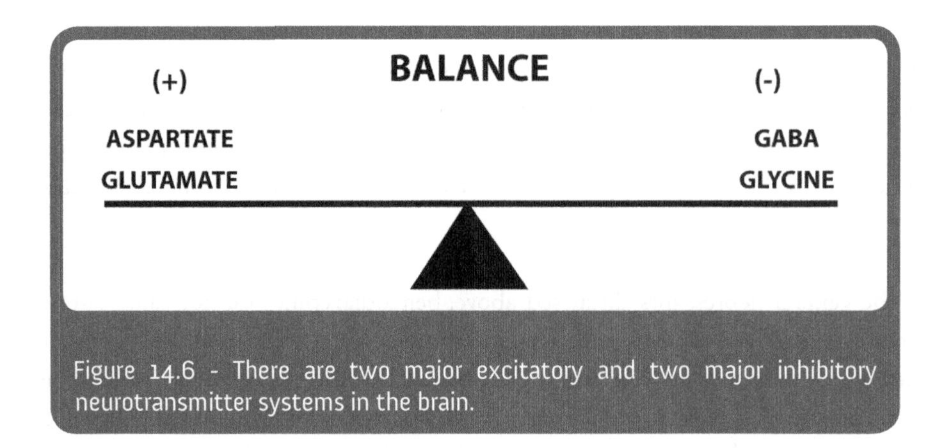

Figure 14.6 - There are two major excitatory and two major inhibitory neurotransmitter systems in the brain.

As one might expect, there are side effects associated with these drugs. The side effects include drowsiness, fatigue, muscular uncoordination, dizziness, blurred vision, and some unpredictable behavioral disturbances. Although respiratory depression is less prevalent than with barbiturates, benzodiazapines and alcohol together can be a very lethal combination. Tolerance does develop to these drugs. Perhaps the most prevalent occurrence is associated with the use of these drugs for anxiety. Withdrawal symptoms become marked when relatively high doses have been taken for some time. Withdrawal is expressed as increased anxiety and increased excitability of the nervous system. Under some circumstances, withdrawal from benzodiazapines can cause seizures.

There is a second neurotransmitter system in the nervous system, which like GABA, always inhibits brain signaling. This system uses the transmitter glycine. Although drugs which manipulate this system are not available for medical use, there is one drug which does quite specifically block the inhibitory signal of glycine. The poison strychnine blocks glycine signals. The result is hyperexcitability of the nervous system often to the point of seizures. Legally, strychnine is available for use in research and as rat poison. However, it is not uncommon for small amounts of it to be added to "street drugs" to add a "kick" to the drug. This illustrates a very important point: most of the drugs acting on the nervous system are powerful poisons. Another potent poison that influences the glycine system is tetanus toxin. Tetanus toxin is produced by the bacterial infection caused by clostridium tetani. This toxin blocks the release of glycine by neurons in the spinal cord. Considering the effect of blocking glycine action, it is no wonder that the disease is sometimes called "lockjaw."

GABA and glycine are purely inhibitory transmitters. When cells containing these transmitters signal, the effect is a decrease in general signaling activity in the nervous system. In balance, there are also two transmitter systems that always stimulate signaling activity in the brain. The two transmitters are glutamate and aspartate (Fig 14.6).

When cells containing these neurotransmitters signal, activity in the brain is stimulated. In general, the cells containing these transmitters do not seem to run between levels of organization. What they appear to do is to sit within a level of organization and act to control the amount of signaling activity that passes through that level on its way to another level. Although specific drugs to manipulate these systems are not available, there is considerable interest in these because of two food additives.

The first is monosodium glutamate (MSG). MSG enjoys extensive use as a food additive to "enhance" taste. Its use for such purpose dates to ancient times. However, as in many other cases, modern society has expanded the use of MSG considerably. Strangely, the mechanism of its effect in enhancing taste is not clear. The two basic ideas are that it adds directly as taste or that it acts as to augment glutamate in the brain. Consistent with this second idea are the reports that "sensitive" individuals react to MSG with increased neural excitability and headaches.

Similarly, aspartame is a food additive used as an artificial sweetener. Aspartame (Nutrasweet) is a combination of two amino acids: aspartate and phenylalanine. Both amino acids are significant in their relationships to neurotransmitters. Phenylalanine is a central molecule in the synthesis of biogenic amines (see below) while aspartate as discussed above is an excitatory neurotransmitter. When aspartame is digested it yields aspartic acid (40%), phenylalanine (50%) and methanol (10%). There are concerns about possible untoward effects of all three of these molecules. There is concern that aspartic acid adds to excitation in the relevant pathways. Of particular interest are possible contributions to hyperactivity in children and seizures in both children and adults. Studies addressing this question continue. Similarly, there are concerns that phenylalanine increases activity in biogenic amine pathways. Finally, there is concern about methanol because it is a poison.

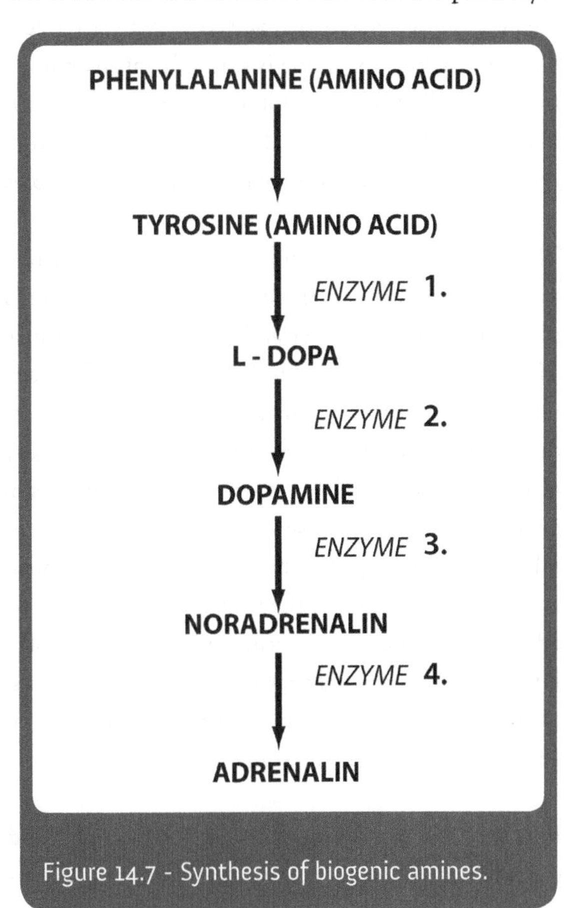

Figure 14.7 - Synthesis of biogenic amines.

Drugs which manipulate neurotransmitter systems and have "mixed" effects: Returning to the story of neurotransmitters, there are many neurotransmitters which are neither purely excitatory nor purely inhibitory. Their effect depends on the cell receiving the signal. They are located in many paths of information flow. They act to excite activity in some paths and inhibit activity in others. From a point of view of drugs, this makes them difficult to predict. Drugs are introduced into the entire body and from there into the entire brain. Since these drugs have actions in both directions in different places, the result is that some functional paths will be excited and others will be inhibited. The consequence is that selective effects are difficult to obtain.

One such group of transmitters involves three biogenic amines: adrenalin, noradrenalin, and dopamine. These three neuorotransmitters are made in a single enzymatic sequence cascade (Fig. 14.7).

Starting with the amino acid phenylalanine, the first enzyme produces a second amino acid, tyrosine. A second enzyme then produces a molecule called L-dopa from tyrosine. A third enzyme then produces the neurotransmitter dopamine from L-dopa. A fourth enzyme then produces the neurotransmitter/hormone noradrenalin from dopamine. The last enzyme produces the hormone adrenalin from noradrenalin.

The concern about phenylalanine produced from aspartame involves the neurotransmitters produced by this cascade. Part of the concern is related to a genetic disease that affects this cascade. Phenylketonuria is an inherited disease in which the first enzyme in the cascade is missing or defective. The consequence is that phenylalanine and several toxic metabolites build up in the individual's tissues. If not treated very quickly after birth severe retardation occurs. As discussed previously, adrenalin and noradrenalin are heavily involved outside the brain in effecting the acute sympathetic stress response. In the brain, noradrenalin as a neurotransmitter is involved in a variety of functions related to both emotions and stress responses. Noradrenergic paths in the limbic system are thought to be involved in activating reward systems, arousal, and mood. In the medulla, noradrenalin is involved in activating acute stress responses with particular effects on increasing blood pressure. Dopamine is similarly involved in a variety of functions in the brain. Dopamine, in an area of the brain called the substantia nigra, is involved in motor control. In the limbic system dopamine is involved in a diffuse cluster of social behaviors. Finally, in the hypothalamus dopamine is involved in the control of the releasing factors that control the output of hormones from the pituitary.

There are a significant number of drugs designed to either stimulate or block either dopamine or noradrenalin as neurotransmitters. The easiest way to understand how these drugs work is to mentally progress through the steps in neurotransmitter action. These neurotransmitters are synthesized and stored in the nerve terminal. There are drugs designed to influence both their synthesis and storage. When activated, the transmitters are released from the nerve terminal. There are drugs designed to influence their release. The released transmitter diffuses to the receiving cell. The transmitter has its effect by binding to a receptor in the dendritic membrane of the receiving cell. This is where complexity begins. There is not a single receptor for either noradrenalin or dopamine. There are several different forms of the receptors for both. There are drugs designed to try and selectively target specific noradrenalin and specific dopamine receptors. These drugs can bind to receptors and mimic the action of the transmitter (agonist), or they can bind and block the transmitter (antagonist). In addition, the effect of the transmitter is turned off by two processes. Some of the transmitter is taken back into the nerve terminal that released it (this process is called reuptake) (Fig. 14.8). There are drugs designed to influence reuptake. Some of the transmitter is destroyed. There are also drugs designed to influence the enzymes that destroy the transmitter.

Finally when considering the effect of any of the drugs influencing the dopamine, noradrenalin/adrenalin cascade it is important to consider selectivity. There is considerable cross-reactivity between noradrenalin and dopamine receptors. There is even more cross-reactivity between subtypes of noradrenalin and dopamine receptors. Selectivity for a particular receptor is at best partial. The drug has its effect on that particular receptor subtype at a lower concentration than its effects on the rest of the receptors. In addition to cross-reactivity between classes of noradrenergic and dopaminergic drugs, there is often cross-reactivity with another transmitter system, serotonin.

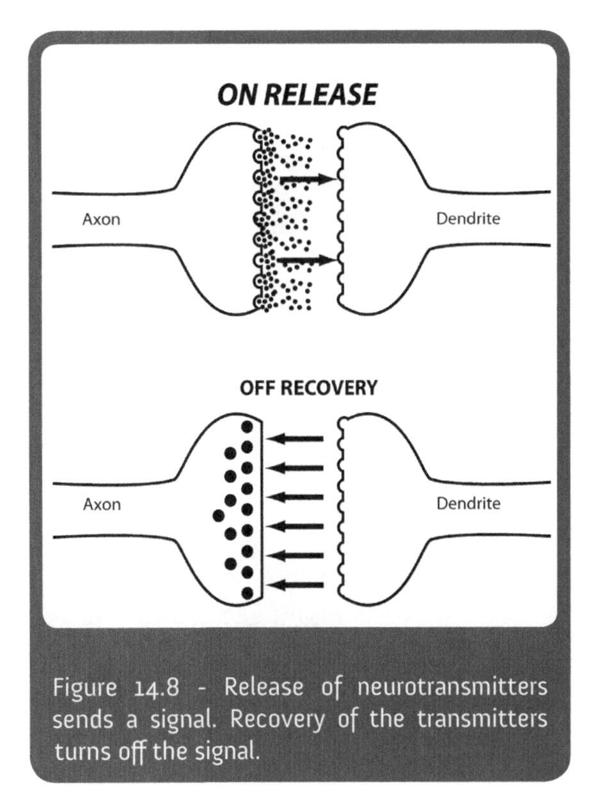

Figure 14.8 - Release of neurotransmitters sends a signal. Recovery of the transmitters turns off the signal.

Serotonin, also known as 5-hydroxytryptamine, is widely distributed throughout the body as a chemical message. However it also occurs as a neurotransmitter in the brain. By far the widest distribution of serotonin is in the lining of the digestive system where it is involved in secretions. It is also found in platelets where it is involved with blood clotting. In the brain, there are serotonin sensitive paths in the limbic system. These affect both mood and drives (hunger, thirst, and sex). There are serotonin paths in the cerebral cortex which are involved in sensory perception. There are serotonin paths in the reticular formation involved with sleep and attention.

Finally, as discussed in Chapter 9, there are serotonin paths in the pineal gland which participate in controlling hypothalamic/pituitary hormone secretions. Like dopamine and noradrenalin, serotonin is released by the nerve terminal and recovered. Also like dopamine and noradrenalin, there are several different serotonin receptors. Dopamine, noradrenalin, adrenalin and serotonin are referred to collectively as biogenic amines.

Amphetamines and cocaine: Perhaps the most famous or infamous drugs influencing cells responding to dopamine, noradrenalin, adrenalin, and serotonin paths are two powerful stimulants, amphetamines and cocaine. These drugs both work by causing the release of biogenic amines from the nerve terminal and then blocking their reuptake. Outside the brain they cause an acute stress response. The heart speeds up, blood flow changes, airways open up, sugar and fat are dumped into circulation, etc. If the dose is sufficient the heart can be accelerated enough to loose coordination (fibrillation). The restriction of blood flow to the surface of the body can be sufficient to cause an individual to run a fever. Inside the brain, the effects are quite widespread. The individual becomes hyperactive. The individual becomes euphoric and excited. The individual often carries out the same behavior over and over. Attention can be focused on a particular sensory input.

With continued use the person can exhibit paranoid schizophrenic behavior. Depending on the dose, the person can have a seizure. Seizure activity is facilitated by the increased body temperature due to the change in blood flow caused by the stress response.

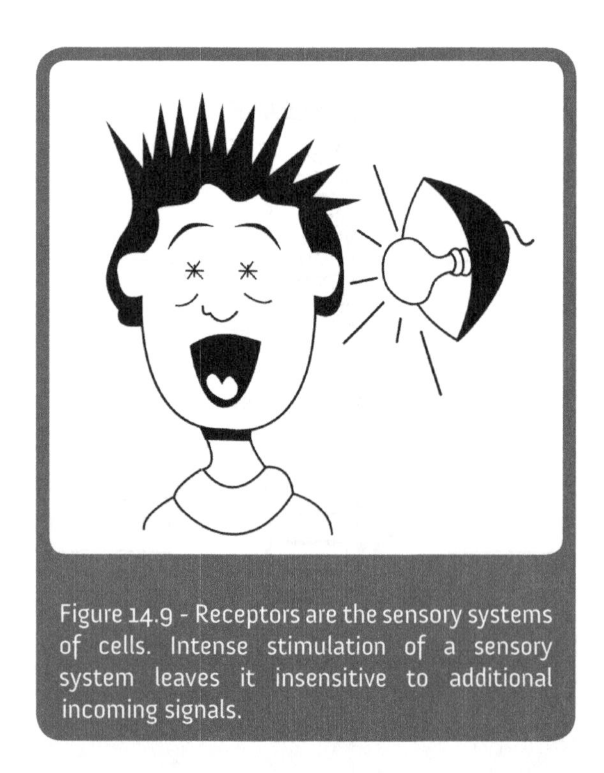

Figure 14.9 - Receptors are the sensory systems of cells. Intense stimulation of a sensory system leaves it insensitive to additional incoming signals.

Dependence on these drugs has an interesting pattern. People tend to repeat dose over and over. In part this is probably an attempt to avoid the "crash" that follows use. The effects of these drugs do not last a long time. When the drug is cleared, all of the various systems that have been activated are left "burned out." By analogy, think of the effect on your vision of a photographic flash set off close in front of your eyes. You are blinded by the flash. These powerful stimulants "blind" the biogenic amine systems to additional input (Fig. 14.9).

There are a few useful, albeit controversial, uses of these drugs. Amphetamines have been used as diet pills. Among all of the other effects, these drugs suppress appetite. The second use is for treating hyperactive children. It is not uncommon for the amphetamine Ritalin (and more recently a similar drug called Adderall) to be given to "hyperactive" children (a condition referred to as Attention Deficit Disorder, ADD). These are children who appear to be unable to focus attention on tasks in school. More often than not, these are boys who do not pay attention. The drug is used because of its effects on the reticular formation. The idea behind the use of amphetamines is that part of the acute stress response is a focusing of attention. Such treatment of these children is controversial because the "diagnosis" is somewhat subjective.

Antidepressants: Depression to some degree influences all humans. Feelings of misery, pessimism, apathy, guilt, inadequacy and indecisiveness are a part of every life at some time. The question of when depression becomes abnormal is one of frequency and intensity. Clinically, depression is divided into two categories. A person diagnosed with unipolar depression fluctuates between depression and normal mood. A person diagnosed with bipolar depression fluctuates between depression and mania. Mania, the opposite of depression, is unfounded feelings of enthusiasm, optimism, fervor, zeal and excitement. When a person is clinically depressed he or she may not have the motivation to get out of bed in the morning. He or she does not eat. He or she generally does nothing. When a person is manic he or she is like a car with its wheels slipping on ice. He or she puts a great deal of effort into doing nothing.

For milder depression, people are treated with drugs to help them deal with a relatively normal life. Severely depressed individuals are treated with more drugs to keep them from killing themselves. It is thought that depression is caused by depletion of one or more of the biogenic amines in certain paths, most probably in the limbic system. It is thought that mania is caused by an excess of one or more of the biogenic amines in those same pathways. All classes of drugs to treat depression except for one are founded on this somewhat diffuse hypothesis.

Tricyclic antidepressants are one of the longest used drugs to treat depression. These drugs work by blocking the recovery of both noradrenalin and serotonin back into the nerve that released them. The result is that when these transmitters are released their effect lasts longer and is enhanced because of build-up at the synapse. The actions of these drugs are a bit more selective than amphetamines and cocaine in that they generally don't influence release, and dopamine is much less affected. It takes several weeks for the antidepressant effects to begin. When the individual first starts taking the drugs he or she is initially lethargic, confused, and uncoordinated in movement. With time these effects decline and his or her mood seems to even out. There are a variety of side effects associated with the use of tricyclic antidepressants. As mentioned above they cause some drowsiness and confusion. They also cause postural hypotension. When a person stands up, blood flow is not adjusted for gravity and flow to the brain precipitously drops. The person faints. They can also cause seizures and mania. In men they prevent the blood flow changes necessary to attaining an erection. Finally, they interfere with parasympathetic control. The result of their effects on the parasympathetic nervous system is dry mouth, constipation, and blurred vision. These are "expensive" drugs in terms of side effects.

The second class of drugs that is used to treat depression is called monoamine oxidase inhibitors. Monoamine oxidases are enzymes that break down monoamines. The biogenic amines are all monoamines. These drugs act inside the nerve to reduce the normal turnover of transmitter. The result is that there is a build-up of transmitter within the nerve. The increased transmitter then leaks out of the nerve, generally increasing the amount of transmitter in the local environment. These drugs increase the amount of noradrenalin, dopamine, and serotonin in the local area around nerves releasing these transmitters. The result is increased excitement, activity, and a feeling of well-being. The problem with these drugs is that the enzyme monoamine oxidase is not exclusive to the brain. This enzyme is intricately involved in the digestion and metabolism of amine-containing molecules. The enzyme is extensively distributed in the digestive system. The enzyme is also used to metabolize amine-containing molecules in the liver. Of particular importance is a group of molecules called tyramines. These are plentiful in cheese and red wine. If such foods are ingested while taking the drug, the result can be severe high blood pressure and liver toxicity.

The most recent additions to the antidepressants are drugs which are relatively selective in blocking the reuptake of serotonin.

Prozac is the most common of these drugs. These have certain advantages over both the tricyclics and the monoamine oxidase inhibitors. They have very low toxicity, thereby reducing the risk of overdose. There is also no problem with food reactions. However, they do have the potential of causing nausea and insomnia. They also suppress hunger, causing anorexia in some individuals. The final side effect is somewhat controversial. There is anecdotal information that these drugs can cause hallucinations and aggressive, violent behavior.

The last antidepressant is lithium. Lithium, like sodium, exists in the body as a positive ion (Li^+). Although its mechanism of action is not understood, it probably at least in part involves substitution of Li^+ for Na^+ in ion-based electrical activity. The effectiveness of lithium is limited by the fact that the range between no effect and toxicity is very narrow. This is of great importance because even a little too much can cause severe neurological problems, seizures, coma, and death. There are many side effects to the use of lithium. Prolonged use of lithium can cause kidney damage. It can also suppress thyroid function.

Drugs to treat mental disorders: The potential power of these neurotransmitter systems is illustrated by the effects of drugs known as anti-psychotic drugs. These drugs are used to treat a group of psychiatric disorders known as psychoses. Psychoses are not a singular entity. Psychosis is defined as a condition where the individual is out of touch with reality. Although reality is defined somewhat differently by each normal individual, these individuals maintain their own reality which is so different from that of the "normal" individual in society as a whole that they are considered "sick." The expression of psychoses is quite variable. Very severely depressed or manic individuals can be considered psychotic. Such individuals can barely move, even to eat (severe depression). In contrast, some psychotic individuals can be so active that they constantly move (severe mania). There is simply a high degree of activity that is unfocused and without sustained purpose even from moment to moment. These people are treated extensively with the antidepressants we just discussed.

The type of psychosis known as schizophrenia usually involves both delusions and hallucinations. For example, a paranoid schizophrenic can believe that others are going to do him or her harm. Such individuals can be quite dangerous. Whatever the expression, schizophrenia is a severe behavioral disorder. There are a number of ideas concerning the cause(s) of schizophrenia. These ideas primarily derive from the action of drugs on behaviors. The two major ideas revolve around serotonin and dopamine. One idea is that a serotonin deficiency in parts of the limbic system causes the group of disorders known as schizophrenia. The origin of this idea stems from studies on the drug LSD. LSD is a serotonin antagonist. It blocks the serotonin receptor. The fact that LSD has strong hallucinogenic activity supports this idea.

The second idea is that an excess of dopamine in these same areas of the limbic system causes schizophrenia. This is supported by observations that drugs such as amphetamines and L-dopa that elevate dopamine cause delusional behavior. The drug L-dopa provides an interesting illustration of the selectivity problem of drugs and the brain. Parkinson's disease is caused by the progressive degeneration of dopamine-containing cells in an area of the brain known as the substantia nigra. There is a depletion of dopamine in this part of the brain. Parkinson's disease is a disorder of movement. It is marked by tremors, muscle rigidity, and difficulty carrying out the decision to move. As discussed previously, dopamine is produced from L-dopa in the cascade which also produces noradrenalin and adrenalin. By providing L-dopa, the level of dopamine in the brain can be increased. A side effect of treatment is schizophrenic-like behavior. Whatever the cause of schizophrenia, most drugs used in its treatment manipulate the serotonin system, the dopamine system, or both.

The phenothiazines are a typical group of drugs often given to individuals who express a psychosis with a delusionary quality. Although other drug groups exist for this purpose, in general all groups are similar in effect. The presumed mechanism of action is that these drugs block the effect of dopamine by binding to the dopamine receptor. In fact, they block all dopamine receptors, some noradrenalin receptors, some serotonin receptors, some acetylcholine receptors, and some histamine receptors. The effect on many patients is to "even out behavior" within the context of some sedation. However, the drug also causes muscle tremors similar to those observed in patients with Parkinson's disease. This is also a dopamine blocking effect.

Phenothiazines seem to reduce the level of signaling activity in the cerebral cortex, and probably reduce the flow of signals between the neocortex (decisions) and the limbic system (emotions). These effects on higher intellectual centers probably are the basis of their therapeutic value. Paradoxically, they can also lower seizure threshold and because of this can promote seizure activity. Because dopamine is involved in many hypothalamic functions, side effects associated with the control of hormones and other basic functions such as hunger are common. In addition, effects on other transmitter systems are observed. These include the "dry mouth" associated with interference with parasympathetic control of salivary secretions. Similarly, postural hypotension is also common. In postural hypotension, normal control of blood pressure is disrupted. The expression of this disruption can be seen when a person stands up. Normally, automatic sympathetic adjustments are made in the vasculature in order to prevent gravity from pulling the blood away from the head towards the feet. When these vascular adjustments do not occur, blood pressure in the head drops and the individual feels dizzy or faints. Finally, the phenothiazines block many reflexes such as the emetic (vomiting) reflex controlled in the brainstem medulla.

Epilepsy: We have discussed seizures as they relate to both taking and withdrawing from drugs. However some people develop seizures that are unrelated to drugs. Often the consequence of some type of brain damage is the development of a condition called epilepsy. Epilepsy is a condition in which in some place in the brain there are a relatively few cells that at times will send out high frequency, relatively intense uncontrolled signals. These signals can then spread and dominate activity in a local area or in some cases dominate much of the brain. Seizures can take many forms. If the uncontrolled electrical activity spreads to motor area the entire body can be overwhelmed by rigid jerking. If the uncontrolled electrical activity spreads to the reticular formation the person may just lose awareness. If the uncontrolled activity spreads to a temporal lobe of the brain the person may just "disappear" into his or her memories. If the activity spreads to a frontal lobe the person may just become confused. If the activity spreads to the limbic system the person may exhibit strong uncontrolled emotional behavior. Every individual case of epilepsy is different depending on how much and what parts of the brain are involved.

Approaches to treating epilepsy involve both blocking initiation of the uncontrolled electrical activity and preventing the spread once it has begun. There are a variety of drugs used for this purpose. Barbiturates, such as phenobarbital, both block initiation and prevent the spread of this activity. As you will recall, barbiturates both stimulate GABA paths and generally depress electrical activity in the brain. Benzodiazepines are also used because they enhance the effects of the inhibitory neurotransmitter GABA. We have already discussed side effects of these drugs.

A second approach is to directly interfere with the movement of ions across the nerve-cell membrane. As discussed in Chapter 8, electrical activity in the body involves the movement of ions across a cell membrane. Regardless of whether the electrical activity is in the brain, the heart, skeletal muscle, or smooth muscle, the controlled flow of ions across a cell membrane mediates the activity. Drugs like phenytoin (Dilantin) and valproate are used to treat epilepsy because they block the channels for the movement of Na^+ ions across the membrane. By blocking ion movement they suppress electrical activity. More recently, the use of drugs that block the flow of calcium (Ca^{++}) across membranes has been investigated for use with epilepsy. The use of these drugs to treat epilepsy illustrates an important point about the coordination of drug treatments. Many drugs used to treat problems with the rhythm of the heart work on the same principle. They also block the flow of ions across the cell membrane. For this reason, it is important to consider effects on the heart when prescribing drugs for the brain and vise versa.

Other recreational drugs: Caffeine, theophylline and theobromine are xanthines that are contained in coffee, tea, some soft drinks and chocolate. People use these drugs to give themselves a "lift." In addition, both theophylline and caffeine are used as drugs. Theophylline is often added to drugs that are used to treat asthma. Caffeine is the active ingredient in "stay awake" pills such as NoDoz.

One of the major effects of xanthines is to block the breakdown of the second messenger, cyclic AMP. A second messenger is simply the second reaction in a receptor cascade. Many biogenic amine effects are mediated by cyclic AMP. As a result, these drugs will prolong the effects of many of these signals. These drugs do not stimulate directly. Rather, they block the turn-off of stimulation once it has occurred. Activity increases because of the build-up of activity. Their effect is to increase activity in many parts of the brain. Activity increases in the reticular activating system. This increases and focuses attention. Activity increases in the respiratory and cardiovascular control centers in the medulla. Heart rate and breathing increase. They also increase sympathetic stress responses by acting directly on the heart, smooth muscle, and skeletal muscle. The use of xanthines in asthma medication prolongs the opening of the airways. In addition, many hormone systems use cyclic AMP as a second messenger. Therefore, they influence many hormonal effects. For example, they are diuretics (promote water loss in the kidney) because of effects on the kidney's water balance mechanism.

A second recreational stimulant also used by millions of people is nicotine. Nicotine is the active drug in cigarettes, snuff, and chewing tobacco. Nicotine directly stimulates one type of acetylcholine receptor. Acetylcholine is another neurotransmitter. It is used for signaling between nerve and muscle throughout the body and in the autonomic nervous system. It is also used as a transmitter in the brain. The primary reason that people use nicotine is that it stimulates a sympathetic acute stress response.

As with many drugs, tolerance and dependence develop with the continued use of nicotine. In some respects, addiction to nicotine is similar to addiction to opiates such as heroin. As with opiate addiction, the withdrawal effects begin as the drug is being cleared from the system. The result is the individual takes another dose. Withdrawal from nicotine is as expected: opposite of its effect. While nicotine stimulates and focuses attention, withdrawal leaves the person lethargic and unable to concentrate.

Marijuana is another commonly used recreational drug. The active compounds in marijuana are a group of related compounds known as cannabinoids. Although most of these compounds have pharmacological activity, some are more potent than others. It has only been fairly recently that some understanding of their mechanism of action has been gained. Initially it was thought that they acted like alcohol as a non-specific general depressant. This is not the case. Because of some of the effects of the cannabinoids, the idea that they acted like hallucinogens (LSD) and blocked serotonin was also considered. Only recently was it discovered that, like the opiates, they act rather specifically on a class of receptors. Cannabinoid receptors have been discovered within the body. Two types of cannabinoid receptors have been identified. One is restricted to tissues of the immune system. The other is distributed in certain areas of the brain. These receptors are found in the cerebral cortex, in the limbic system and in the cerebellum. Like endorphins and enkephalins and the opiate receptor, an endogenous messenger has also been identified. The endogenous transmitter is an oily molecule called anandamide.

Cannabinoids have a number of effects on mood, sensation, and behavior. They cause the person to have a general feeling of well-being. They heighten awareness of sound and to a lesser extent to sight. They stimulate hunger. They also interfere with both learning and motor coordination. They induce an acute sympathetic stress response, speeding up the heart, opening the airways, raising blood sugar and fat, etc. They also have two effects of potential medical importance.

They reduce intraocular pressure. This is of importance to treating glaucoma. In glaucoma, intraocular pressure is excessive and threatens vision. They also reduce feelings of nausea by suppressing central emetic centers. This effect, along with the stimulation of hunger, has potential value to cancer and AIDS patients. Interestingly neither tolerance nor dependence develops with marijuana smoking. Marijuana is not biologically addictive.

Messing with the brain can be dangerous: At present, understanding brain function and the development of neuropharmacology is at an infant stage. In some respects, we know enough to be constructive at times and dangerous at other times. In a sense, the brain contains a myriad of "buttons to be pushed" which will accomplish specific effects. At present, the problem is sorting out all of the "buttons" and figuring out how to selectively push them without disturbing the balance too much.

chapter fifteen

nature and nurture

The life of an individual: Every individual starts as a single cell, which is the product of the mixing of genes by a male and a female. That male and female derived their genetic mix as the result of a previous male and female mix. In one sense, the most important time in the life of an individual is the moment of conception. The egg gets together with one out of about 500 million sperm and the genetic composition of the single cell that will become the individual is determined. That single cell contains the potential that

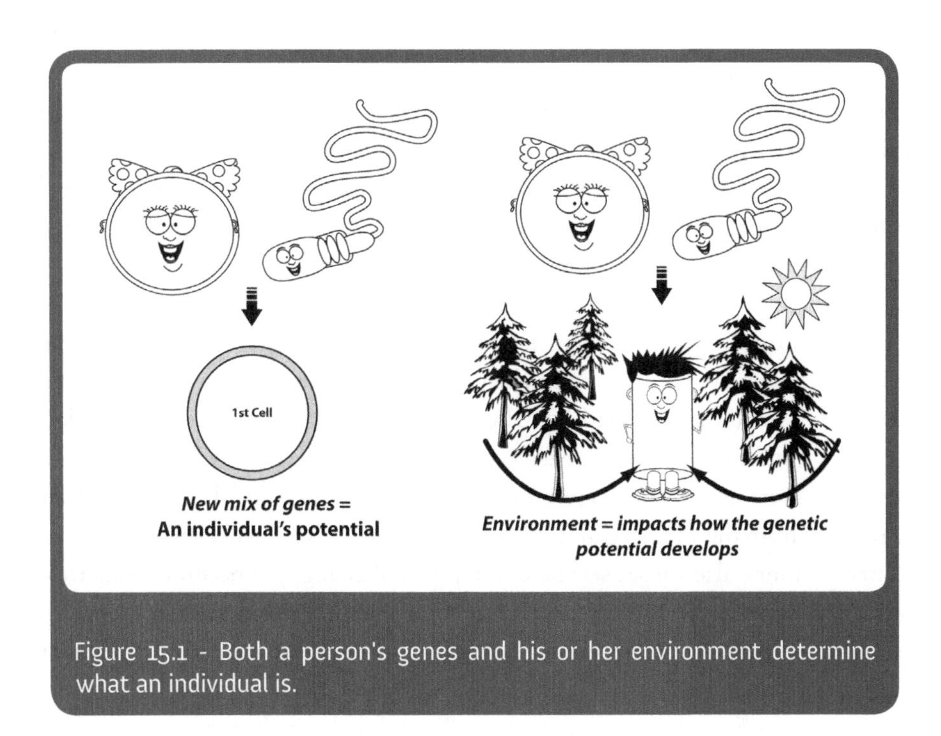

Figure 15.1 - Both a person's genes and his or her environment determine what an individual is.

defines what that individual CAN BECOME. What that individual ACTUALLY BECOMES is a function of environmental circumstances acting on the genetic mix (Fig. 15.1).

About seven days following conception, a small ball containing about 200 cells will implant in the uterus. There are two types of cells in this ball. One cell type will become the placenta while the other cell type is composed of the stem cells that will form the individual. These hundred or so stem cells will begin to divide and to specialize into the 200 or so different types of cells that form all of the organ systems of the adult. This specialization is essentially complete by the end of the first three months (the first trimester). At the end of this period the heart, limbs, teeth, palate and even the ears are completely formed. Although much of the central nervous system had been formed, the completion of the brain and spinal cord will not be finished until close to the end of the nine months. If something, whether it be a drug like thalidomide or a virus like Rubella (German measles), is going to cause a major structural defect it will happen during the first trimester.

The second three months is the time when many organ systems like the heart start to function (the second trimester). At the end of the second three months the individual weighs about one pound. If highly specialized health care is available then the individual may be viable outside the mother's body at this time. It is viable only with considerable medical assistance because the various organ systems are not prepared for life outside the mother. Of particular importance is the ability to breathe. Breathing is a complicated process. Breathing involves not only the lungs but also the brain and the musculature. The musculature is there but must be connected up properly with the nervous system. The nervous system is still developing not only until birth, but after birth as well. The nervous system must also be connected up correctly with the sensors that tell the brain that it is time to take a breath. Breathing is a complicated "emergency." The capability to survive outside of the womb is what develops during the last three months (the third trimester). Emergencies like breathing require the coordinated action of several systems. Coordination requires connections and integration at many levels. Breathing requires coordination and connections at many levels.

Perhaps one of the best illustrations of the importance of the last three months of pregnancy is fetal programming. If an individual is born prematurely, he or she usually has respiratory distress syndrome (RDS). This means that the individual has difficulty breathing. The cause of RDS is that the genes that make the surfactant that allows gas exchange in the lungs have not been expressed. Under normal circumstances, shortly before the baby "decides" to be born, its hypothalamic/pituitary/adrenal axis "turns on." The result is that a glucocorticoid, cortisol, is released from the adrenal gland. As discussed previously, glucocorticoids are a stress hormone. However, this same hormone also causes several developmental changes to occur. During this time, cortisol has many targets, the purpose of which is to prepare the individual to live on its own. Cortisol induces genes in many tissues that are important for living outside the womb. For example, up until birth the individual has been obtaining glucose from its mother. Now the individual's body will need to provide a constant source of glucose. Failure to do so even for a short period of time will result in brain damage, coma, and death. There are only two sources of glucose; eating or gluconeogenesis in the liver and kidney. The pre-birth output of cortisol from the adrenal induces the expression of the enzymes in the liver and kidney that are necessary for making glucose from amino acids and glycerol. The second emergency at the beginning of life outside the womb is obtaining oxygen and removing the waste product carbon dioxide. After birth, the mother no longer does this for the baby. The pre-birth output of cortisol from the adrenal also induces the genes in the lungs that are necessary for making the surfactant that reduces surface tension and allows for gas exchange. If for some reason the individual is born before the brain initiates the output of cortisol from the adrenal then breathing is problematical.

During the course of the pregnancy, the individual is actually protected from the stress hormone cortisol produced by the mother. Cortisol is normally circulating in a circadian pattern in the mother's blood such that it is high when the mother is awake and dealing with stressors and low when the mother is asleep. However, the developing individual is not exposed to these fluctuations in this stress hormone. The placenta expresses an enzyme that destroys cortisol and protects the unborn individual from exposure. When modern medicine figured out the role of cortisol in preparing the individual to breathe at birth, a solution was found to the problem of RDS in premature babies. There are many synthetic glucocorticoids that are used for a variety of purposes associated with suppressing the immune system and inflammation. As discussed previously these range from organ transplant to autoimmunity and asthma. Some of these synthetic glucocorticoids such as dexamethasone and betamethasone have a fluoride atom in their structure. The presence of the fluoride atom protects them from the enzyme in the placenta that destroys other corticosteroids such as cortisol. These corticosteroids can pass through the placenta into the developing baby. The solution was to begin treating a woman who starts premature labor with one of the fluorinated corticosteroids. If the premature labor results in birth the baby has been "primed for life." It can breathe and make its own sugar. It is not uncommon for premature labor to start and stop any number of times during the last three months. The result is that a woman can be treated several times in anticipation of pre-term birth. Even if the pregnancy continues to full term the result of this treatment is often a "low birth weight" baby.

Besides intentional treatment with corticosteroids to prevent respiratory distress syndrome in a premature baby there are a variety of circumstances that can lead to a "low birth weight" baby. These circumstances include exposure to a variety of environmental pollutants, maternal malnutrition, excessive stress in the mother and maternal use of a variety of drugs such as nicotine. A developing picture of low birth weight babies is one of an increased probability of developing metabolic syndrome as an adult. Why should low birth weight lead to type 2 diabetes, hypertension and arthrosclerosis as an adult? The current hypothesis is that such individuals have a fetal programmed "thrifty" metabolism. If an individual is subjected to unusual stress in the womb then many aspects of general metabolism are reprogrammed to be more efficient. The individual is reprogrammed to consume more and store more. Throughout the course of evolution, the ability to adapt in the womb to life in a sparse environment outside the womb had survival value. In modern Western societies with a grocery store on every corner the result can be obesity and metabolic syndrome.

Although the rapidity of change during postnatal development (development after birth) is far less than the prenatal development period, substantial change is still occurring. A major aspect of this change is the organization of the central nervous system. This is a combination of the brain "learning" how to control the body and body "teaching" the brain how to control it. Sight organizes the visual system and sound organizes the auditory system. Movement of the skeleton by the musculature progresses from random movement of the body to purposeful movement. All of this development is the formation of connections between nerve cells. It also involves death of nerves. Nerve cells that don't form a meaningful relationship with another nerve cell or muscle fiber die. At the end of this organization process, when the individual is but a few years old, there are fewer cells in the central nervous system then when they were born but the ones that are left have highly organized relationships with each other.

The other part of postnatal development is growth. A young adult is 20 to 30 times larger than the newborn. Growth itself results in some degree of vulnerability to an individual. This vulnerability is illustrated by polio (a viral disease of the nervous system that often left an affected individual unable to walk). Polio was also known as "infantile paralysis." Although adults such as President Franklin Roosevelt could get the disease, the impact on young, growing children was much greater. Motor neurons were a major location for the polio virus. When a motor nerve was infected it was not controlling the fibers of its motor unit, hence paralysis. The skeletal system develops in coordination with the muscle applying force. A spine grows straight because of the balance of muscular forces being exerted on it. If growth is occurring in the context of imbalanced forces curvatures can occur that get locked in place by the growth.

If you want to know what can go wrong with you, look to your parents and grandparents. One of the most important pieces of information one provides to his or her physician is his or her family medical history. Clearly some medical problems are caused by the direct, or Mendelian, inheritance of defective genes. Some of these diseases are autosomal recessive and you must inherit two defective copies, one from each parent. Phenylketonuria (PKU), in which the enzyme phenylalanine hydroxylase is defective, is an autosomal recessive disease. Some inherited diseases are co-dominant, in which the people with two defective alleles are just "sicker" than those with only one defective allele. Sickle cell anemia is an example of co-dominance. Other inherited diseases are autosomal dominant, in which case you only need to inherit one copy of the gene to develop the problem. Huntington's is an example of an autosomal dominant disease.

Beyond simple straight-forward inheritance, genes have a more subtle relationship to disease. Each individual, which is a unique genetic mix, responds differently to factors such as lifestyle and the environment. For example, fair-skinned individuals have a higher probability of developing melanoma. Melanoma is a cancer of the pigment producing cells of the skin. Such individual's ancestors evolved in northern climates where there is not a lot of sun, hence the light skin. A significant contributing factor in the development of melanoma is repeated sunburns.

Conversely, dark-skinned people's ancestors evolved in areas of the earth around the equator where there is a lot of sun. Dark skin evolved as a protection against the sun. However, the darker skin can lead to the development of rickets in dark-skinned children growing up in northern cities. Rickets can be caused by a vitamin D deficiency. Vitamin D is necessary for absorption of calcium. Vitamin D can be obtained from the diet, but it also is produced by the interaction of sunlight with the skin. The sun protection afforded by dark skin also inhibits the production of vitamin D. Therefore, dark-skinned children must obtain vitamin D in their diet. That is why vitamin D was added to whole milk.

There are many traits that are not the product of a single gene, but rather of many genes, all of which influence the development of that trait. These are traits that appear in the population as a continuum of graded responses, rather than "all or none." We have already used the example of height as an example of such polygenic ("many genes") inheritance. With polygenic traits, environmental factors very often have a substantial imput on the final expression of that trait. There are many conditions or diseases that, while not directly inherited in a simple Mendelian fashion, still have a strong genetic component. Examples include type 2 diabetes, obesity, many types of heart disease, high blood pressure, many types of cancer, depression, schizophrenia, and many others. Such conditions tend to "run in families."

One aspect of the study of the human genome is the study of variations in single genes. Much of this work focuses on the study of single nucleotide polymorphisms (SNPs). SNPs are small variations in the nucleotide sequence of the gene. SNPs have been identified in both the coding and non-coding sequence of genes. Such variations occurring in the coding sequence may change an amino acid in the structure of the expressed protein and either increase or decrease activity. When such variations occur in non-coding sequence they can alter the regulation of the expression of the gene. Variations in genes can change the probability of developing certain diseases. For example, certain inheritable mutations in specific genes have shown to greatly increase the risk of developing breast cancer. One such gene is the Breast Cancer 1 Gene (BRCA1). BRCA1 is a gene involved in identifying and correcting mistakes in copying DNA during cell division. Inheritable mutations in BRCA1 are associated with an increased incidence of both breast and ovarian cancer in women and prostate cancer in men. Increasingly, inheritable mutations in genes associated with the development of different kinds of cancer are being identified.

Similarly, inheritable mutations in a number of different genes are being identified that result in an increased probability of developing other diseases such as type 2 diabetes. The presence of variants in genes that are associated with the development of a particular disease does not mean the individual will absolutely develop the disease, it simply means he or she has a higher probability. Whether or not a person carrying a particular variant of a gene will develop the associated disease depends to a great degree on lifestyle, like the example of skin cancer and sun exposure.

As people get older, they develop more medical problems. Aging is not an accident; it is a design of nature.

One of the most important observations relevant to understanding aging was made in the 1960's by Leonard Hayflick. He developed data demonstrating that there is a limit to the number of times a normal cell can divide in culture. In human cells this Hayflick Limit is about 52 divisions. The importance to aging is that when one compares the Hayflick Limit across species this limit is proportional to the lifespan of the species of animals. Animals with short lifespans, such as mice, have a lower Hayflick Limit than animals with longer lifespans, such as humans. These and other similar observations suggest that senescence is "built into life."

With all of modern medicine, humans have not pushed very far beyond the Biblical life span of three score and 10 (70) years.

However, there are many hypotheses as to why some individuals age faster than others. Many of these hypotheses derive from the idea that with time there is an accumulation of something "bad." Different hypotheses focus on different aspects such as the accumulation of damage, of errors, of metabolic waste, of mutations and of antibodies against self (autoimmunity). The importance of these ideas is that understanding the factors that contribute to aging may allow mitigation. In the narrowest sense, the quality and quantity of modern human life is becoming focused on health care and medicine. Increasingly medical care and life in general is becoming about drugs. It is therefore important to develop some understanding of how drugs work and where they come from.

Where did drugs come from and where are they now? From the beginning of recorded history, humans have been looking for potions to protect themselves from or to cure the ills of life, potions to increase strength and potency, and potions to relieve the discomforts of existence. Various extracts of plants and animals were found, by trial and error, to have curative powers. If a substance seemed to work, and if it didn't do too much damage, then it became a "magic potion" of the spiritualist or herbalist. These potions had a mystical quality, because their use had no foundation in understanding of how they really worked in the body. A very good example of such a drug that is still in use today is aspirin. Aspirin is a type of chemical compound known as a salicylate. A similar compound, salicin, is contained in the bark of the willow tree. The use of brews containing willow bark to treat mild pain and fever dates back thousands of years. Similarly, the powers of the opium poppy, coca leaves, and the marijuana plant all date back thousands of years. Thus, from the search for cures to the problems of life sprang many potions which manipulated the mind and body.

In modern times, the potions of the spiritualist have been replaced by the drugs of science. However, nature still remains the source of many drugs. Drug companies spend millions of dollars searching the flora and fauna of the world for compounds with medicinal properties. For example, the cancer drug Paclitaxel (Taxol) was found in the bark of a type of yew tree.

In addition, molecular biology has made all hormones and chemical messages of the body potential drugs. Modern molecular biology has allowed us to determine the nucleotide sequence of these individual chemical messages and insert them into bacteria that can be grown up in the lab and made to produce those proteins for us. By this process of cloning the gene sequence for a protein into bacteria, we can turn those bacteria into "factories" to produce that particular protein. This has made such drugs both cheaper and safer. Before cloning, such drugs were generally isolated from animal sources, and the differences in the animal product and human product often made the drug less effective and could cause unwanted side effects.

The first hormone that was cloned and used as a drug was insulin in the 1970's. Before that time insulin dependant diabetics injected themselves with pig or cow insulin. Synthetic erythropoietin (Procrit and Epogen), a hormone that stimulates red blood cell formation, is available to treat anemia. Interferons, which are a type

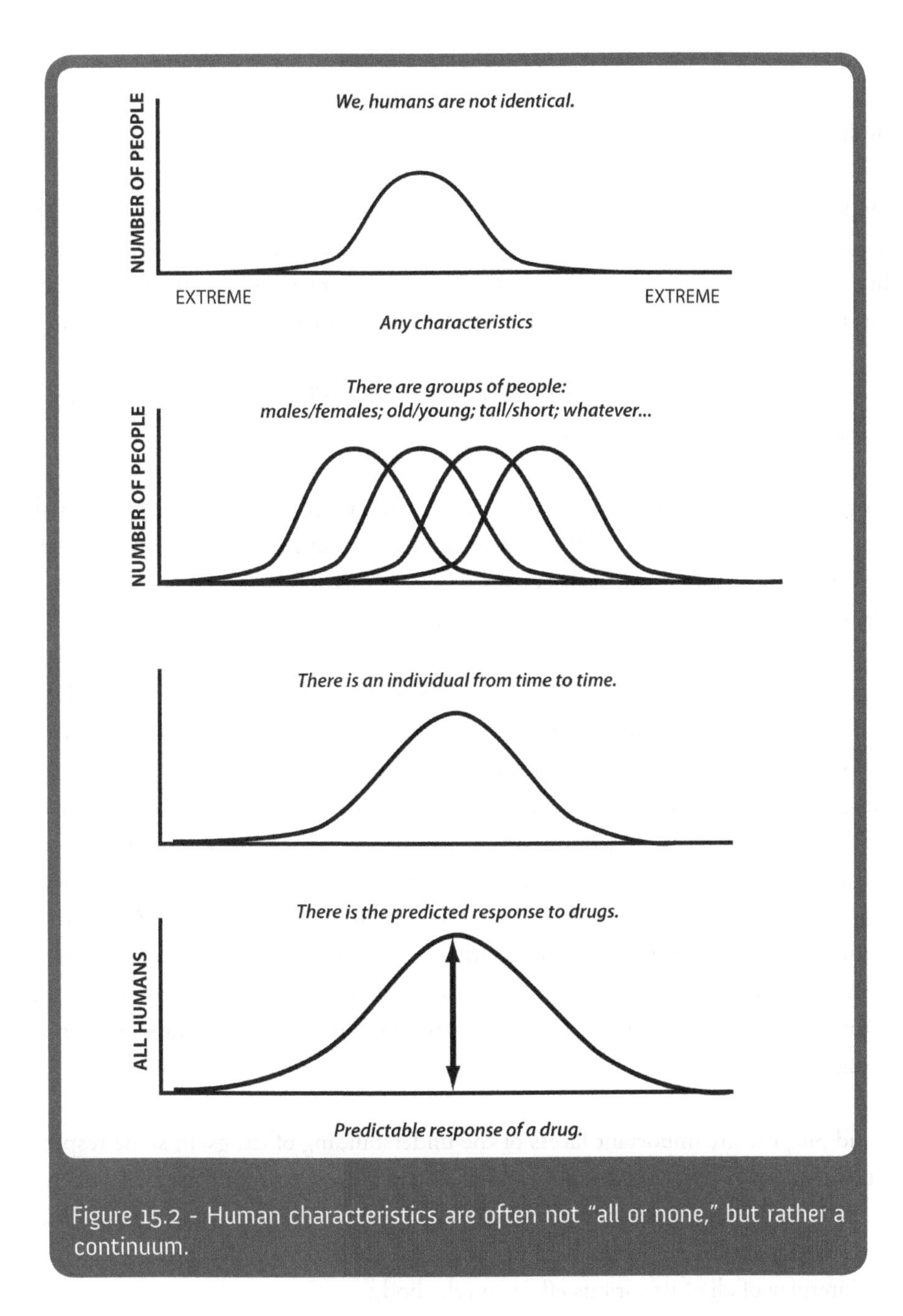

Figure 15.2 - Human characteristics are often not "all or none," but rather a continuum.

of chemical message of the immune system, are being tested for their value in treating immune related diseases ranging from types of leukemia to multiple sclerosis.

One would think that with a foundation in science, most drug effects would be predictable. However, even with current knowledge, drugs still have an unpredictable quality. The effect of any drug on any given individual at any given time is still often enigmatic and unpredictable. The lack of predictability stems from the fact that the elements of life are quite variable (Fig. 15.2).

Statistics is the tool used to introduce an element of predictability into the use of drugs. The completely average individual in the average situation will respond to a particular drug in the average manner. The lack of predictability stems from the non-existence of the absolutely average individual and the absolutely average situation. Who is average when one considers all of the myriad characteristics imparted by the genes?

Who has had a completely average development and maturation? Who always eats the average diet, and who is under the average amount of stress all of the time?

At present, drugs are unpredictable because of individual differences and because we are still ignorant of many aspects of human function. One of the goals of medicine that may be obtainable in the 21st century is individualized medicine. The genome projects, the study of SNPs, Bioinformatics and the vast computing power available is making this goal potentially attainable.

What is a drug? Like many words, "drug" has taken on many meanings and associations with use. We have drug abuse, wars on drugs, designer drugs, drug testing, drug addiction, and drug side effects. We also have orphan drugs, experimental drugs, drugstores, prescription drugs, and over-the-counter drugs. The wide use of the word drugs in a wide range of different contexts is informative to the degree that it indicates that drugs are a pervasive part of life. It is also important because it alludes to the power or potency of drugs. Drugs are substances that do have unusual power over the body. In small quantities drugs are able to modify the functioning of the body, and in larger quantities virtually all drugs can kill. However, the lexicon does little to define the large group of substances referred to as drugs. Antibiotics, vaccines, birth control pills, antacids, aspirin, antihypertensives, antiarrhythmics, antiepileptics, antidepressants, opiates and stimulants are all drugs. Drugs are also substances that occur in commonly consumed substances such as tea, coffee, soft drinks, tobacco and chocolate.

Although use and purpose are important facets of the understanding of drugs, in some respects the lexicon misrepresents the nature of drugs. For example, aspirin is classified as a non-narcotic analgesic. Aspirin relieves low-level pain. Aspirin also reduces fever, reduces inflammation and inhibits blood clotting. As a result, its use and purpose are situational to the individual. However, one cannot put such a drug to a particular use without considering the interplay of all of its various effects on the body.

In the broadest sense, a drug is simply a molecular structure whose effect on the body is great even in small amounts. Drugs are molecular structures that magnify their presence in the body by influencing critical events within a large sequence of events. Like millions of dominos set up to fall one after another, life is a continuous complex cascade of molecular events fueled by energy. Death occurs when the cascade stops. If you modify the event of the fall of one domino in the cascade, then the thousands of subsequent events represented by the fall of all of the dominos down the line is changed. Such is the nature of drugs: they modify molecular events causing cascading consequences.

By this definition, hormones are the natural drugs made by the body. If you add to them some minerals, all vitamins, a large variety of natural molecules made by plants and animals, and molecules made by human beings in laboratories, you have a complete collection of all drugs. Each drug is a particular molecular structure which modulates the functioning of the body.

Drugs in the body are molecules with a purpose. The purpose is the reason for taking the drug. Drugs have target functions in selected groups of cells. The target cells are carrying out the function that the drug is designed to alter. However, when a drug enters the body it distributes amongst many environments. Some drugs have very wide distributions and some have relatively narrow distributions. For example oily drugs like steroids move into membranes and fat very easily, while a charged drug like penicillin does not.

In addition, different amounts of different drugs will enter different environments. How much drug enters any given environment will depend on the character of the drug and the character of the various environments available to the drug. Molecules in all environments have a tendency to bind or stick together in both specific and non-specific ways. When a drug molecule enters an environment and sticks to another molecular structure it is both prevented from interacting with other molecules and it modifies the behavior of the molecule that it stuck to. For example many drugs, such as penicillin, bind to molecular structures in circulating blood. This binding can greatly influence how long the drug stays in the body and how much drug is available to affect cells outside of the blood. It is also involved in causing allergic reactions to drugs.

Finally, the body has ways of destroying drugs. Many drug molecules are destroyed before they even have a chance of coming near their target. How much effect a drug will have will depend on how much of the drug interacts with that target at any one time. This will depend on the distribution of the drug. How long the drug will continue to have an effect will depend on how long some of the drug stays in the body. This will depend on how long the drug stays in each of the various environments it entered and how fast the body destroys the drug. How much of any dose of a drug that participates in causing the desired effect of the drug, as well as how much will participate in causing side effects, will always be determined by where the molecules go and how long they stay in the various places (Fig. 15.3).

Figure 15.3 - Drugs in the body.

How do drugs have their desired effects on the body? There are many different kinds of drugs designed to accomplish many different purposes. Some drugs like steroids are copies of natural chemical messages. These drugs work by substituting or adding a man-made message to the message systems of the body. The information of the various chemical message systems is carried by how much of the message is present at different times. The addition of a man-made message to the existing chemical message will change the directions being given to the body. For example, estrogen/progesterone birth control pills change the message and tell a woman's brain that she is pregnant when she is not. This basically instructs the brain not to send the message to the ovaries to develop an egg to be fertilized. Substitute messages can also be used in conditions of disease such as type 1 ("juvenile") diabetes. In type 1 diabetes, the cells of the pancreas that make insulin have been damaged by the immune system. As a result, they do not produce enough insulin. An individual with type 1 diabetes gives himself or herself a shot of insulin after he or she eats to provide the "store nutrients" signal to most cells of the body. Finally, substitute messages can be used to turn up or down the activities of groups of cells whose function is deemed detrimental to the individual at a particular time. For example, large amounts of the chronic stress message can be used to turn off the body's normal immune defense mechanisms and prevent the rejection of a transplanted organ such as a heart or kidney.

Just as there are drugs designed to mimic normal biological chemical messages, there are drugs designed to block the messages of the body. These drugs usually block the reception of the signal by binding to the target structure and thus preventing some or all of the body's own message from being received. For example, one way to control blood pressure is to block the signals for raising blood pressure. Often drugs that block the signal for raising blood pressure are given to individuals at risk of having a stroke or heart attack.

Drugs can also work by binding to specific molecules that carry out specific reactions.

There are many large molecules made by the body that carry out specific reactions on other specific molecules. Such reactions are usually part of a sequence of reactions designed to carry out some function. Like the dominos, if you block one of the reactions in the sequence, you can block the function from being carried out. For example, many cancer chemotherapeutic drugs block some reaction in the sequence of reactions necessary for reproducing genetic material. If a cancer cell cannot reproduce its genetic material, it cannot reproduce itself. If these cells cannot reproduce themselves then the cancer cannot grow and spread.

Finally, drugs can work by simply dissolving into particular environments in the body. When they dissolve into the particular environment, they change the properties of the environment. Changing the environment then alters the functioning of the other molecules in the environment. For example, some local anesthetics have their effect by dissolving into the environment of the barrier membrane of nerve cells. The effect is to block the signal from crossing that local area on the long nerve cell. The pain signal cannot cross that point on the nerve and the signal never reaches the brain.

Why are different drugs given in different ways? In order to work, a drug must first enter the body. There are several different ways in which drugs are given. How a particular drug is administered will depend upon where one wants the drug to go and what one wants it to do.

If a drug is required to treat a condition in a local surface area of the body it may be applied topically (directly to the area). Often drugs used topically are in the form of creams, lotions, or sprays. Whether or not a drug is given this way may depend upon whether or not there is a break in the surface barrier or whether or not the drug is absorbed through the skin layers. In more recent times, surface application has developed as a method of systemic dosing. Drugs with high penetrance are applied in "patches" against the skin. The drug slowly leaves the patch, enters the skin and is absorbed into circulation. This method provides a good way for delivering a relatively constant amount of drug over a reasonably long period of time. Nicotine can be delivered this way as a drug to aid in stopping smoking, and female sex steroids for birth control or treatment of post-menapausal symptoms are also available as patches.

Another common mode of administration for many drugs is oral administration. The drug is taken in through the digestive system. Drugs administered orally may be in the form of pills, capsules or as a liquid. They enter the digestive tract in some form and are absorbed into the blood. Absorption occurs in the small intestine. This is a very simple and effective method for administering many drugs. The person can easily and painlessly self-administer the drug.

However, there are some drugs that cannot be administered orally. Some drugs are destroyed by the digestive system. For example a diabetic cannot take insulin orally. Insulin is a protein. If administered orally, by the time a protein drug reaches the small intestine it would be just a collection of amino acids instead of the drug. In other cases, some drugs may be harmful to either the digestive tract itself or to the beneficial bacteria that reside there. The form in which a drug is presented can be very important in oral administration. Sometimes drugs are complexed to salts or other compounds, some are coated or they are placed in special protective capsules. Structural characteristics of drugs can be changed either to increase the solubility of the drug, increase its absorption, or in order to protect it from the rather harsh environment of the digestive system.

Oral administration is a relatively slow way to get a drug into the body. It takes a certain period of time for the drug to travel through the digestive system and be absorbed into the bloodstream. In addition, since the drug enters the hepatic portal system it must first pass through the liver before entering general circulation. Often a significant percentage of the drug is picked up and destroyed by the liver. Oral administration is not used in situations where it is important that the drug rapidly enter the body. In addition, the state of the digestive system is important. How much, what, and when a person has eaten can influence how a drug is absorbed or what effect it has on the digestive system itself. That is why specific directions for taking a drug orally are often supplied with the drug. For example, a drug may be prescribed to be taken after a meal, or on an empty stomach. It may be recommended that a drug be taken only with certain foods such as a glass of milk. It may be suggested that a drug not be taken with a certain food. The reason for these types of recommendations is

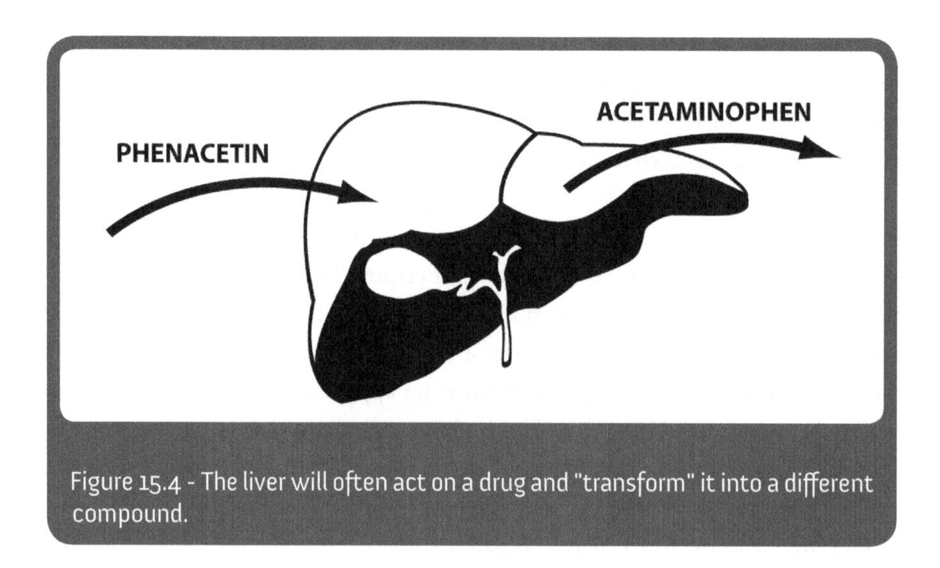

Figure 15.4 - The liver will often act on a drug and "transform" it into a different compound.

that certain drugs and food interact with each other. They influence each other's transit through the digestive system into circulation.

Another way to administer a drug is by injection. There are several different ways to use injections. A drug may be injected subcutaneously (s.c.). In s.c. administration, the drug is injected under the skin. It then must diffuse into the vasculature in order to enter the blood and travel through the body. Another way that drugs are sometimes although much less commonly injected is intraperitoneally (i.p.). The drug is injected into the body cavity that contains the liver, stomach, kidneys and intestines. Drugs given in this manner are absorbed into the capillaries around the digestive system. The hepatic portal system then carries the drug to the liver. The drug then must pass through the liver screening system before it is distributed throughout the body. Often a significant amount of the drug is destroyed by the liver and never reaches the body. Yet another common way to introduce a drug by injection is intramuscular (i.m.). An intramuscular injection is an injection given directly into a muscle. Muscle is a highly vascularized tissue, and the drug diffuses fairly rapidly into the blood. However, there are some drugs that cannot be given this way. Because of their chemical nature, some drugs may be damaging to the muscle tissue. The final way that a drug may be injected is intravenously (i.v.). An i.v. injection introduces the drug directly into the blood. It is by far the most rapid way to administer a drug. In this mode, the drug is carried rapidly throughout the body. For all intents and purposes, the drug is present in the body virtually immediately. For the same reason, it is also the most potentially dangerous way to give a drug. Because the drug enters the system immediately, any possible untoward reactions will also be immediate.

The body acts on the drug: The way in which a drug is administered as well as its chemical form not only determines where it goes in the body, but also how long it stays in the body. The liver and the kidney are the two organs which are primarily responsible for clearing drugs from the body. How rapidly the liver and kidney eliminate a drug will depend on both its form and how quickly it reaches these organs. If a drug travels in circulation bound to another molecule or a cell, then it will leave the body slower because only "free" drug will be filtered. In addition, the liver often acts on a particular drug and turns it into a chemically different compound from the one that was originally given (Fig. 15.4). This process is called biotransformation. That is the original compound is "transformed" into a different compound which may be more or less potent, or may have a totally different effect. Recall that Prontosil was transformed by the liver into a sulfonamide. In some cases, this is part of the design of the drug.

A "prodrug" is given with the intent that the liver will transform it into the active molecule. Recall that phenacetin is a prodrug that is transformed by the liver into acetaminophen. However, in some cases the liver can transform a molecule into a toxic molecule. Finally, it is important to remember that the liver is a major site of drug interactions. What happens to a drug in the liver will depend on what other molecules are being processed by the liver at the same time. For example, there are reports of liver damage when acetaminophen and alcohol are taken together.

When a drug is taken is also important: The dosage schedule of a drug also has a big influence on how well a drug accomplishes its intended purpose. The pattern of presentation of the drug to the body, as well as the absolute amounts, can be very important in determining drug action. For example, intermittent use of some drugs can be very unwise. This is especially true of antibiotics. Intermittent use of antibiotics leads to the development of antibiotic resistance. The microbial population can evolve during the course of an infection. The intermittent use of an antibiotic makes it a force for selection of resistant microbes. How you stop taking a drug can also be important. Especially with drugs acting on the nervous system, rapid withdrawal can have devastating effects on the body. If a drug causes tolerance, then the body has adapted to its presence. The proper way to stop such drugs is to slowly reduce the dose. This allows the body to adapt to the absence of the drug. For all of these reasons, it is important that directions as to when to take a drug be carefully followed.

How new drugs are developed: Initially drug development meant simply finding extracts of plants, animals or minerals that seemed to have some effect on the human body. Throughout history, drug development was always more or less of a "trial and error" endeavor. However, with the accumulation of more and more knowledge, drug development has moved towards being a more exact science.

With recent progress in science, we know more about the details of how the human body operates, and more about how many diseases operate. With such knowledge, drug development has become more of a technological problem. An attempt is made to tailor drugs to a specific adjustment problem in the body. The biggest difficulty in drug research usually is finding out in biochemical and/or molecular terms what causes a particular disease. The idea for a drug to treat a particular disease or condition stems from understanding both normal and abnormal human biochemistry.

Drug development is a big business. A "good" drug to treat some common malady is literally worth a fortune. However, developing and obtaining permission to sell a drug can also cost a fortune. For this reason companies are very careful in deciding on where to invest their research and development money. If a disease is rare, then the chances of a company investing money in its treatment are low. If a disease is common, many companies may be investing hundreds of millions of dollars searching for and developing improved drug treatments.

The testing and marketing of drugs in this country today is regulated by the federal government under the purview of the FDA (Food and Drug Administration). Although the functioning of this agency may not be without fault its existence, as well as the passage of certain laws governing the marketing of new drugs, has done much to improve drug safety over the years. The goal in developing drugs is to come up with a product that is both safe and effective. However, both "safe" and "effective" are relative terms. Drugs are more or less safe and more or less effective.

A network of drug research is carried out in drug companies, hospitals and universities around the world. Once an idea for a drug has been developed, and the substance is at hand, it must be tested. The first level of testing in this country is pre-clinical. Pre-clinical studies involve both chemistry and biology. The chemical compound must be thoroughly characterized and precisely defined as a molecule. The biological characterization first involves an evaluation of its action on cells in culture. The goal is to ascertain how the molecule influences life processes at the most basic level.

Biological evaluation then involves evaluation of the molecule in living mammals. Although no other mammal is identical to humans, for the most part they work very similarly. It is usual to screen in at least two or three different species.

For example, rats, mice, guinea pigs, rabbits and dogs are often used for such tests. These tests are extremely important because they define the general behavior of the drug in a complex organism which is similar to humans. These tests answer questions about such factors as toxicity, distribution, and metabolism. They also address questions concerning the effect of the drug on the animals, and define the amount of drug that causes the effect. The two most important effects to be characterized with respect to dose are those related to the purpose of the drug and to death: what doses are necessary to have the desired effect, and how much is necessary to kill the animal. Virtually all drugs will kill at some dose.

Although these studies can supply valuable information, a drug cannot be totally tested by such methods. There are several problems with such testing. One problem is that most human diseases don't have an appropriate animal model. However, the major problem is that these tests are not being done in humans, and no animal model or system is exactly like a human. Not all aspects of how the drug will behave in the human body can be ascertained by such animal studies. However, a lot can be learned from such pre-clinical studies. These screens will at least initially indicate the rough limits of safety of the drug. They will also indicate if the drug might be effective as a treatment for the particular malady. A well designed drug screen will at least give an indication that a "good" drug may work and that a "bad" drug will not work. Equally, if not more important, a good drug screen will provide information about possible toxicity problems with a particular compound such as mutagenesis (ability to cause irreversible damage to genetic material), teratagenesis (ability to cause birth defects in a developing fetus), and carcinogenesis (the ability to promote cancer). Just as many potential drugs never make it to pre-clinical testing, many potential drugs flunk pre-clinical testing. They are not worth more investment.

Once pre-clinical testing is completed, the company can submit the results to the FDA. The purpose of the application is to obtain approval to begin studies with humans (clinical testing). In order to gain this approval the company must submit extensive pre-clinical data along with documentation on how it was obtained. Lying to the FDA is a crime. Second, the company must specify in great detail the precise conditions under which humans will be given the drug. If approved, clinical testing in humans can begin.

These human tests involve two important aspects. The first is the use of placebos. Placebos are simply "sugar pills," given to half of the people in the study. The subjects don't know whether or not they are getting the drug or just the placebo. This is designed to factor out any psychological effects of the individual receiving the drug thinking that the drug has some effect. The second is the doubleblind aspect of the design. In addition, neither the clinical investigators nor the evaluators of the study know which individuals have actually received the drug. This is designed to prevent any personal bias associated with how the investigator "wants" the results to come out from influencing the interpretation of results.

Human testing is initially divided into three phases. Phase I involves a small group of volunteers (from a few dozen to a few hundred). They are given small doses of the drug for an initial evaluation of how it will affect humans.

These studies generally last for a relatively short period of time. These usually progress from relatively short exposures to no more than six months. These are normal volunteers, and the drug is evaluated for any toxic sidereactions. It is the "go slow" part of clinical testing.

If all goes well with Phase I, then the company can return to the FDA for approval to start Phase II testing. As with Phase I approval, Phase II approval requires well-documented results and a thorough description of the study to be done. In this phase, a much larger number of healthy volunteers is used, along with a small number of "diseased" individuals. At the same time, long-term chronic toxicity studies are run with lab animals. Phase I has already shown that the drug is not likely to kill anyone; Phase II will hopefully tell if a drug is potentially beneficial and identify problems with chronic toxicity.

If Phase II gives positive results, then the company can return to the FDA for approval to begin Phase III. In Phase III the number of people with the disease is expanded. This usually involves several university hospitals across the country. Also, other specialists are brought in who can better understand particular side effects of a drug being tested. It is in this stage of testing that the "risk/benefit ratio" of a drug is assessed. This ratio describes the therapeutic versus toxic effects of the drug. All drugs have some unwanted and sometimes harmful side effects. Whether or not a drug is deemed "safe and effective" will depend upon the severity of the disease or condition it is designed to alter and how well the drug accomplishes that purpose, versus the

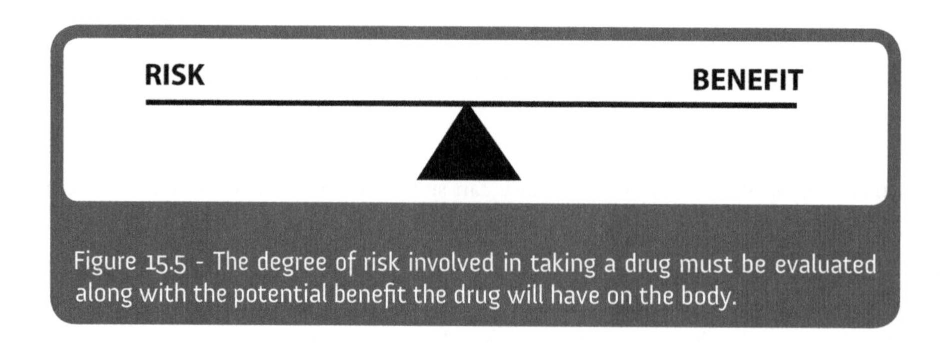

Figure 15.5 - The degree of risk involved in taking a drug must be evaluated along with the potential benefit the drug will have on the body.

detrimental side effects of using the drug. In other words, drugs with more severe side effects can be tolerated with more severe diseases (Fig. 15.5).

If all three of these phases of testing give positive results, the company can return to the FDA for permission to market the drug for a specific purpose. The FDA does not approve a drug per se; rather, it approves a drug for a specific purpose. Herein lies a problem. Once the drug is on the market for one purpose physicians can prescribe it for anything they deem appropriate.

The limitations on its use are greatly diminished. However, drug testing does not end here. The drug is "on probation." Phase IV of testing occurs once the drug is on the market. After a drug is on the market, more information is acquired on that drug through sources such as reports from physicians. This testing and reassessment of the effects of a drug go on for many years after the drug is first marketed. The company is required to report this information to the FDA as well. Thus, when you use a drug that has been on the market for a relatively short period of time, you are a "walking laboratory." You are taking part in Phase IV of drug screening. Even after the drug has passed Phase IV, the company is still required to inform the FDA of any reports of negative effects. A general rule of thumb is that the longer a drug has been on the market, the safer it is, simply because more information is known about both its immediate and long-term effects.

Informed consent: Humans use many drugs. Drugs are in food. Drugs are purchased in a drugstore. Drugs are prescribed by physicians. Drugs are purchased on the street. In all cases, it is the individual who puts or allows the drug to be put into his or her body. A "drug" that has no danger has no effect, and thus is not a drug. Drugs have risks and drugs have values. Both the risks and the potential beneficial effects differ between different individuals, because individual humans ARE different from each other (Fig 15.6). The individual assumes the risk, obtains the value, and pays for any mistakes. Because the ultimate result of the drug, harmful or beneficial, resides with the individual, the responsibility for taking a drug also resides with the individual. Such responsibility requires understanding the nature of the decision being made each time that a drug is taken.

UNPREDICTABILITY OF DRUGS IS DERIVED FROM INDIVIDUAL DIFFERENCES
Who is average?
12

Figure 15.6

index

A

absorption 98, 99, 100, 110, 113, 145, 200, 201, 215, 224, 254, 261
acetaminophen 225, 227, 263
acetyl CoA 118
acid 21, 22, 99, 100, 101, 102, 119, 194, 226, 238
Acquired Immune Deficiency Syndrome. See AIDS
ACTH 186, 192, 212, 213, 222, 229
active immunization 181
active transport 100
acupuncture 223
adenosine triphosphate. See ATP
ADH 78, 214
adrenal gland 116, 118, 165, 186, 207, 210, 212, 213, 214, 222, 238, 252
adrenalin 43, 97, 119, 120, 210, 211, 214, 238
adrenalin gland 211
afferent 132, 134
AIDS 167, 186, 187, 194, 195, 196, 202, 249
alcohol 51, 90, 101, 110, 120, 152, 164, 194, 227, 235, 238, 240, 249, 263
aldosterone 78
allergy 91, 96, 97, 199
alpha islet 43, 115
ALS 160
alveolar 55, 92, 93, 95
amino acid 24, 28, 29, 40, 41, 62, 63, 115, 117, 118, 120, 186, 207, 212, 242, 255
amniotic fluid 164
anabolic steroids. See Androgens
anaphylaxis 182, 200
androgen 53, 135, 156, 160
anemia 108, 150, 185, 196, 202, 254, 256
antacids 101, 258
antagonist 168, 242, 246
antibiotic 5, 40, 46, 91, 110, 193, 197, 198, 199, 200, 201, 202, 263
antibodies 159, 176, 182, 183, 186, 187, 256
antibodies 29, 175, 185, 188, 190
antidiuretic hormone. See ADH
antihistamines 40, 97, 102, 189
antioxidants 110
antiseptics 194, 195
aorta 72, 75
arteries 21, 50, 51, 68, 69, 70, 73, 74, 75, 76, 77, 79, 80, 81, 95, 119, 120, 131, 133, 138, 213, 215
artificial insemination 166

aspartame 241, 242
aspirin 225, 226, 227, 229, 256, 258
asthma 55, 97, 248, 253
atoms 1, 2, 16, 17, 18, 19, 20, 21, 24, 25, 26, 27, 29, 31, 37, 105, 106, 109, 118, 194
ATP 9, 31, 35, 36, 37, 38, 44, 46, 58, 61, 64, 83, 100, 105, 106, 107, 108, 110, 111, 112, 113, 115, 119, 125, 145, 226
autonomic 86, 98, 133, 134, 138, 208, 209, 221, 228, 248
autosomes 151, 153
autosomes 148, 150

B

bacteria 5, 21, 24, 25, 27, 29, 36, 37, 45, 46, 105, 106, 109, 110, 179, 194, 196, 197, 198, 200, 201, 256
beta islet cells 6, 113, 120, 121, 123, 140
bilateral 137, 221
bile salts 99, 100
biotransformation 263
birth control pill 51, 138, 168, 258, 260
blood-brain barrier 40
blood group 183, 185
blood pressure 12, 60, 67, 70, 71, 72, 74, 78, 79
B-lymphocyte 175, 176, 177, 179, 185
body temperature 13, 56, 57, 82, 83, 84, 85, 86, 87, 89, 90, 118, 205, 216, 225, 226, 243
bone 7, 10, 60, 61, 81, 108, 110, 136, 165
bone marrow 6, 39, 64, 90, 175, 176, 196, 199
brain 11, 12, 13, 38, 40, 51, 55, 156, 157, 181
breast 135, 155, 158, 160, 164, 165, 166, 168
breathing 9, 10, 54, 55, 81, 92, 93, 96, 105, 111, 138, 209, 229, 248, 252, 253
Broca's area 143
bronchial 92

C

caffeine 65, 97, 101, 120, 248
calcium 24, 41, 61, 105, 108, 109, 110, 145, 165, 248, 254
cancer 39, 104, 110, 129, 168, 169, 171, 172, 174, 175, 178, 186, 187, 188, 189, 190, 191, 192, 193, 195, 196, 203, 204, 254, 255, 256, 260, 264
capillaries 39, 40, 68, 69, 92, 159, 174, 262
carbohydrates 24, 30, 31, 98, 99, 113, 119, 121
carbon dioxide 10, 22, 30, 31, 37, 54, 64, 68, 69, 83, 91, 93, 94, 95, 96, 115, 138, 163, 213, 226, 253
carbonic acid 22, 23, 94, 95
carbonic anhydrase 94, 95
cell division 6, 7, 28, 38, 39, 155, 167, 178, 186, 189, 197, 255
cells 4, 5, 6, 7, 11
cellular reproduction 38
cell wall 44, 46, 197, 200
cerebellum 145, 249
cerebral cortex 137, 140, 141, 142, 143, 144, 145, 220, 221, 223, 228, 239, 243, 247, 249
channels 33, 34, 37, 41, 42, 61, 78, 85, 109, 248
charge 4, 16, 17, 18, 19, 20, 21, 22, 23, 24, 25, 26, 29, 32, 33, 34, 41, 42, 61, 99, 100, 101, 106, 107, 108, 109, 120, 126, 216, 259
chemical message 42, 43, 53, 57, 62, 63, 64, 86, 90, 97, 101, 103, 108, 117, 124, 125, 132, 135, 138, 139, 157, 159, 160, 171, 172, 173, 174, 176, 178, 179, 180, 186, 189, 190, 192, 196, 210, 213, 219, 220, 222, 226, 234, 243, 256, 257, 260
chewing 98
chloride 22, 23, 106, 126
cholesterol 79, 99, 116, 202
choroid plexus 40
chromosome 27, 28, 36, 148, 149, 150, 151, 152, 153,

154, 185, 197
X chromosome 152
Y chromosome 152
chromosome 21 144, 151
circular muscle 51
cloning 152, 256
clotting 59, 110, 226, 243, 258
cocaine 51, 90, 120, 243, 245
collagen 109
congestive heart failure 74
constipation 101, 102, 245
contraction 53, 70, 72, 74, 75, 76, 79, 81, 98, 103, 108, 130, 166, 210, 216
convection 88
copper 105
coronary artery 75
corpus callosum 142
corticosteroids 91, 189, 253
cranial nerves 131, 134
cytotoxic T-cells 176, 178, 179

D

daughter cells 39, 152, 163, 175, 176, 177
demineralization 108, 110, 165
dendrites 125, 126, 128, 129, 130, 220
deoxyribonucleic acid. See DNA
deoxyribonucleotides 27, 29, 38, 196
dependence 229, 230, 231, 237, 238, 244, 248, 249
depolarization 129, 136
DES 168
diabetes 63, 118, 120, 121, 165, 166, 168, 186, 192, 253, 255, 260
diaphragm 53, 59, 92
diethylstilbestrol. See DES
digestive system 10, 47, 53, 57, 58, 59, 60, 77, 84, 97, 98, 99, 100, 101, 102, 103, 109, 133, 173, 201, 204, 210, 213
disinfectants 194, 195
diuretics 78, 248
DNA 24, 27, 28, 29, 36, 37, 38, 40, 108, 148, 149, 151, 153, 175
dopamine 6, 242, 243, 245, 246, 247
Down's syndrome 151, 155
drug 258, 263
drug resistance 196, 200, 201, 263

E

efferent 132, 133, 134
egg 153, 154, 155, 157, 160, 161, 162, 163, 167, 168, 251, 260
ejaculation 167
electrical potentials. See potentials
embryonic stem cells 6, 38
endorphins 213, 218, 222, 223, 228, 231
energy 1, 7
enkephalins 222, 223, 228, 231
enzyme 21, 22, 29, 38, 44, 58, 94, 98, 99, 100, 101, 116, 117, 133, 187
esophagus 98, 99, 101
estrogen 116, 157, 158, 160, 161, 162, 163, 165, 166, 168, 260
evolution 4, 5, 11, 125, 126, 137, 172, 175, 187, 189, 192, 201, 207, 253
excitatory 129, 132, 225, 230, 236, 240, 241
exercise 9, 38, 65, 77, 79, 108, 111, 112, 118, 119, 120, 121, 160, 213, 214, 218

F

Fallopian tubes 163, 167
fat 7, 8, 9, 25, 26, 60, 61, 62, 63, 64, 75, 97, 98, 99, 100, 101, 110, 112, 113, 115, 118, 119, 120, 121, 164, 186
fatty acid 98, 100, 118, 119, 120

fear 119, 140, 141, 206, 207, 216, 218, 229
feast 103, 111, 113
fertility 148, 167
fetal alcohol syndrome 164
fetus 156, 163, 164, 183, 185, 264
filtration 77, 133
first trimester 156, 163, 252
FSH 157, 158, 159, 160, 161, 162, 163, 167, 168
FSHRF 157, 158, 159, 160, 162
fungi 36, 44, 171, 172, 174, 193, 194, 200, 201, 202

G

GABA 236, 239, 240, 247
gall bladder 99
Genes 27, 28, 121, 144, 148, 149, 150, 151, 152, 154, 155, 183, 187, 197, 203, 251, 252, 254, 255, 258
germ cells 153
glial cells 38, 129
glucagon 43, 115
glucocorticoids 116, 117, 118, 120, 164, 165, 168, 186, 189, 192, 193, 207, 211, 212, 216, 222, 252, 253
glutamate 236, 240, 241
glycerol 98, 100, 118, 252
glycine 240
glycogen 31, 32, 62, 63, 113, 114, 115, 116, 118, 119, 120, 121, 210, 213
goiter 108, 138, 139
gravity 70, 75
growth factors 43

H

HCG 163
HDL 120
heart 67, 68, 69, 70, 71, 72, 73, 74, 75, 76, 77, 79, 80, 81, 82
heart attack 51, 75, 77, 226, 238, 260
heat capacity 85, 86
heat transfer 57, 62, 85, 86
helper cells 38, 129
hemispheres 142
hemoglobin 22, 23, 29, 34, 86, 93, 94, 95, 96, 108, 121, 150, 202
Hemolytic disease 183, 185
hepatic portal vein 58, 100, 113, 114, 121
high density lipoprotein. See HDL
histamine 57, 64, 97, 101, 102, 189, 226, 247
HIV 186, 187, 195, 196, 202
hormones 43, 53, 64, 135
human chorionic gonadotropin. See HCG
Human Immunodeficiency Virus. See HIV
humidity 88, 89, 92
hydrogen ion 21, 22, 23, 93, 95
hypothalamus 85, 86, 135, 138, 157, 162
hypothermia 8, 12, 90

I

ibuprofen 225, 227
immune system 39, 64, 172, 178, 181, 192
immunodeficiency 186, 192, 202
implantation 155, 160, 162, 163, 167, 168
impotency 160
infection 5, 44, 45, 46, 64, 91, 171, 191, 226, 240, 263
inflammation 96, 97, 173, 174, 188, 189, 226, 227, 253, 258
information ports 42, 43, 53, 113, 126
inhibitory 129, 130, 132, 166, 222, 225, 229, 230, 236, 239, 240, 241, 247
insulin 6, 11, 63, 113, 114, 115, 118, 120, 121, 123, 140, 165, 168, 186, 212, 216, 256, 260, 261
insulin resistance 121, 165, 168, 216
interpretive areas 143
iodine 24, 105, 108, 109, 138, 139, 194
ions 22, 23, 24, 26, 29, 32, 33, 34, 41, 60, 61, 68, 77,

78, 95, 96, 99, 101, 106, 107, 126, 145, 226, 248
ions 21
iron 2, 16, 24, 118, 164
Iron 108
IUD 167, 168

K

ketone bodies 64, 118, 119, 216
kidney 40, 41, 58, 60, 62, 77, 78, 79, 84, 101, 106, 109, 110, 112, 113, 117, 119, 121, 133, 134, 135, 160, 168, 186, 196, 199, 201, 202, 207, 208, 210, 212, 213, 227, 236, 237, 248, 252, 262
Klinefelter's syndrome 155

L

lactation 165
lactic acid 38, 112, 119
LDLs 120
L-dopa 242, 246
learning 141, 144, 148, 218, 222, 249, 253
LH 157, 158, 159, 160, 161, 162, 167, 213, 229
light/dark cycle 116
limbic system 137, 140, 141, 143, 148, 207, 209, 214, 220, 222, 229, 239, 242, 243, 246, 247, 249
lipid 25, 26, 27, 36, 61, 62, 63, 116, 121
lipopolysaccharide. See LPS
lipoproteins 120
liver 10, 31, 38, 40, 41, 43, 58, 59, 60, 61, 62, 63, 65, 99, 100, 103, 108, 110, 112, 113, 114, 115, 116, 117, 118, 119, 120, 121, 198, 202, 207, 216, 227, 238, 261, 262, 263
local anesthetics 223, 224, 225, 260
LPS 173, 181
lungs 22, 53, 54, 55, 64, 68, 71, 72, 81, 91, 92, 93, 94, 95, 96, 97, 111, 120, 132, 189, 207
luteinizing hormone. See LH
lymphatics 10, 82, 101, 113, 115, 202
lymphocyte 174, 175, 176, 179, 185, 187, 195

M

macrophage 55, 57, 58, 59, 90, 97, 103, 173, 174, 177, 178, 179, 185, 187, 226
magnesium 24, 105
malaria 44, 202
manganese 105
manipulating reproduction 166
medulla 137, 138, 140, 145, 209, 215, 220, 222, 226, 228, 229, 242, 247, 248
melatonin 157
membrane 4, 29, 35, 36
memory 44, 124, 141, 144, 180, 228
meningitis 5, 46, 174, 180, 197
menstrual cycle 162
menstruation 157, 161, 162, 163, 168
metabolism 108, 109, 110, 119, 120, 121, 135, 225, 253, 264
milk 110, 135, 165, 166
minerals 24, 58, 97, 98, 100, 105, 106, 107, 109, 110, 164, 259, 263
molecules 1, 2, 15, 16
morning after pill 168
motor nerve 92, 98, 136, 188, 195, 254
MSG 241
muscle 10, 29, 37, 38, 53, 60, 61, 62, 63, 65, 77, 80, 81, 83, 84, 85, 89, 92, 98, 108, 112, 113, 114, 115, 117, 118, 119, 120, 121, 132, 134, 143, 145, 160, 200, 205, 207, 210, 211, 212, 213, 215, 216
muscular dystrophy 154, 160
mutation 7, 153, 187, 196, 202, 255, 256
myelin 129, 187
myoglobin 118

N

naproxen 225, 227
narcotic analgesics 228
natural selection 172
negative feedback 138, 139, 158, 159, 160, 161, 162, 168
neocortex 140, 141, 208, 214, 247
nerve cell 11, 53, 101, 106, 125, 126, 127, 128, 129, 130, 133, 135, 144, 187, 195, 220, 223, 234, 239, 248, 254
neuron 125, 126, 128, 129, 130, 132, 144, 240, 254
neurotransmitter 43, 125, 126, 128, 129, 136, 145, 189, 213, 222, 234, 235, 236, 239, 240, 241, 242, 243, 247, 248
nicotine 65, 120, 248, 253, 261
nitrogen 4, 15, 16, 17, 24, 27, 54, 62, 91, 109, 117
non-narcotic analgesics 225
non-polar 26, 33
noradrenalin 228, 238, 242, 243, 245, 246, 247
nucleus 16, 17, 18, 19, 20, 21, 26, 27, 28, 29, 36, 46, 64, 117, 125, 149, 187
nutrient 61, 105, 171, 208

O

oil 4, 9, 25, 26, 194, 195
opiates 213, 218, 228, 229, 230, 231, 248, 258
opium 228, 256
oral contraceptives 110
osteoporosis 108, 165
ovaries 153, 156, 157, 158, 160, 161, 162, 163, 167, 260
ovulation 162, 168
oxygen 1, 2, 4, 9, 10, 15, 18, 19, 20, 21, 22, 23, 24, 27, 29, 30, 31, 37, 38, 54, 55, 64, 68, 69, 91, 93, 94, 95, 96, 108, 111, 112, 118, 119, 213
oxygen debt 9, 38, 112, 119
oxytocin 165, 166, 214

P

pacemaker cells 72
pain 75, 85, 132, 141, 213, 218, 219, 220, 221, 222, 223
pain signal 219, 220, 221, 222, 223, 224, 225, 228, 260
pancreas 6, 11, 43, 63, 99, 100, 113, 115, 121, 123, 140, 260
parasites 44, 172, 174, 193, 194, 202
parasympathetic 133, 134, 137, 209, 245, 247
Parkinson's disease 6, 239, 246, 247
passive immunization 182, 183
patches 261
pattern translation 136
penicillin 40, 200, 201, 259
perception 102, 213, 223, 229, 243
pH 21, 22, 101, 205
phagocytosis 173, 179
phospholipid 32, 33, 36, 41, 92
phosphorus 15, 24, 105, 108
photosynthesis 2, 31
pineal gland 157, 158, 243
pituitary gland 12, 78, 116, 131, 135, 138, 139, 157, 159, 186, 208, 209
placenta 6, 162, 163, 164, 165, 185, 252, 253
pneumothorax 93
polar 4, 25, 26, 100, 125
portal system 103, 135, 159, 261, 262
potassium 24, 105, 106
potential energy 2, 3, 29, 30, 31, 33, 37, 62, 63
potentials 33, 34, 106, 126, 128, 129, 145, 223, 224
pregnancy 108, 148, 155, 156, 160, 162, 163, 164, 165, 166, 167, 168, 185, 190, 208, 252, 253
primary sensory 143
primary sex characteristics 148

progesterone 138, 158, 160, 161, 162, 163, 164, 165, 166, 168, 260
prolactin 166, 214
prostaglandins 165, 225, 226, 227
protein 24, 28, 29, 33, 41, 99, 117, 149, 192
pseudohermaphrodite 156
puberty 156, 157, 158, 165

R

radiation 30, 31, 57, 87, 195, 203
receptors 42, 43, 53, 87, 126, 128, 175, 176, 220, 234, 242, 244
red blood cell 22, 64, 93, 94, 95, 108, 183, 185, 202, 256
reflex 98, 132, 145, 209, 220, 229, 247
relative resistances 74, 76
relaxin 165
releasing factor 135, 136, 139, 157, 158, 159, 242
representational 143, 145, 221
reproduction 64, 65, 135, 147, 148, 152, 153, 154, 156, 157, 158, 160, 166, 214
resistance 73, 74, 75, 76, 77, 79, 80, 81, 89, 182, 196, 200, 213
reticular activating system 141, 213, 220, 221, 248
Rh antigen 185
ribonucleic acid. See RNA
risk/benefit 164, 182
RNA 24, 28, 29, 40, 45, 46, 117, 186, 187, 193, 195, 196, 198
RU486 168

S

saliva 98, 247
salt 8, 12, 23, 24, 60, 79, 107, 109, 261
secondary sex characteristics 157, 159, 160
segment 132, 136
seizure 234, 243, 247
selection 5, 172, 192, 198, 263
seminal fluid 167
sensory 43, 132, 141
sex 143, 147, 148, 154, 155
sex chromosomes 151, 153, 155
sexual reproduction 153, 156
side effect 67, 101, 169
signal 11, 12, 43, 125, 126, 128, 129, 130, 131, 136, 209, 210
sinuses 59, 80, 81
skin 38, 39, 53, 56, 57, 62, 77, 83, 84, 85, 87, 88, 89, 90, 91, 104, 110, 132, 145, 171, 173, 189, 194, 195, 201, 203, 204, 205, 206, 223, 254, 255, 261, 262
small intestine 21, 34, 58, 59, 98, 99, 100, 101, 103, 110, 113, 114, 121, 135, 188, 199, 238, 261
smoking 55, 96, 164, 189, 203, 261
smooth muscle 43, 51, 68, 98, 102, 103, 133, 210, 229, 236, 248
soap 194, 195
sodium 16, 23, 24, 60, 73, 77, 78, 79, 105, 106, 126, 145
sperm 153, 154, 155, 157, 158, 159, 160, 166, 167
spinal cord 11, 39, 54, 85, 93, 124, 131, 132, 133, 136, 137, 145, 219, 220, 221, 222, 223, 224, 228, 229
starvation 7, 118, 119, 186, 192, 215, 216
statistics 169, 258
stem cells 6, 7, 38, 39, 51, 56, 64, 104, 129, 206, 252
steroids 116, 117, 156, 158, 160, 186, 238, 259, 260
stomach 21, 22, 34, 98, 99, 101, 102, 201, 226, 238, 262
stress 64, 65, 119, 124, 206
strychnine 130, 240
Substance P 222
substances 40, 258
substantia nigra 6, 242, 246

sugar 8, 9, 10, 24, 26, 29, 31, 32, 37, 43, 59, 62, 63, 77, 98, 99, 105, 110, 111, 112, 113, 114, 115, 116, 117, 118, 119, 120, 121, 207, 210, 213, 216
sun 2, 3, 10, 12, 30, 31, 87, 110, 203, 207, 254, 255
superfemale 155
surfactant 92, 194, 252, 253
sweat 57, 85, 88
sympathetic 133, 134, 137, 207, 208, 209, 210, 211, 213, 215, 220, 221, 229
sympathetic response 79, 236, 238, 242, 248, 249
synthesis 36, 63

T

temperature 7, 8, 13, 16, 51, 83, 85, 86, 90, 229, 243
testosterone 116, 138, 156, 157, 158, 159, 160
tetanus toxin 240
thalidomide 169, 252
threshold 144, 222, 223, 228, 247
thyroid gland 108, 139, 140
thyroid hormone 108, 120, 139, 214, 216
T-lymphocytes 175, 176, 178, 187
tolerance 176, 187, 188, 214, 229, 230, 231, 237, 238, 240, 248, 249, 263
toxic substances 41, 60, 100, 110
toxin 182, 197, 240
transplantation 183, 185, 189, 193, 202
tubal ligation 167
tumors 168, 188, 189
Turner's syndrome 155

U

ulcers 101
unconsciousness 84, 141, 229, 237
uterus 6, 155, 160, 161, 162, 163, 164, 165, 166, 167, 168, 252

V

vagina 163, 166, 201
vagus nerve 134, 135
vas deferens 167
vasectomy 167
veins 21, 69, 80, 81, 202
ventricle 68, 69, 71, 72, 81
virus 44, 45, 90, 91, 171, 172, 174, 178, 179, 181, 182, 186, 187, 189, 192, 193, 194, 195, 196, 197, 203, 254
vitamin 40, 46, 47, 58, 97, 98, 100, 105, 109, 110, 120, 164, 198, 238, 254, 255
voltage 34, 106
volume 8, 32, 73, 74, 75, 76, 77, 78, 79, 80, 81, 82, 164, 181, 182

W

waste 10, 35, 49, 83
water 4, 5, 7, 8, 9, 10, 15, 16, 19, 20, 21, 22, 23, 32
Wernicke's Area 143
white blood cell 64, 174
withdrawal 44, 105, 230, 231, 238, 240, 248, 263

X

xanthines 97, 248
X chromosome 151, 154, 155

Y

Y chromosome 151, 154, 156
yeast infection 44, 202

Z

zinc 105

CPSIA information can be obtained at www.ICGtesting.com
Printed in the USA
LVOW01s0905210814

400126LV00003B/13/P

9 781621 312109